KAIST Research Series

More information about this series at http://www.springer.com/series/11753

Chong-Min Kyung
Editor

Theory and Applications of Smart Cameras

CISS
Center for Integrated Smart Sensors

Springer

Editor
Chong-Min Kyung
Department of Electrical Engineering,
Center for Integrated Smart Sensors
KAIST
Daejeon
Korea, Republic of (South Korea)

ISSN 2214-2541 ISSN 2214-255X (electronic)
KAIST Research Series
ISBN 978-94-017-9986-7 ISBN 978-94-017-9987-4 (eBook)
DOI 10.1007/978-94-017-9987-4

Library of Congress Control Number: 2015942807

Springer Dordrecht Heidelberg New York London

Printed on acid-free paper

Springer Science+Business Media B.V. Dordrecht is part of Springer Science+Business Media (www.springer.com)

Contents

Part I
Fundamental/Energy-related Issues of Smart Cameras

CMOS Image Sensor for Smart Cameras

JongHo Park

Abstract A smart camera is a vision system with special features implemented to achieve its specific purpose. A smart camera which can be used for security or surveillance purpose requires high dynamic range of the sensor to cover broad illumination range of the scene. A stick- or badge-type smart camera operates as a stand-alone device so that the power consumption is one of the most important parameters. For applications such as nondestructive inspection using infrared (IR), sensitivity of the image sensors should be improved to obtain suitable SNR for reliable output. This chapter describes basic imaging principles and dynamic range expansion methods of the CMOS image sensors.

Keywords CMOS image sensor (CIS) · Charge coupled device (CCD) · Active pixel sensor (APS) · Wide dynamic range (WDR) · Correlated double sampling (CDS)

1 Imaging Principles

Image sensors in CMOS technology are implemented using an array of smart photo sensors called active pixel sensor (APS). The design of the APS is flexible and various types of pixels have been developed for various applications including WDR imaging [1]. The performance of CMOS APS in areas of high-end digital imaging has been proven to be comparable to their CCD counterparts due to the ability of on-chip image processing [2]. Image processing to a certain degree could be performed within the pixel itself by integrating signal processing circuitry in each pixel [3]. Further image processing can be done in the subsequent circuit stages before image information is read out [4].

J. Park (✉)
Center for Integrated Smart Sensor, ITC Building (N1), KAIST, Daehak-ro 291, Yuseong-gu, Deajeon 305-701, Republic of Korea
e-mail: parkjh20@kaist.ac.kr

C. Kyung (ed.), *Theory and Applications of Smart Cameras*,
KAIST Research Series, DOI 10.1007/978-94-017-9987-4_1

1.1 Solid State Imaging Devices

1.1.1 Charge Coupled Device

A Charge coupled device (CCD) is the most important technology for image sensors. The CCD offers guaranteed image quality because it uses optimized photodetectors for achieving low noise, low dark current, and high sensitivity [5]. The basic concept of CCD is a simple series connection of MOS capacitors. The individual capacitors are physically located very close to each other, which compose an analog shift register driving by two, three or four phase clocks for charge transfer. Figure 1 shows the simplified structure of an interline-transfer CCD. A charge transfer must occur at high enough rates to avoid image corruption by leakage, but slow enough rates to ensure high charge transfer efficiency [5].

The limitation of CCD technology comes mainly from insufficient charge transfer efficiency [6]. To avoid insufficient charge transfer, high-speed and high-voltage clock control schemes in CCD increase the system complexity and the power consumption. Another major drawback in CCD technology is that peripheral circuits such as ADC cannot be integrated on the same chip. Nevertheless, understanding CCD technology is very important because CMOS image sensors (CISs) have been developed based on CCD.

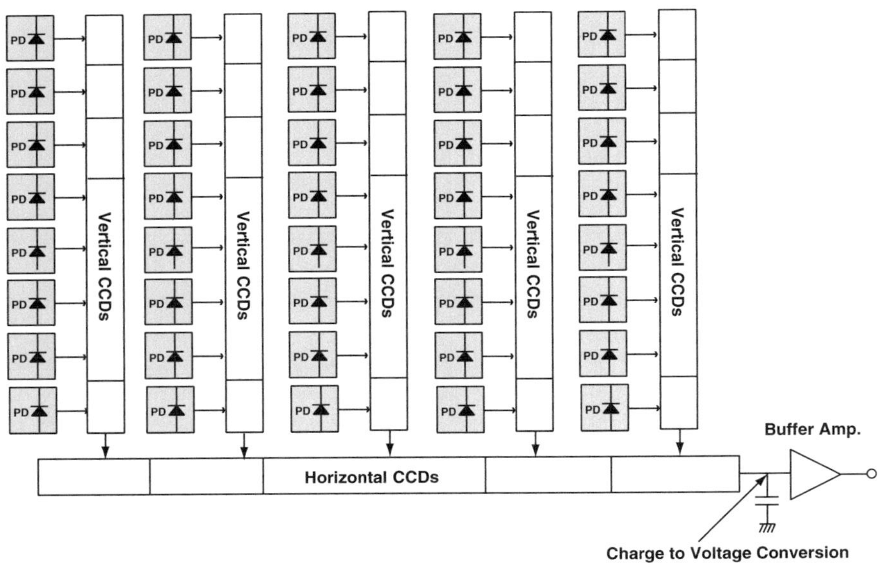

Fig. 1 Simplified interline-transfer (IT) CCD structure

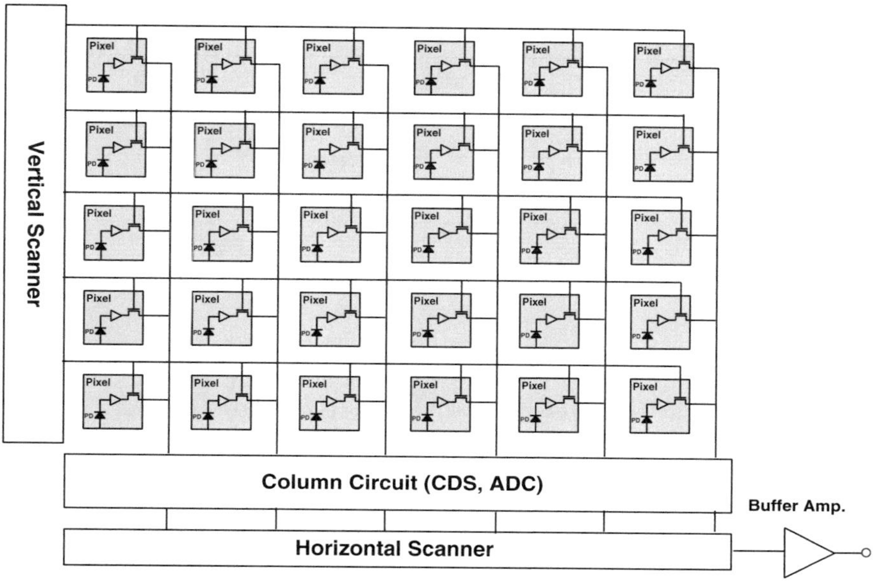

Fig. 2 Structure of conventional CMOS image sensors

1.1.2 CMOS Image Sensors

Unlike CCD technology, CISs are fabricated with normal CMOS technology so that pixel array and peripheral circuits such as ADC and image processing circuits can be integrated on the same chip. A conventional CIS architecture with an APS is shown in Fig. 2. The CIS consists of pixel arrays, column-parallel analog front end (AFE) circuits, and peripheral circuits. Since the pixels are selected in a row by row fashion, each pixel has a row selection transistor connected to a row selection signal from a vertical scanner. The vertical and the horizontal scanners are used to sequentially access pixels. Column parallel readout circuits such as correlated double sampling (CDS) circuit and ADC perform noise cancelation [7] and analog-to-digital conversion before reading out the signal. Pixel outputs selected by row selection signal are sampled to the column circuits for analog signal processing in parallel. After analog signal processing, the horizontal scanner generates access signals to read out digitized output sequentially.

1.2 Preliminaries for CMOS Image Sensors

1.2.1 Readout Timing for CMOS Image Sensors

In conventional CISs, one frame image output is read out in a frame period. Figure 3 shows a signal accumulation and a readout timing for conventional CISs.

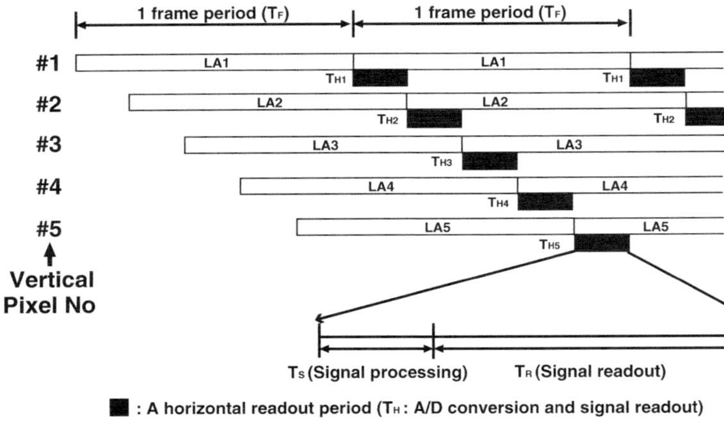

Fig. 3 Timing diagram showing the integration and the readout timing in a frame period

In this diagram, it is assumed that the sensor has 5 vertical pixel arrays for simplification, and long accumulation time signals, LA, are read out in a frame period T_F. A horizontal readout time T_H can be written by $T_H = T_F/N_V$ where, N_V is the total number of vertical pixel arrays. T_H includes the time for column signal processing and the readout time of the column signal. If the time for column signal processing such as CDS and A/D conversion is T_S and the time required to scan out all column signals is T_R, then $T_H = T_S + T_R$. Here, one pixel readout time T_P is defined by $T_P = T_H/N_H$, where N_H is the number of horizontal pixel arrays.

1.2.2 Column Noise Canceler for CMOS Image Sensors

Pixel output itself contains various noise sources from photodiode and source follower amplifier. A fixed pattern noise (FPN) and a part of temporal noise are suppressed or canceled by column circuits shown in Fig. 4a. In this circuit diagram, the input labeled V_{in} shows pixel output. A transistor M_3 is a current source circuit which is common to all source followers in each column.

A timing diagram describing switching conditions for each phase is shown in Fig. 4b. V_{in} from the pixel can be represented in two levels, V_{reset} and V_{signal}, respectively. During T_1 period, the noise canceler samples V_{reset} into capacitor C_1, and the difference between V_{reset} and V_{signal} is output in T_2. The output is given by

$$V_{\text{out}} = \frac{C_1}{C_2}(V_{\text{reset}} - V_{\text{signal}}) + V_{\text{ref}} \tag{1}$$

Removing a FPN and a correlated temporal noise by differentiating two outputs is called a CDS [8]. A reset noise which has a correlation between the two levels in 4-transistor APS is mostly removed by applying CDS, while 1/f noise is partially

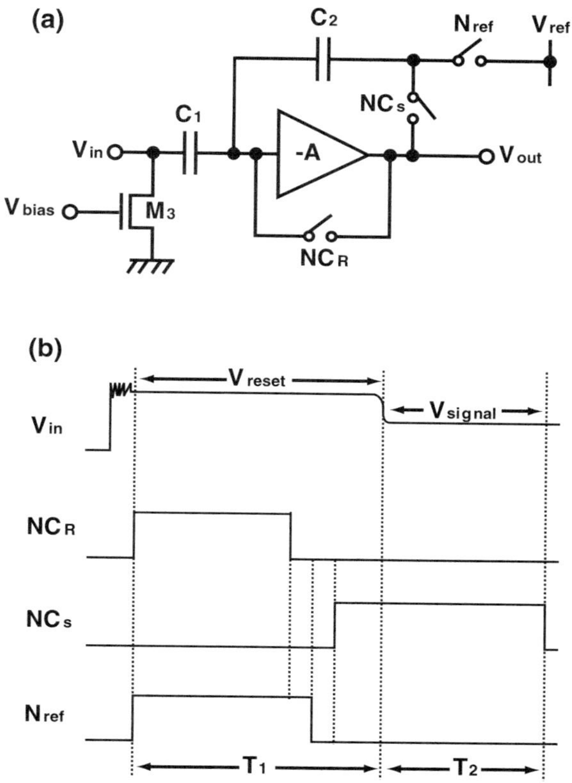

Fig. 4 Column noise canceler: **a** Schematic diagram, **b** timing diagram

removable by applying CDS [9]. In addition, if the system gain given by the ratio of two capacitors is set to be higher, an input-referred noise can be reduced by bandwidth limitation effect [10].

1.3 Pixel Structure for CMOS Image Sensors

1.3.1 Three Transistor APS

Although special functionality such as high-speed and low-noise imaging can be achieved by adding dedicated circuits on CIS, the pixel is the most important building block which affects overall performance of CIS.

There are two types of APS. One is a 3-transistor APS which has three transistors in a pixel and the other is a 4-transistor APS. Figure 5a shows a schematic of 3-transistor APS including a simple column noise canceler. The pixel consists of a reset transistor, a source follower transistor, and a selection transistor as well as a photodiode.

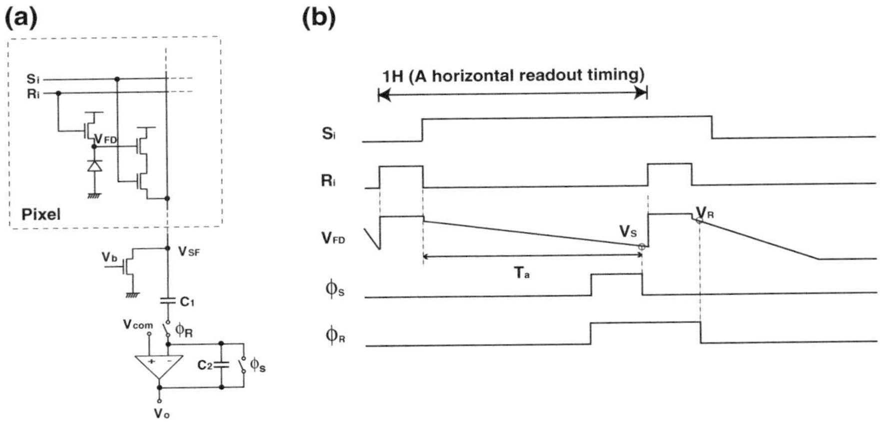

Fig. 5 Three transistor APS: **a** Pixel structure, **b** timing diagram

Incident photons into the photodiode generate a proportional amount of signal charge that is accumulated in the photodiodes. The potential of the photodiode is varied by the accumulated signal charges, and applied to the in-pixel source follower. The selection transistor controlled by the signal S_i is used as a switch to read out pixel output row by row.

The timing diagram and corresponding voltage of V_{FD} are shown in Fig. 5b. The signal level, V_S is read out first through the in-pixel source follower when the S_i turns on. The ϕ_s is used to sample the signal level V_S to the noise canceler and reset the capacitor C_2. After reading the signal level V_S, the photodiode is reset by R_i. A reset level, V_R is sampled to the noise canceler again by ϕ_R. The difference between the signal level and the reset level is obtained as the final output. However, the reset noise still remains in 3-transistor APS because of uncorrelated V_S and V_R as shown in Fig. 5b.

The photodiode accumulates signal charges which form a photocurrent I_{ph}. The potential of photodiode which is initially set to VDD decreases to V_S during the accumulation time T_a. If I_{ph} is constant, V_S is given by

$$V_S = \frac{I_{\mathrm{ph}} \times T_a}{C_{\mathrm{FD}}} \tag{2}$$

where, C_{FD} is a capacitance of photodiode. The signal level V_S is proportional to the accumulation time T_a.

The detailed pixel structure and a potential distribution of the 3-transistor APS are shown in Fig. 6. In order to suppress dark current, the photodiode used in 3-transistor APS can be partially pinned because the potential of the photodiode is directly read out through the source follower shown in Fig. 6b. However, partially pinned photodiode shows high dark current which is a significant drawback of the 3-transistor APS for high performance imaging applications [11].

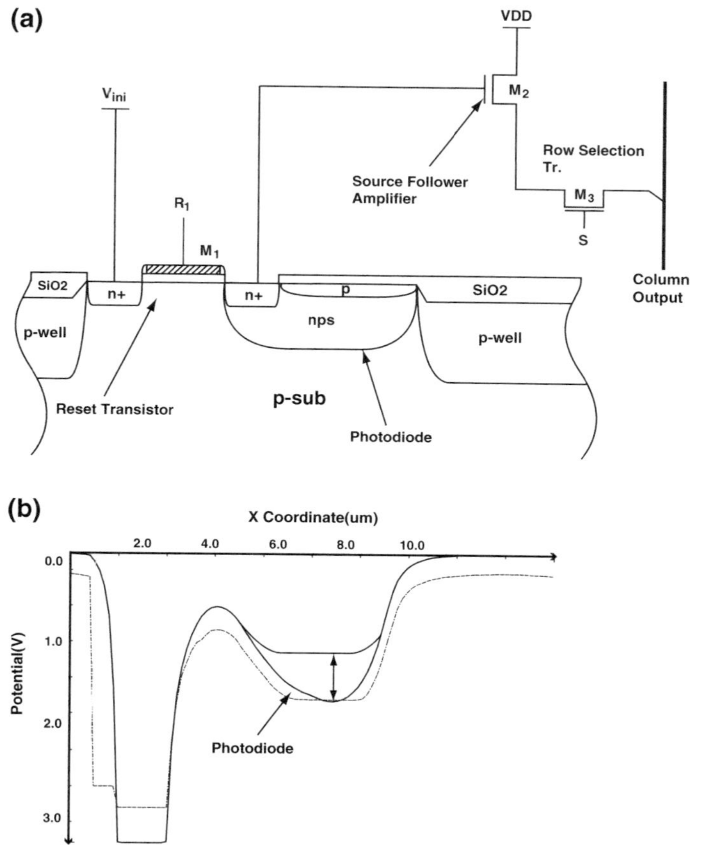

Fig. 6 Detailed structure and potential profile for three-transistor APS: **a** Pixel structure, **b** potential profile

1.3.2 Four Transistor APS

Recently, a four-transistor APS shown in Fig. 7a is mostly used as pixel because of its excellent performance. Unlike 3-transistor APS, potential variation of the photodiode due to signal charges in four-transistor APS is not directly read out. The signal charges in photodiode are transferred to a floating diffusion node, V_{FD} before reading out. Therefore, the conversion gain can be increased independently because the capacitance of the floating diffusion, V_{FD} can be designed as small as possible. Moreover, strong correlation between the reset and the signal levels is achieved by separating the regions of signal collection and detection.

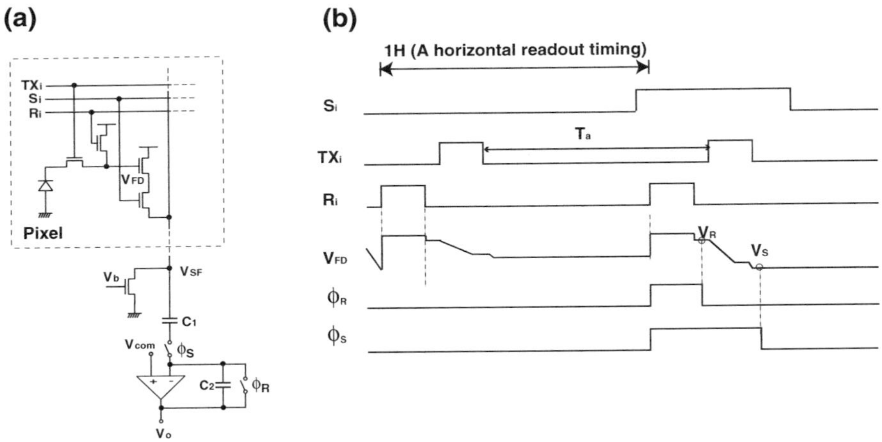

Fig. 7 Four-transistor APS: **a** Pixel structure, **b** timing diagram

The timing diagram of four-transistor APS is shown in Fig. 7b. The signals S_i, TX_i, and R_i represent control signals applied to the pixel, and ϕ_R and ϕ_S are clock signals for reset and sample in column circuit. Initially S_i goes high to turn on the select transistor. The floating diffusion node V_{FD} is then reset via R_i. The reset level is read out by a noise canceler with ϕ_R. Signal charges colected in the photodiode are then transferred to the FD node V_{FD} by TX_i. The potential variation of V_{FD} is proportional to the amount of accumulated signal charges in the photodiode. The signal level is read out to the noise canceler by the signal ϕ_S. The difference between reset level and signal level is obtained as the final output.

The detailed pixel structure of the four-transistor APS is shown in Fig. 8a. The photodiode can be fully pinned by holes because the charge integration and detection are performed in different device, which is separated by a transfer transistor M_2. Dark current is significantly reduced by using a fully pinned photodiode. The signal charges accumulated in the photodiode are perfectly transferred to the floating diffusion. Therefore, an image lag due to any remaining signal charges in the photodiode does not exist in four-transistor APS. The reset noise is cancelled by the noise canceler because of strong correlation of the reset and signal outputs [12, 13]. The sensitivity of the 4-transistor APS can be increased by the small capacitance of charge detection node (V_{FD}) without reducing the size of photodiode. Figure 8b shows a potential distribution of the four-transistor APS when the signals V_R and V_{TX} are set to low. A potential variation of the floating diffusion node is read out through a source follower.

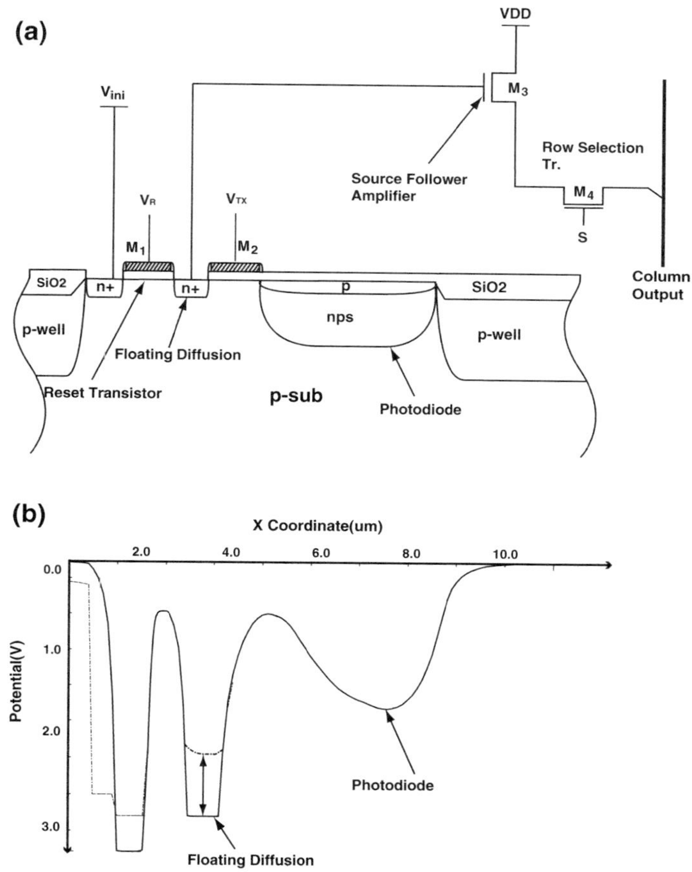

Fig. 8 Conventional four-transistor APS: **a** Pixel structure, **b** potential profile

1.4 Dynamic Range of Image Sensors

In image sensors, maximum detectable signal is called the full well capacity in both of CCD and CIS. The ratio of the maximum non-saturating signal to the smallest detectable input signal is called dynamic range (DR) [14] which can be expressed as

$$\mathrm{DR[dB]} = 20\log\frac{V_{S,\max}}{V_{S,\min}} \tag{3}$$

where, $V_{S,\max}$ is the largest non-saturation signal which can be obtained from the pixel and $V_{S,\min}$ is the smallest detectable input signal determined by sensor's noise level.

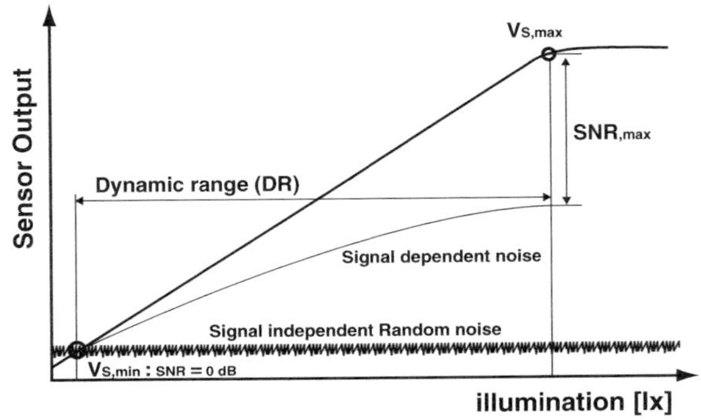

Fig. 9 Definition of the dynamic range for image sensors illustrated by sensor's output characteristic

The maximum non-saturation signal has a close relationship with the maximum detectable charge $N_{S,\max}$, the sensitivity of photodiode, S_{PD}, and the accumulation time, T_a. The relationship can be expressed by $V_{S,\max} \propto N_{S,\max}/S_{\text{PD}} \times T_a$. Therefore, the dynamic range for bright region can be increased by following methods.

1. Increasing the maximum number of electrons $N_{s,\max}$ by increasing the size of photodiode.
2. Increasing $V_{S,\max}$ by reducing integration time T_a.
3. Adjusting the photodiode sensitivity, S_{PD} which is inversely proportional to $V_{S,\max}$.
4. Employing a nonlinear detector to prevent the saturation of pixel output.

The dynamic range of dark region can be expanded by improving sensor's noise performance, i.e., lowering the detectable low illumination. Random noise such as reset noise, readout noise, and shot noise of dark current limits the minimum detectable input signal. Effective suppression of these noise should be considered to expand the dynamic range in both directions, i.e., toward both bright and dark regions. Figure 9 illustrates the definition of the dynamic range from the sensor's output characteristics. The minimum detectable illumination is defined by the signal to noise ratio (SNR) being equal to 0 dB where the sensor's output is the same as noise level.

2 Dynamic Range Extension

Important specifications of wide dynamic range (WDR) image sensors can be described as high SNR, low noise for low illumination as well as high dynamic range (DR). Various methods to achieve both of high SNR and high DR have been reported for the last decade.

Representative WDR methods can be clarified according to response types of CIS. One is a nonlinear response type and the other is a linear response type.

2.1 Nonlinear Response Type

2.1.1 Logarithmic Photodetector

A logarithmic compression is a well-known method to expand DR of image sensors [15]. Figure 10a shows a circuit diagram of the logarithmic photodetector. The gate of the transistor M_1 is connected to the drain which differs from three-transistor APS. The transistor M_1 operates in subthreshold region because a photocurrent that flows through diode-connected transistor is typically several tens to hundreds of femto (10^{-15}) ampere. The photocurrent I_{ph} can be written as

$$I_{\text{ph}} = \frac{W_{M1}}{L_{M1}} I_{D0} e^{\frac{V_D}{U_t}\frac{1}{n}} \tag{4}$$

where I_{D0} is a reverse leakage current of photodiode, U_t is a thermal voltage defined by kT/q, and n is a subthreshold factor affected by fabrication process. Therefore, the node voltage V_D is proportional to $\log(I_{\text{ph}})$ as shown in Fig. 10b.

In the logarithmic photodetector, the pixel output is directly converted to the corresponding voltage from the photocurrent, which is logarithmically proportional to the photocurrent. Unlike integration-type APS, no operation for charge integration is required in logarithmic photodetector due to this mechanism. An important advantage with nonintegration type operation is that the pixel can be accessed randomly.

Although the logarithmic photodetector can extremely expand the dynamic range for bright region, it also has many drawbacks. First, a MOSFET working in the subthreshold region shows a large variation of threshold voltage, which leads to

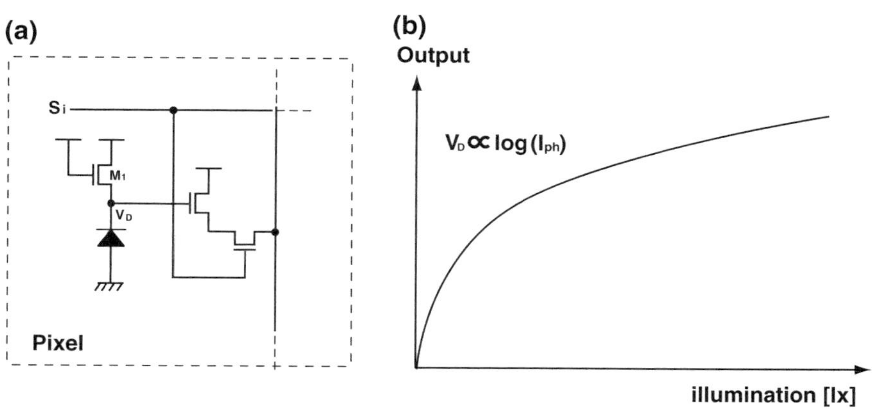

Fig. 10 Logarithmic photodetector: **a** Schematic diagram, **b** output characteristics

a high pixel FPN. A slow response time due to very low photocurrent causes an image lag, which is not acceptable for high-speed imaging. A nonlinear output also makes the signal processing like a color correction difficult.

2.1.2 Linear-Logarithmic Method

An image sensor showing both linear and logarithmic responses has been reported [16, 17], which can address the image lag of the logarithmic detector due to the slow response time. Figure 11a shows a pixel schematic of the linear and logarithmic sensor. The difference between the two technologies is that the gate of transistor M_1 can be set to an arbitrary voltage level compared to logarithmic pixel. Initially, the node V_D is reset through M_1 by applying the signal V_{trans}, and the signal accumulation is performed after the reset operation. The potential V_D changes linearly until V_D goes to $(V_{trans} - V_{th,M1})$ because M_1 is turned off. When V_D reaches $(V_{trans} - V_{th,M1})$, the transistor M_1 enters into subthreshold region. Therefore, the pixel's response becomes logarithmic according to the subthreshold operation of the transistor M_1, which results in the dynamic range expansion by preventing saturation of pixel output. The response of the linear-logarithmic sensors is shown in Fig. 11b.

The lin-log sensor can expand the dynamic range for a bright region maintaining the three-transistor properties in the low-light region. However, nonlinear image sensors still suffer from difficulties in reducing random noise such as dark current and reset noise. The other important issue of the lin-log sensor is that the transition of linear and logarithmic is strongly affected by the pixel to pixel variation of the transistor properties such as threshold voltage.

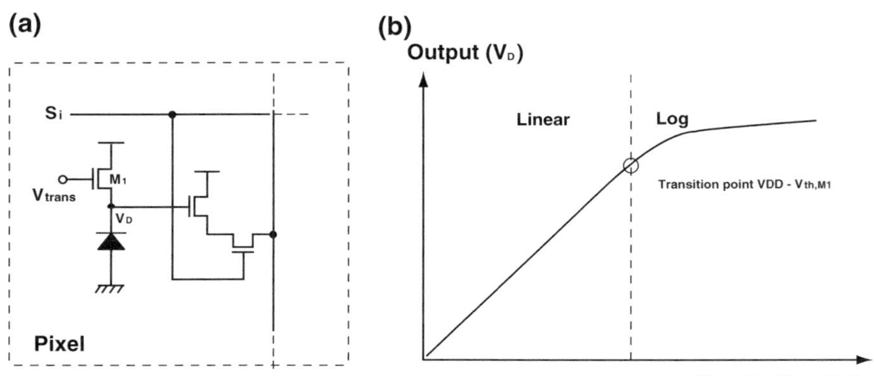

Fig. 11 Linear-logarithmic sensor: **a** Schematic diagram, **b** output characteristics

2.2 Linear Response Type

2.2.1 Well Capacity Adjusting Method

Controlling the well capacity of photodiode can be used to prevent the saturation of pixel output, which expands the dynamic range with partially linear output [18]. Figure 12a shows the equivalent circuit for the pixel with transistor M_4 for well capacity adjusting. The pixel structure is the same as four-transistor APS, but transistor M_4 is controlled with a certain reference voltage according to the illumination. In operation, the photodiode is reset by a lateral overflow gate M_3. Then the potential barrier between the charge sensing node, the cathode of photodiode and the drain is lowered by reset, and the lower barrier allows the excess charges to flow into the drain. After the reset, the potential of lateral overflow gate is raised to the given level. Signal charges proportional to the illumination begin to be accumulated in the photodiode. If the illumination is sufficiently high and the potential of photodiode reaches a certain voltage, the accumulated charges in the photodiode are drained through the lateral gate M_3. At the end of integration, pixel output is read out through the source follower M_1 by turning on the selection transistor M_2.

Example barrier curves and their corresponding output curves are shown in Fig. 12b. The barrier height $b(t)$ represents the charge capacity of the sensing node and is the product of the sensing-node capacitance and barrier potential, which is a function of the lateral overflow gate voltage $b(t)$. Therefore, the DR expansion ratio is controllable by $b(t)$.

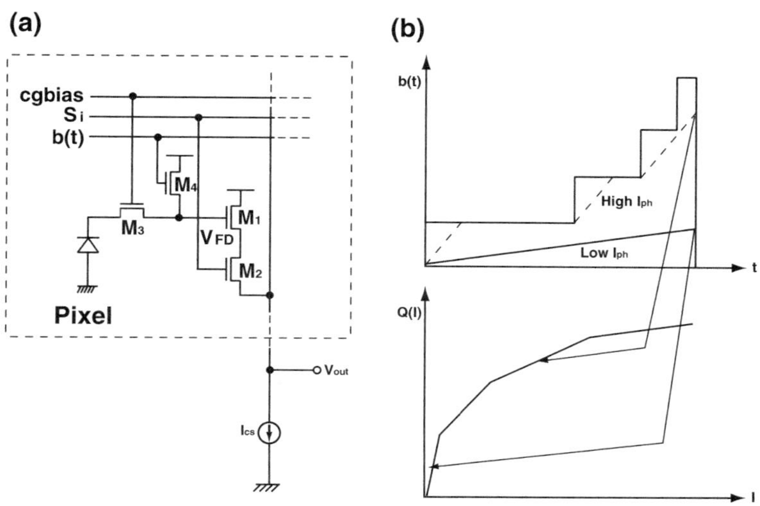

Fig. 12 Well capacity adjusting method: **a** Pixel schematic, **b** barrier curve $b(t)$ and compression curve $Q(t)$

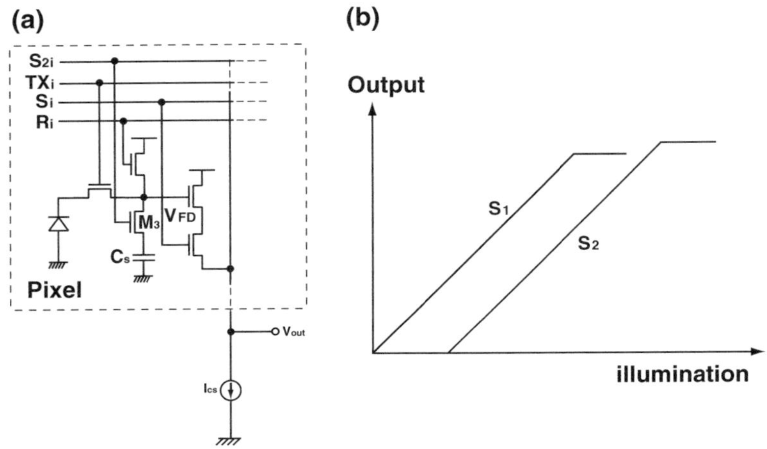

Fig. 13 DR expansion method using lateral overflow integration capacitor: **a** Pixel schematic, **b** output characteristic

2.2.2 Method to Use a Lateral Overflow Integration Capacitor

A DR expansion method using a lateral overflow capacitor has been proposed by Tohoku university [19]. This sensor shows a high DR of 100 dB without degradation of image quality at low-light conditions. Figure 13a shows a pixel schematic for the proposed sensor. A transistor M_3 and a capacitor C_S are the key elements newly added to the conventional four-transistor APS. Overflowed extra charges from the photodiode are accumulated in the capacitor C_S. The charge accumulation behavior before saturation is the same as the conventional four-transistor APS. When accumulated charges in the photodiode exceed the well capacity of the photodiode, overflown photo-electrons are now accumulated in the floating diffusion and the overflow capacitor C_S through M_3. The overflowed charges can be used for signal charges, which means expansion of full well capacity.

The leakage current at the floating diffusion and the overflow capacitor was minimized by dedicated pixel design and process control. Figure 13b shows the photoelectric conversion characteristics of the proposed DR extension method with linear responses.

2.2.3 Dual Sampling Method

A dual sampling method having the DR of 109 dB using two different exposures has been proposed by [20]. Two successive frames with different exposure times are output and synthesized as a WDR image. The dual exposure method can also support linear output which is preferred for signal processing in many applications.

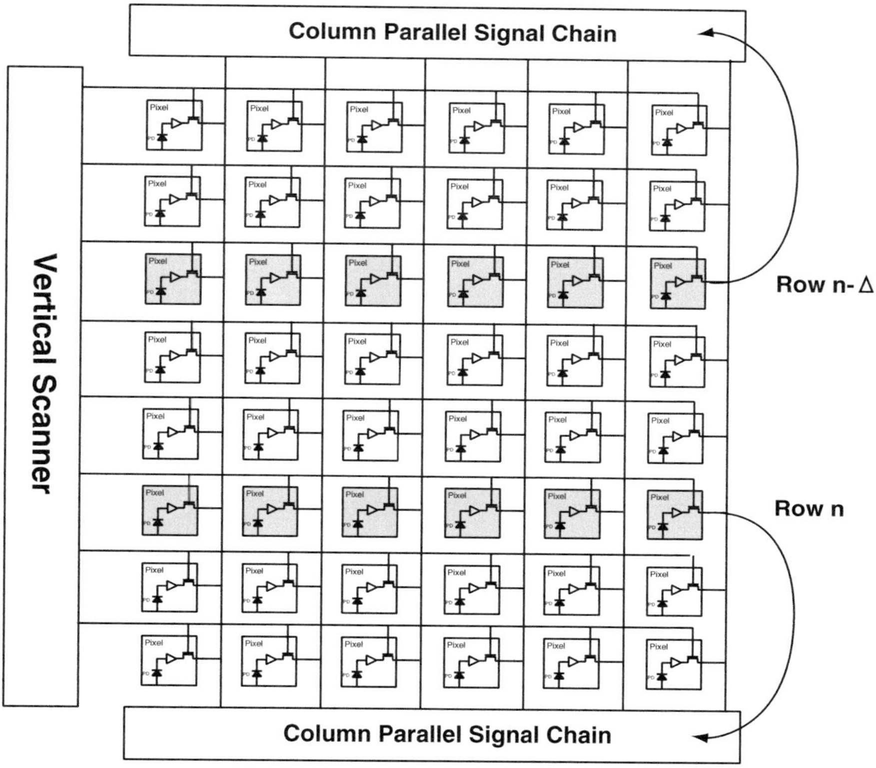

Fig. 14 Block diagram of the dual sampling method

The structure of the dual sampling method is shown in Fig. 14. In this architecture, two blocks of column circuits at the top and bottom of the pixel array are employed, which enable two row's simultaneous readout. The pixels for row n are selected and sampled into the capacitors of the column circuits on the lower side. At the same time, the pixels for row $(n - \Delta)$ are also selected and read out from the circuits on the upper side. The integration time of pixels being read out through the circuits on lower side is given by

$$T_{\text{long}} = (N - \Delta) \times T_{\text{row}} \ln \frac{f_2}{f_1} \tag{5}$$

and the integration time for the pixels being read out though the circuits on the upper side is

$$T_{\text{short}} = \Delta \times T_{\text{row}} \tag{6}$$

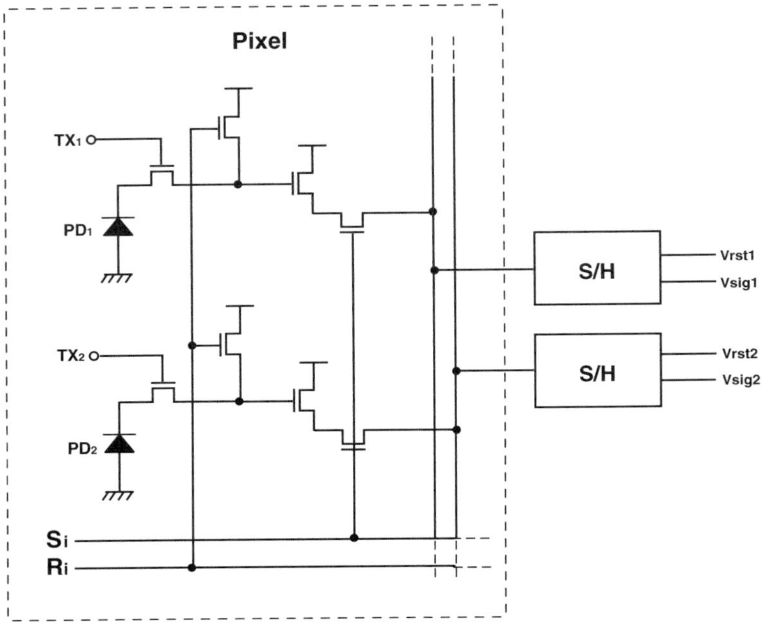

Fig. 15 Pixel structure using double sampling method

Therefore, two outputs which have different exposure times, long exposure T_{long} and short exposure T_{short} are obtained simultaneously. The DR extension is given by an exposure time ratio, $T_{\text{long}}/T_{\text{short}}$.

2.2.4 Double Sampling Method

A special pixel having two photodiodes in each pixel has been proposed by [21]. Figure 15 illustrates pixel structure of double sampling method, which contains signal processing circuits on both sides of a pixel array. The upper circuits cover normal light conditions from low light to medium light levels on PD_1 with a long exposure time. The lower circuits cover the pixel output PD_2 with short exposure time. The dynamic range is extended from the two pixel outputs.

2.2.5 Burst Readout Multiple Exposure

Another WDR image sensor using the burst readout multiple exposure (BROME) has been proposed by [22]. The WDR image sensor with the dynamic range of 117 dB and a linear response has been implemented.

Signal accumulation and readout timing for the sensor is shown in Fig. 16 [22]. In this diagram, it was assumed that the sensor has n vertical pixel arrays and four

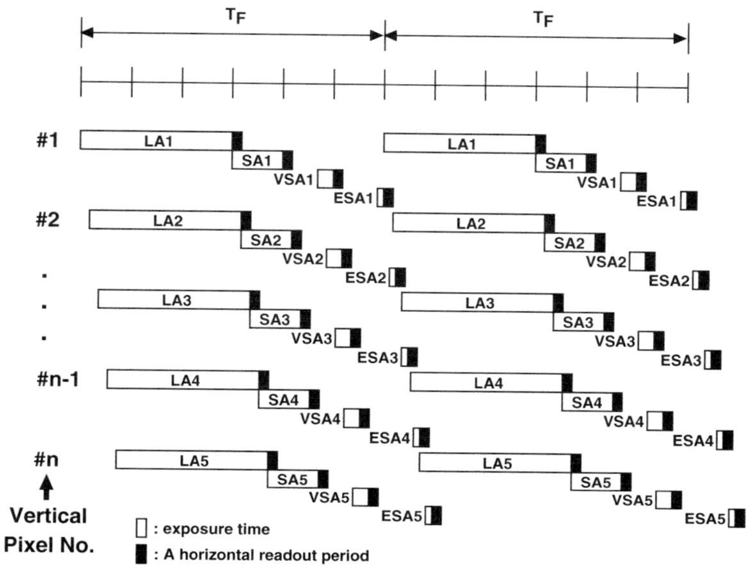

Fig. 16 Dynamic range expansion by multiple exposures

different exposure times. The long, short, very short, and extremely short accumulation are denoted as LA, SA, VSA, and ESA, respectively. A long and three short exposure signals are read out in a frame period, T_F. The LA signals occupy three time slots for a signal accumulation. SA signals are accumulated with overlapping the time slots for reading the LA signals. The VSA and ESA signals are accumulated with overlapping the time slots for reading the SA and VSA signals, respectively. In Fig. 16, the unit time for one slot is one-sixth of one frame period. If the image sensor operates at 30 frames per second, the unit time is $1/180\,\text{s} \cong 5.5\,\text{ms}$.

To expand the dynamic range further to higher illumination, an extremely short exposure time signal using inverted reset-signal sampling is proposed in [23]. Extremely high dynamic range of 153 dB with linear response was achieved using BROME and a low-noise circuit.

Acknowledgments This work was supported by the Center for Integrated Smart Sensors funded by the Ministry of Science, ICT & Future Planning as Global Frontier Project. (CISS-3-4)

References

1. Hynecek J (2002) High dynamic range active pixel CMOS image sensor and data processing system incorporating adaptive pixel reset, U.S. Patent, 0113886 A1
2. Meynants G, Dierickx B, Scheffer D (1998) CMOS active pixel image sensor with CCD performance. Proc SPIE 3410:68–76

3. Mendis SK, Kemeny SE, Gee RC, Pain B, Staller CO, Kim Q, Fossum ER (1997) CMOS Active pixel image sensors for highly integrated imaging systems. IEEE J Solid-State Circuits 32(2):187–197
4. Yasuda T, Hamamoto T, Aizawa K (2003) Adaptive-integration-time image sensor with real-time reconstruction function. IEEE Trans Electron Devices 50(1):111–120
5. Theuwissen AJP (1995) Solid-State Imaging with Charge-Coupled Devices. Kluwer Academic Publishers, Dordrecht
6. Hardy T, Murowinski R, Deen MJ (1998) Charge transfer efficiency in proton damaged CCD's. IEEE Trans Nuclear Sci 45(2):154–163
7. Yonemoto K, Sumi H (2000) A CMOS image sensor with a simple fixed-pattern-noise-reduction technology and a hole accumulation diode. IEEE J Solid-State Circuits 35(12):2038–2043
8. Hynecek J (2002) CDS noise reduction of partially reset charge-detection nodes. IEEE Trans Circuits Syst 49(3):276–280
9. Tian H, Gamal AE (2001) Analysis of 1/f noise in switched MOSFET circuits. IEEE Trans Circuits Syst 48(2):151–157
10. Kawai N, Kawahito S (2004) Noise Analysis of high-gain, low-noise column readout circuits for CMOS image sensors. IEEE Trans Electron Devices 51(2):185–194
11. Furumiya M, Ohkubo H, Muramatsu Y, Kurosawa S, Okamoto F, Fujimoto Y, Nakashiba Y (2001) High-sensitivity and no-crosstalk pixel technology for embedded CMOS image sensor. IEEE Trans Electron Devices 48(10):2221–2227
12. Tian H, Fowler B, Gamal AE (1999) Analysis of temporal noise in CMOS APS. Proceedings of SPIE, vol 3649. San Jose, pp 177–185
13. Tian H, Fowler B, Gamal AE (2001) Analysis of temporal noise in CMOS photodiode active pixel sensor. IEEE J Solid-State Circuits 36(1):92–101
14. Rodericks B, Hoffberg M (2002) Wide dynamic range digital imaging system and method. PCT Patent, WO 02/103391 A1
15. Kavadias S, Dierickx B, Scheffer D, Alaerts A, Uwaerts D, Bogaerts J (2000) A logarithmic response CMOS image sensor with on-chip calibration. IEEE J Solid-State Circuits 35 (8):1146–1152
16. Takanayagi I (2006) Wide dynamic range linear-and-log active pixel. U.S. Patent, 0036785
17. Hara K, Kubo H, Kimura M, Murao F, Komon S (2005) A Linear-logarithmic cmos sensor with offset calibration using an injected charge signal. In: Proceedings of ISSCC, pp 354–603
18. Decker S, McGrath RD, Brehmer K, Sodini CG (1998) A 256 × 256 CMOS Imaging array with wide dynamic range pixels and column-parallel digital output. IEEE J Solid-State Circuits 33:2081–2091
19. Sugawa S, Akahane N, Adachi S, Mori K, Ishiuchi T, Mizobuchi K (2005) A 100 dB dynamic range CMOS Image sensor using a lateral overflow integration capacitor. In: Proceedings of ISSCC, pp 352–353
20. Yadid-Pecht O, Fossum ER (1997) Wide intrascene dynamic range CMOS APS using dual sampling. IEEE Trans Electron Devices 44(10):1721–1723
21. Nakamura J (2002) Wide dynamic range pinned photodiode active pixel sensor (APS), U.S. Patent, 0096124 A1
22. Mase M, Kawahito S, Sasaki M, Wakamori Y, Furuta M (2005) A Wide dynamic range CMOS image sensor with multiple exposure-time signal outputs and 12-bit column-parallel cyclic A/D converters. IEEE J Solid-State Circuits 40(12):2787–2795
23. Park JH, Mase M, Kawahito S, Sasaki M, Wakamori Y, Ohta Y (2005) A 142 dB dynamic range CMOS image sensor with multiple exposure time signals. In: Proceedings of ASSCC, Taiwan, vol A2L-3, pp 85–88

Architectural Analysis of a Baseline ISP Pipeline

Hyun Sang Park

Abstract An ISP is an entity that performs various image-processing algorithms on a raw image from an image sensor. A number of functions are incorporated in an ISP, and they are combined together similarly but differently among ISP implementers. ISP functions are divided into pixel-based and frame-based ones, and are dedicated to one of three color domains in Bayer, RGB, or YCbCr. Although it is an essential component for a camera system, surprisingly, its architecture has not been analyzed in the context of standards. The purpose of this chapter is to remove ambiguity when analyzing an ISP architecture or designing a new ISP architecture. At the end of this chapter, a baseline ISP pipeline is presented, which is tentatively built to conform to the existing standards.

Keywords Image signal processor · Image pipeline · Image sensor · Bayer sensor

1 Introduction

The functions implemented in ISP can be categorized into two groups. The first includes pixel-based functions. It makes the result by utilizing an input pixel and its surrounding pixels. It is also regarded as a spatial filter because its output is generated by exploiting spatial information. The second contains frame-based functions. To obtain the processed result, these functions require the whole pixels of an image. Frame-based functions are further divided by how many images are exploited to get the outcome.

One is to refer to global features of the whole frame of a single image. The image quality of an image needs to be consistent over all portions of the image. The method for extending the dynamic range of an image can be included in this category. There are many other algorithms such as auto-white balance,

H.S. Park (✉)
Division of Electrical Electronics and Control Engineering, Kongju National University, Gongju, South Korea
e-mail: vandammm@kongju.ac.kr

C. Kyung (ed.), *Theory and Applications of Smart Cameras*,
KAIST Research Series, DOI 10.1007/978-94-017-9987-4_2

auto-exposure, contrast enhancement, which extract the global features from the given single image. The other functions that require a plural number of image frames often utilize temporal correlation among them. Some algorithms to reduce noise or distortion are included in this group. They analyze the temporal correlation between frames, and include following algorithms such as temporal noise reduction, rolling-shutter removal, image stabilization, and so on.

Frame-based functions are not handled in traditional ISPs except for auto-exposure control, auto-white balance, and auto focus (also known as 3A or 3-auto). For example, if noise is to be reduced by considering temporal correlation, at least two image frames have to be stored to check if it can be regarded as noise or not. Basically, an ISP has been developed to be embedded in an image sensor. Because of this requirement, it cannot work with functions requiring the frame memory. The 3A algorithm doesn't need the frame memory because the global features that 3A requires can be extracted while scanning the current frame. Although they are regarded as frame-based ones, they could be considered as basic components in the traditional ISP architecture since they do not need the frame-memory itself. In general an ISP can be implemented in three ways.

1.1 Embedded ISP in an Image Sensor

It is what is called the baseline ISP, which has a cascaded pipeline architecture composed of spatial filters and point functions. Allowed frame-based functions are limited to only 3A algorithms, which do not require any frame-memory.

1.2 Discrete ISP Package

In the early era of ISP commercialization, a baseline ISP itself was built solely as a discrete chip. These days it is often produced in a multi-chip package with a stacked SDRAM as frame memory. Because it embeds the frame memory inside, it can support frame-based functions such as image stabilization, temporal noise reduction, wide dynamic range, and so on. However, it still has difficulties in handling those algorithms derived from computer vision technology, which also utilize the images stored in the frame memory but require large number of floating-point operations and complicated control flow. Adopting power-consuming CPU and/or GPGPU is not considered yet.

1.3 Embedded ISP Inside an AP

There are powerful programing units like CPU/GPGPU inside an application processor (AP). Besides, the application processor provides abundant memory

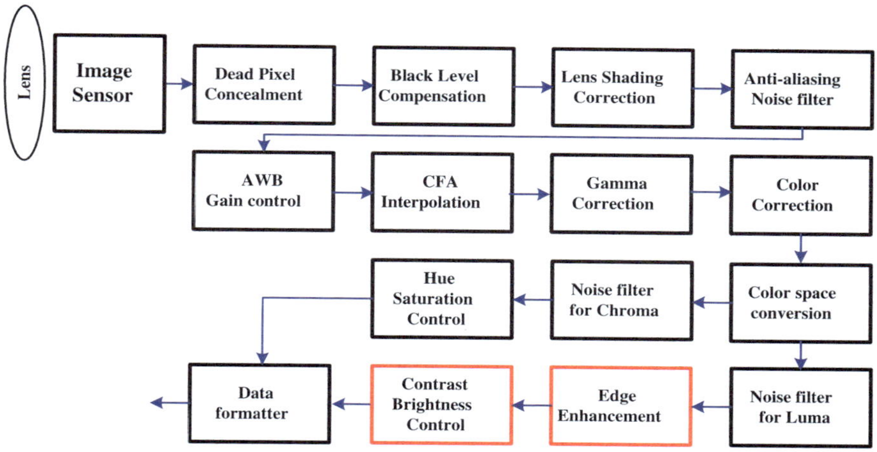

Fig. 1 The definitive form of a baseline ISP pipeline

space as well as bandwidth. So the pixel-based functions can be processed with a legacy baseline ISP, while the frame-based functions can be processed by programing GPU/GPGPU. This form of the ISP implementation consumes much energy since it uses the power-hungry memory device and the hot computing units. Nevertheless, it can provide the best quality of an image for end-user satisfaction.

The pipelined chain of an ISP is not standardized, such that each implementer has devised lots of very similar, yet different ISP pipelines. In this section, the baseline ISP in Fig. 1 will be discussed in the context of known standards.

2 Primary ISP Architecture for Bayer Image Sensors

The ISP itself is not a subject under standardization, but the standardization of digital video has been built continuously for a long time. Rec. ITU-R Rec. 601 [1] and Rec. ITU-R BT. 656 [2] (also known as CCIR601/656) constituted in 1982 claims the standardization of basic component of an ISP for the first time.

The camera module in Fig. 2 consists of an optical module, an image sensor, and an ISP. The ISP here contains three components: quantization, color space

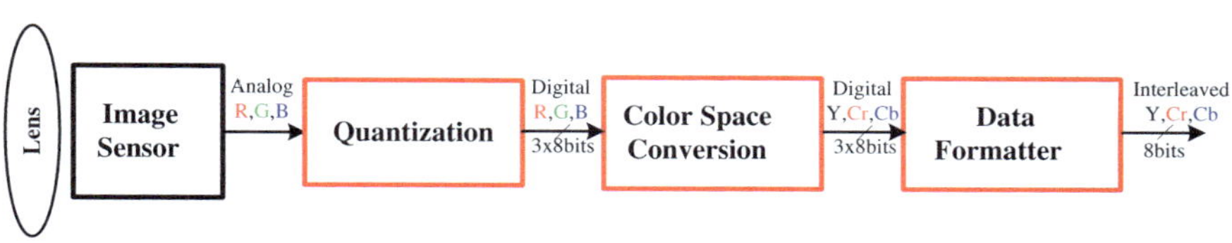

Fig. 2 Camera module architecture in Rec. ITU-R BT.601 and Rec. ITU-R BT.656

conversion, and data formatter. The image sensor is assumed to produce analog R, G, and B signals at every pixel position. In Rec. ITU-R BT.601, the first two functions are standardized, and in Rec. ITU-R BT.656 the last function is standardized.

The title of Rec. ITU-R BT.601 is "Studio encoding parameters of digital television for standard 4:3 and wide-screen 16:9 aspect ratios" and defines common regulations on digitization of digital video for SDTV (Standard Definition Television). Video in this standard has the resolution of 720 × 480 or 720 × 576 at the sampling frequency of 13.5 MHz. This recommendation standardizes how to obtain the corresponding digital video data. When analog R, G, and B signals—E_R, E_G, and R_B—of the 1.0 volt dynamic range are given, 8-bit digital RGB signals are quantized as below. They will have 219 values which reside between 16 and 235.

$$\begin{aligned} E_{R_D} &= \mathrm{int}(219E_R) + 16 \\ E_{G_D} &= \mathrm{int}(219E_G) + 16 \\ E_{B_D} &= \mathrm{int}(219E_B) + 16 \end{aligned} \tag{1}$$

Y, C_R, and C_B signals are calculated from these digital R, G, and B signals. The formula to convert R-G-B into Y-C_R-C_B is defined a little bit differently according to the recommendations. For example, Rec. ITU-R BT.709 [3] and Rec. ITU-R BT.2020 [4] specify the digital video format for HDTV (High Definition Television) and UDTV (Ultra Definition Television) in a very similar way to Rec. ITU-R BT.601. Although these recommendations standardize the digital video formats at difference resolutions, their color space conversion to the Y-C_B-C_R space is not the same. That is, there is no color compatibility between them. In case of inverse transformation from Y-C_R-C_B made by a different regulation to R-G-B, there may be some differences among reconstructed R-G-B data. So any ISP implementer should obey the formula specified in the appropriate recommendation. Equation (2) is what is recommended in Rec. ITU-R BT. 601. Each arithmetic operation in Eq. (2) is designed to be implemented by integer operations. Allowing the use of integer operations gives consistent calculation results among different implementations in hardware or software.

$$\begin{aligned} Y &= \frac{77}{256}E_{R_D} + \frac{150}{256}E_{G_D} + \frac{29}{256}E_{B_D} \\ C_R &= \frac{131}{256}E_{R_D} - \frac{110}{256}E_{G_D} - \frac{21}{256}E_{B_D} + 128 \\ C_B &= -\frac{44}{256}E_{R_D} - \frac{87}{256}E_{G_D} + \frac{131}{256}E_{B_D} + 128 \end{aligned} \tag{2}$$

In Rec. ITU-R BT.601, subsampling is performed horizontally with C_B and C_R components after color conversion. There exist some subsampling formats on Y-C_R-C_B signals, such as 4:4:4, 4:2:2, 4:1:1, or 4:2:0. These subsampling formats are available only in the Y-C_R-C_B color space, not for legacy RGB color spaces. The

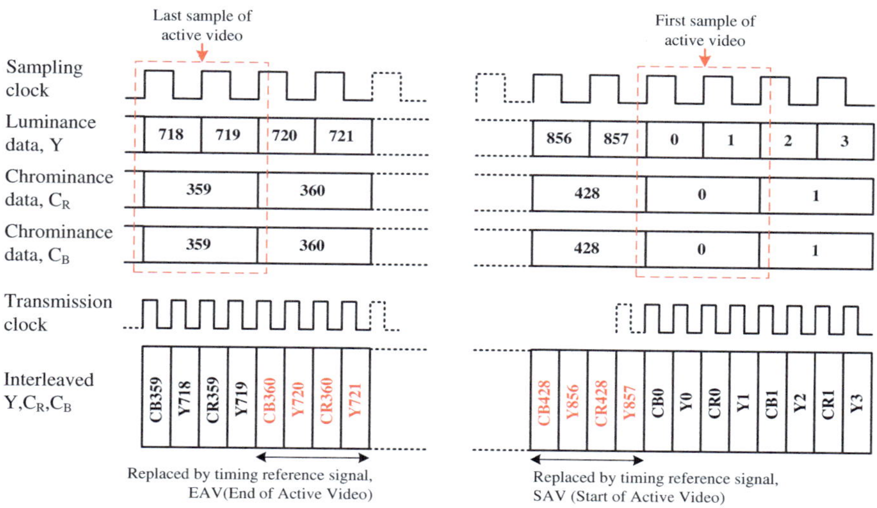

Fig. 3 Interleaved Y-C_R-C_B data format by Rec. ITU-R BT.656

subsampling on C_B-C_R components is desirable when effective data reduction is required without loss of visual quality degradation. There are large correlations between R, G, and B signals, but a few between C_R and C_B signals. Rec. ITU-R BT.601 only regulates 4:4:4 and 4:2:2 chroma subsampling formats. The 4:4:4 chroma subsampling format represents that there is indeed no subsampling. The 4:2:2 chroma subsampling format allows the subsampling on C_B and C_R chroma signals with 2:1 horizontally. The corresponding subsampled Y-C_R-C_B data are then interleaved into a single data stream according to Rec. ITU-R BT.656. The module to do subsampling and to interleave Y-C_R-C_B signals is Data Formatter in Fig. 2. The formatted data by Data Formatter will have the form like that in Fig. 3. In Rec. ITU-R BT.656, only the chroma 4:2:2 subsampling format is allowed. Thus the standardized digital video produced by conventional camera modules always supports the chroma 4:2:2 subsampling format.

Data formatter of an ISP needs the output speed to be twice as fast as any other part in the ISP, instead of using a number of data signals. An ISP usually has two clock domains. For example, a camera module made by Rec. ITU-R BT.601 and Rec. ITU-R BT.656 takes 13.5 MHz as the sampling frequency and 27.0 MHz as the output data frequency, respectively. The transferred signals through Rec. ITU-R BT.656 are only video, and no timing reference signals that define horizontal/vertical blanking periods are explicitly transferred. Instead those timing reference signals are derived from video data, where some reserved codewords are inserted at appropriate locations within the data stream.

In Fig. 3, a line of an image frame consists of 858 luminance (Y) data. Among them, the number of valid video data is 720. The interval of producing invalid data is called the horizontal blanking period. In Rec. ITU-R BT.656, four successive words just before and after the valid data are replaced by a reserved codeword sequence such that correct timing reference signals can be derived. The four codewords substituted at the end of a valid line are called EAV (End of Active Video) and those before the beginning of a valid line are called SAV (Start of Active Video). Each codeword in SAV or EAV has either 8-bit or 10-bit, but 8-bit is preferably used in industries. SAV and EAV have the sequence of 'FF-00-00-XY' in hexadecimal numbers.

The first three codewords constitute a synchronization code to inform the receiver of the existence of timing reference. Because the synchronization code is used to synchronize the communication between a transmitter and a receiver, the synchronization code itself cannot happen by chance in video data. Otherwise erroneous synchronization will happen, which will result in a failed reconstruction of an image. The emulation of the synchronization code will not be made in practice. If Rec. ITU-R BT.601 is used in producing the digital video data, no '00' or 'FF' is allowed to be generated. In practical implementation of an ISP, however, it is often necessary to consider at the transfer stage not to emulate the synchronization code because the ISP may use all 256 values that an 8-bit code can have. There are three timing information signals such as F, V, and H, where they are transferred with protection bits at the last codeword of SAV or EAV. Their bit positions and meanings are given in Table 1.

As described above, the simplest form of an ISP is composed of quantization, color space conversion from R-G-B to Y-C_R-C_B, and data formatter. All of these steps are standardized in Rec. ITU-R BT.601 and in Rec. ITU-R BT.656, respectively. Because the quantization step is mostly embedded in an image sensor, the minimum number of components for the simplest ISP is only two, and the corresponding ISP is shown in Fig. 4.

In Fig. 4, it assumes that the image sensor produces digital R, G, and B data for each pixel. Unfortunately, no image sensors produce the R, G, and B data altogether at the same pixel position, unlike display devices where three or four color sub-pixels exist within a pixel. Foveon [5] invented the image sensor that samples R, G, and B data altogether at any pixel position. However, the practical sensor

Table 1 Timing reference code configuration

Data bit number	First word (FF)	Second word (00)	Third word (00)	Fourth word (XY)
7 (MSB)	1	0	0	1
6	1	0	0	F
5	1	0	0	V
4	1	0	0	H
3	1	0	0	P3
2	1	0	0	P2
1	1	0	0	P1
0	1	0	0	P0

Table 2 Protection bits in SAV and EAV

F	V	H	P3	P2	P1	P0
0	0	0	0	0	0	0
0	0	1	1	1	0	1
0	1	0	1	0	1	1
0	1	1	0	1	1	0
1	0	0	0	1	1	1
1	0	1	1	0	1	0
1	1	0	1	1	0	0
1	1	1	0	0	0	1

F = 0 during field 1; 1 during field 2
V = 0 elsewhere; 1 during field blanking
H = 0 in SAV; 1 in EAV
P0, P1, P2, P3: protection bits (see Table 2)

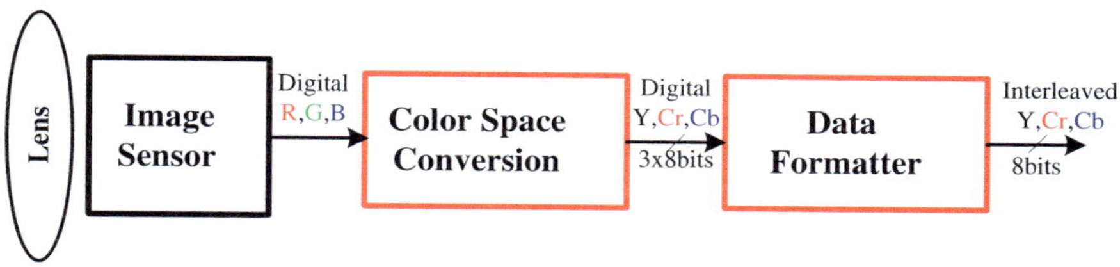

Fig. 4 Simplest ISP architecture

samples one of color components for a pixel as shown in Fig. 5 [6]. This differentiation between two light-related devices comes from the fact that the allowable pixel sizes they use are quite different.

Let's compare the size of two different optical devices shown in Fig. 6: a display panel and an image sensor supporting the same FHD (Full High Definition, 1920 × 1080) resolution, assuming the size of the display panel is 5 inches, and that of the image sensor is 1/3 inch. The display panel is easier to implement, compared to the image sensor because the effective pixel area of the display panel is $(3 \times 5)^2$ times as large as that of the image sensor. The pixel area of an image sensor needs to be as large as possible for improving SNR (Signal to Noise ratio). However, there is another constraint on the pixel size, which claims that an image sensor should be made as small as possible such that it can be packaged inside a compact smartphone of a small form factor. As the sensor shrinks, the SNR becomes lower. So there must be some compromise between the pixel size and the image sensor resolution. If sub-pixels constituting a pixel, e.g., *R*, *G*, *B* sub-pixels, are to be defined as in the display panel, the effective area of each sub-pixel will be much smaller. This is why sub-pixels of a pixel cannot be implemented on the same plane in the image sensor. Thus, the appropriate compromise between the high SNR and the small form factor is to adapt the spatial subsampling strategy such as Bayer array.

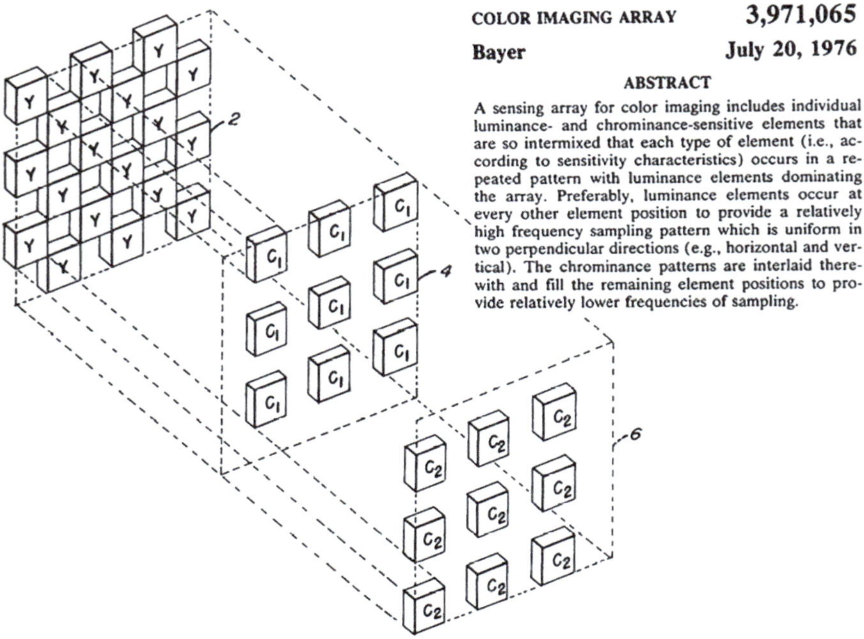

Fig. 5 Color imaging array by Bayer

Fig. 6 Typical color filter pattern in image sensors (*left*) and display panels (*right*)

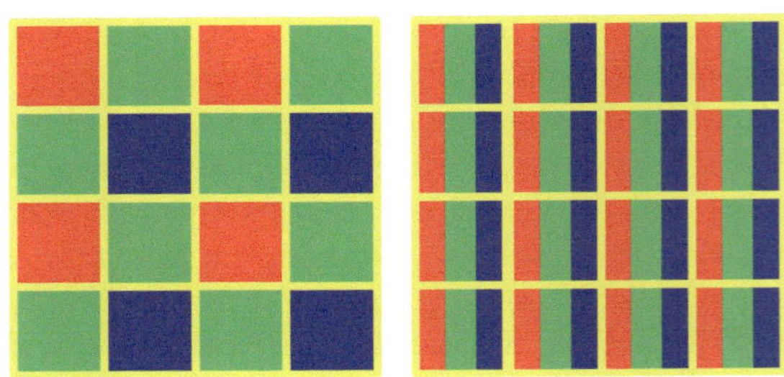

A sensor array made by spatial color subsampling is called color filter array (CFA) or Bayer array. According to the patent by Bayer, only the principle of color subsampling is provided, and which color is sampled is not explained. Thus, colors can be sampled in a variety of ways, and these combinations of sampling are also called Bayer pattern. Some typical examples of Bayer patterns are shown in Fig. 7.

Image sensors with Bayer pattern have high sensitivity with low implementation cost, but the process of restoring deficient color components is additionally required. This process is called interpolation or demosaicing. There are lots of ways [7] in demosaicing color filter arrays. One of the simplest is the 1-st order interpolation, i.e., bilinear interpolation. Among Bayer patterns mentioned in Fig. 7, the

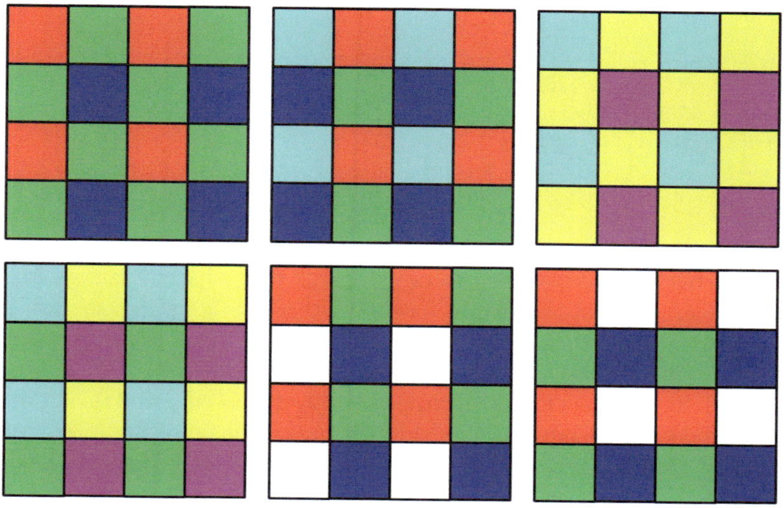

Fig. 7 Bayer patterns

RGB pattern is widely used since it allows better color reproduction. So, the subsidiaries of bilinear interpolation are to be described with this pattern. In Fig. 8, the shaded pixels represent practically sampled ones in the sensor array, while the other unshaded pixels represent those to be restored by bilinear interpolation in Eq. (3).

$$\begin{aligned} R11 &= R11 \\ R12 &= \frac{R11 + R13}{2} \\ R21 &= \frac{R11 + R31}{2} \\ R22 &= \frac{R11 + R13 + R31 + R33}{4} \end{aligned} \tag{3a}$$

$$\begin{aligned} G22 &= \frac{G12 + G21 + G23 + G32}{4} \\ G23 &= G23 \\ G32 &= G32 \\ G33 &= \frac{G23 + G32 + G34 + G43}{4} \end{aligned} \tag{3b}$$

$$
\begin{aligned}
B22 &= B22 \\
B23 &= \frac{B22 + B24}{2} \\
B32 &= \frac{B22 + B42}{2} \\
B33 &= \frac{B22 + B24 + B42 + B44}{4}
\end{aligned}
\tag{3c}
$$

Because bilinear interpolation averages two or four adjacent data of the same color attribute, the interpolated values may be what do not exist in the real scene. Since different interpolation equations are used for color components, the associated color built by combining them can show undesirable color where there are edges with high gradient. These unwanted color artifacts are called pseudo-color or color noise. Figure 9 shows artifacts produced by bilinear interpolation. The periodic noise pattern, which is called zipper noise (or maze noise), is shown with additive pseudo-color. The main role of color interpolation is to suppress such pseudo-color and zipper noise. The zipper noise can be reduced greatly by interpolating pixels along the distinct edges as shown in Fig. 9c.

Edge-directed interpolation is an adaptive approach, where the adjacent pixels around each pixel are analyzed to decide if there exists a horizontal or vertical edge. There are lots of ways to decide the direction of edges. In [19], the simplest form of edge direction detection is presented. Let $G22$ be interpolated using its neighboring

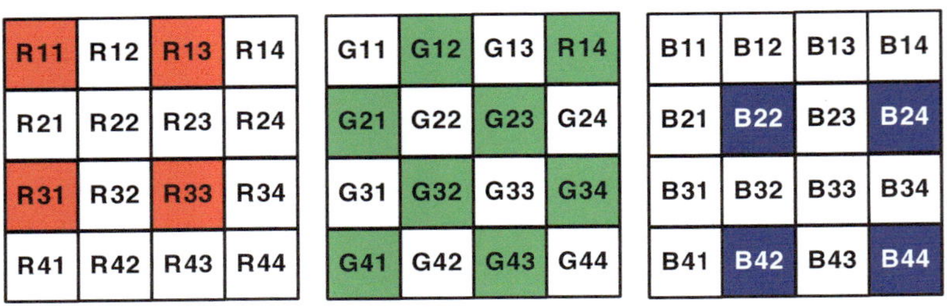

Fig. 8 Pixels to be interpolated by bilinear interpolation

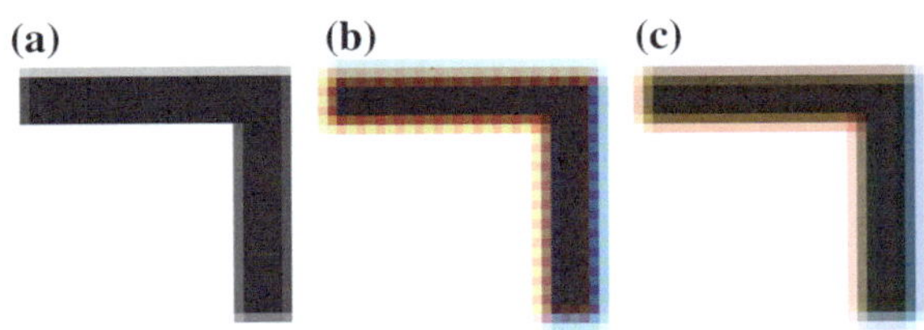

Fig. 9 Artifacts by color filter interpolation. **a** Original; **b** bilinear interpolation; **c** edge-directed interpolation [19]

G pixels in Fig. 8. The horizontal and vertical gradients are defined as $\Delta H = |G21-G23|$ and $\Delta V = |G12-G32|$ respectively. If $\Delta H > \Delta V$, the edge direction is vertical, then $G22 = (G12 + G32) \gg 1$. If $\Delta H < \Delta V$, the edge direction is horizontal, then $G22 = (G21 + G23) \gg 1$. Otherwise, $G22 = (G12 + G21 + G23 + G32) \gg 2$. In this way, the G image is interpolated first, and then the other color planes are acquired by utilizing the G image.

Impulsive noise is easy to remove by legacy noise reduction filters that utilize median filter. The basic assumption about noise is that noise is statistically independent and has very high-frequency components. Zipper noise looks like high-frequency noise, but it is difficult to remove by a legacy noise reduction filter because its frequency components are in the mid-to-high ranges. This is a contradiction to the basic assumption on noise. In Fig. 10 filtering results are presented by applying median filter and mean filter to remove zipper noise. The results show that the zipper noise is very difficult to remove by basic noise reduction tools. Thus, it is desirable to suppress the zipper noise in the interpolation stage instead of using noise reduction filter after color interpolation. Besides, an additional filter for removing pseudo-color is also necessary because it is hard to remove pseudo-color only with interpolation. Figure 9c also shows pseudo-colors along the edges after edge-directed interpolation

Figure 11 is the block diagram of an ISP evolved to compensate for artifacts raised by using a Bayer sensor. Anti-aliasing filter means a low-pass filter adapted before the color sampling to avoid aliasing. The ideal anti-aliasing filter must be an optical low-pass filter (OLPF) because the signals before the spatial subsampling are purely optical. However, OLPF cannot be considered here because it must be considered during the camera module design stage. Nevertheless, the first function of an ISP needs to be a noise reduction filter in the Bayer domain. The purpose of placing the noise filter here is not to prevent aliasing, but to prevent noise propagation through color interpolation. Anyway, this function is often called anti-aliasing noise filter for convenience. By adopting a cost-effective Bayer sensor, thus, an ISP should add following functions as shown in Fig. 11. Among them, the CFA interpolation is mandatory and all noise filters are optional.

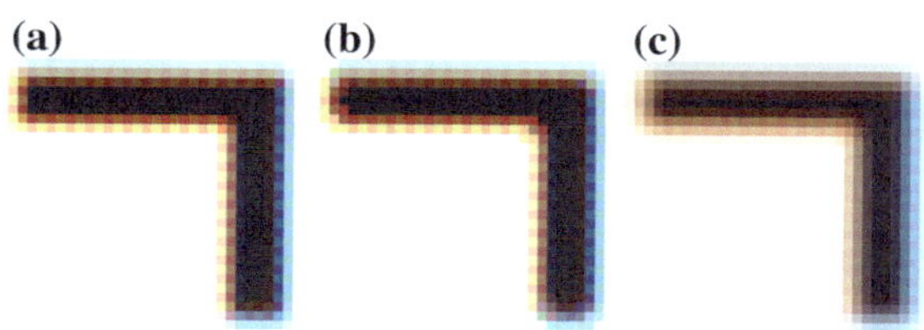

Fig. 10 Applying conventional low-pass filter to reduce zipper noise. **a** zipper noise; **b** median filtering; **c** mean filter

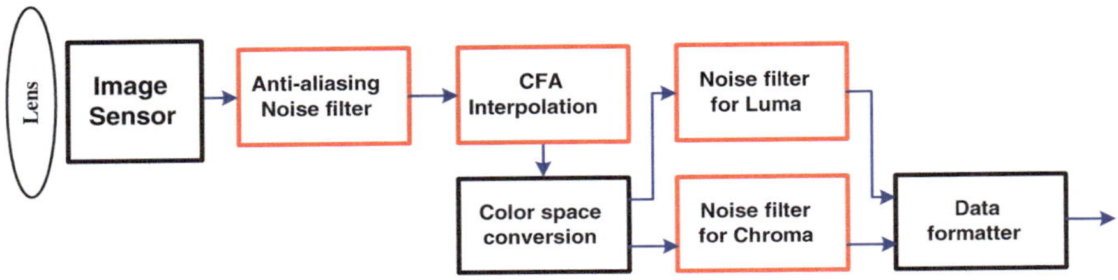

Fig. 11 ISP architecture to recover artifacts from a Bayer image sensor

2.1 Anti-aliasing Noise Filter

Noise needs to be removed in the Bayer domain. Salt-and-pepper noise produced during the manufacturing of image sensor has to be removed before color interpolation. Otherwise the noise will be expanded through color interpolation kernel.

2.2 Color Filter Array Interpolation

This is the process to restore the original color components from the sampled ones. It results in zipper noise and pseudo-color. The zipper noise can be suppressed considering edge direction during color interpolation process.

2.3 Noise Filter for Luma

In an anti-aliasing noise filter, it is not possible to exploit correlation with the adjacent pixels because they are of different color attributes. After interpolation, it is easier to remove Gaussian noise by considering correlation with adjacent data. This noise filter is a legacy noise filter [8] that has been developed for a long time.

2.4 Noise Filter for Chrominance: C_B and C_R

This is a filter for removing pseudo-color caused by subsampling and interpolation process. Because human eyes are very sensitive to rapid color changes, it is necessary to build a natural image by suppressing excessive color changes.

3 ISP Architecture for Color Reproduction

The process for restoring 'natural' color is necessary because the response of silicon to light is quite different from that of human eyes. Color is the response of light receptors of the human eyes to light spectrum. In the retina where there are rods and cones, rods sense brightness and cones sense chromaticity, respectively. Rods are extremely sensitive to light, and can be triggered by as few as six photons [9]. At very low light conditions, visual experience is solely decided by rods. Cones require significantly brighter light than rods. There are three different types of cones, distinguished by their response pattern with different wavelengths of light. Colors can be defined and quantified by the degree with which these cells are stimulated.

A color space is a 3-dimensional representation system into which a perceived color is translated [10]. The whole colors are represented by three-dimensional coordinates in a color space. There are many color spaces such as CIERGB, CIEXYZ, CIELAB, CIELUV, and so on. An RGB color space [11] is any additive color space based on the RGB color model. The most popular RGB color space is sRGB [12], which is used in consumer electronics including digital cameras, video cameras, televisions, projectors, and computer monitors. The RGB color space is defined in Rec. ITU-R. BT. 709.

A particular RGB color space is defined by the three chromaticities of the red, green, and blue additive primaries, and can produce any chromaticity inside the triangle whose vertices are defined by those primary colors. The primary colors are specified with reference to their corresponding chromaticity coordinates (x, y) in the CIE 1931 color space [13]. To completely specify an RGB color space, a white point and a gamma correction curve need to be additionally defined. In Table 3, the three primary colors and white points for popular RGB color spaces are summarized [11].

It should be noted that a gamma correction curve is mandatorily included for specifying a color space. Our "nonlinear" eyes do not perceive light like "linear" image sensors. They are more sensitive to changes in dark tones, and less in bright

Table 3 RGB color space parameters

Color space	Gamut	White point	Primaries					
			Red		Green		Blue	
			x	y	x	y	x	y
sRGB, HDTV	CRT	D65	0.64	0.33	0.30	0.60	0.15	0.06
PAL/SECAM	CRT	D65	0.64	0.33	0.29	0.60	0.15	0.06
NTSC(1987)	CRT	D65	0.63	0.34	0.31	0.595	0.155	0.07
UHDTV	Wide	D65	0.708	0.292	0.170	0.797	0.131	0.046
CIE(1931) RGB	Wide	E	0.7347	0.2653	0.2738	0.7174	0.1666	0.0089

tones, compared to silicon image sensor. It is because human eyes have evolved to enable our vision system to operate over a wide range of luminance. Gamma correction or gamma encoding is the name of a nonlinear operation used to code and decode luminance or tri-stimulus values in image sensors or display systems. Gamma correction is defined by the following power-law expression:

$$V_o = V_i^{\gamma} \tag{4}$$

The input and output values are nonnegative real numbers and are typically in the range of [0,1]. A gamma value which is smaller than 1 (i.e. $\gamma < 1$) is called an encoding gamma, and is used to compress the dynamic range of input values. The nonlinear characteristics of the human eyes to the brightness change can be observed from the exemplary patches in Fig. 12. Figure 12 shows what happens after quantizing continuous tones in an explicitly linear way or a perceptually linear way. Figure 12b shows quantizing into 32 levels by uniform quantization step and Fig. 12c by nonlinear quantization step based on a gamma curve. The quantization step size in Fig. 12c is numerically nonlinear but is perceived linear to the human eyes, while the quantization step size in Fig. 12b is arithmetically linear but looks nonlinear. Thus the true linear response of an image sensor should be perceived linear for human eyes. The nonlinear tone mapping process for human eyes is called gamma correction, which is included in defining a color space. The definition of gamma curve in the sRGB color space is like Eq. (5) below.

$$V_o = \begin{cases} 1.099V_i^{0.45} - 0.099, & 0.018 \leq V_i \leq 1 \\ 4.500V_i, & 0 \leq V_i < 0.018 \end{cases} \tag{5}$$

To support a particular RGB color space, both tone mapping and color mapping are required. The former is gamma correction and the latter is color correction. Gamma correction is nonlinear but color correction is linear, which are generally implemented by a 3 × 3 matrix multiplication. In color correction, the result of color mapping should not be affected by the brightness level of the captured scene. To

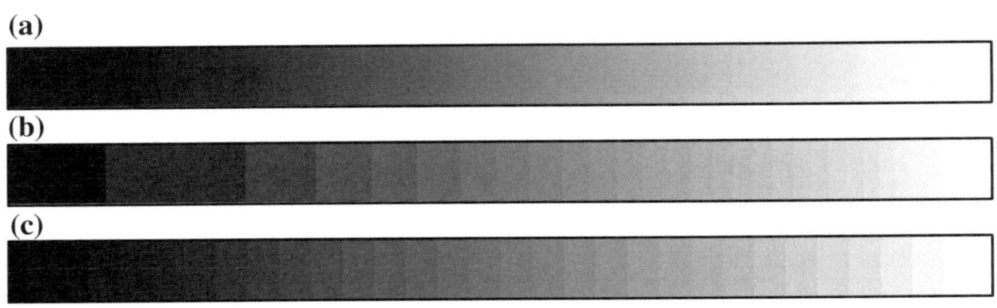

Fig. 12 Linear and nonlinear quantization of continuous tones. **a** Continuous tones from 0 to 1023; **b** linearly quantized tones into 32-levels; **c** nonlinearly quantized tones into 32-levels according to a gamma curve

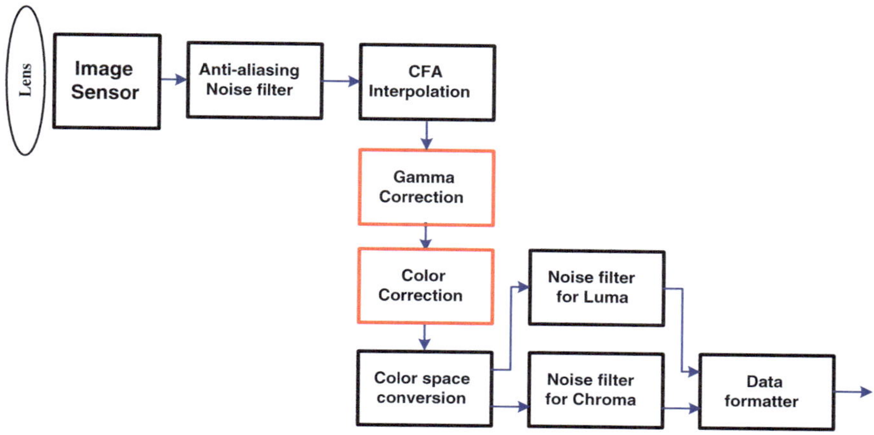

Fig. 13 ISP architecture to support a particular RGB color space

maintain consistency of color correction regardless of brightness, the color correction process needs to be linear. An ISP pipeline containing these two functions to support a particular color space is depicted in Fig. 13.

There is no restriction as to where stage gamma correction is placed. Gamma correction can be located before color interpolation or after color correction. It is also possible to place it even after color space conversion. The purpose of doing gamma correction is a nonlinear tone mapping. As long as this purpose is achieved efficiently, the place of performing gamma correction in the ISP pipeline is not so important. In reality, many ISP implementers do not use gamma correction in the same way. If efficient hardware implementation is pursued, implementing gamma correction in the Bayer domain or in the Y-C_B-C_R domain may be more efficient than in the RGB domain. Figure 14 shows two modified ISP chains with different location for gamma correction.

A white point is also included in the definition of an RGB color space. It is used to standardize the light spectrum and is abbreviated as D65 or E as tabulated in Table 3. The spectrum of a standard illuminant can be converted into tri-stimulus values by integrating it over all wavelength spectrums. The set of resultant three tri-stimulus coordinates of an illuminant is called a white point. CIE Standard Illuminant D65 [14] is a commonly used standard illuminant defined by the CIE. It describes standard illumination conditions at open-air in different parts of the world. D65 is intended to represent average daylight, and has a corresponding color temperature of approximately 6500 K. The power spectrum of illuminant D65 is shown in Fig. 15. CIE standard illuminant D65 should be used in all colorimetric calculations requiring representative daylight, unless there are specific reasons for using different illuminant. Illuminant E [15] is an equal-radiator; it has a constant distribution inside the visible spectrum. That is, it is a theoretical illuminant that gives equal weight to all wavelengths.

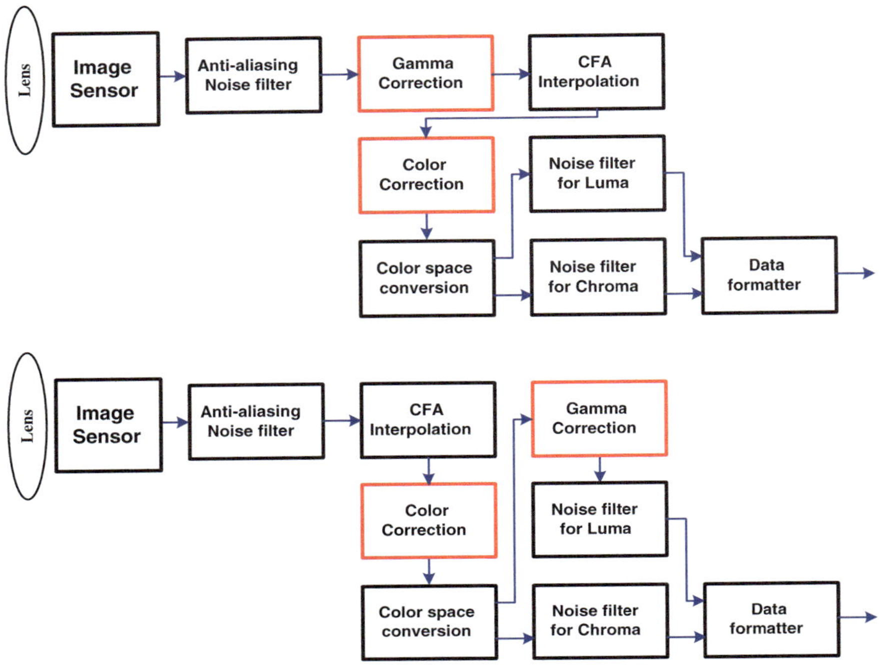

Fig. 14 ISP architecture variants for gamma correction. **a** ISP architecture variant with gamma correction at the Bayer domain; **b** ISP architecture variant with gamma correction at the $Y\text{-}C_B\text{-}C_R$ domain

Fig. 15 Spectral power distribution of Illuminant D65

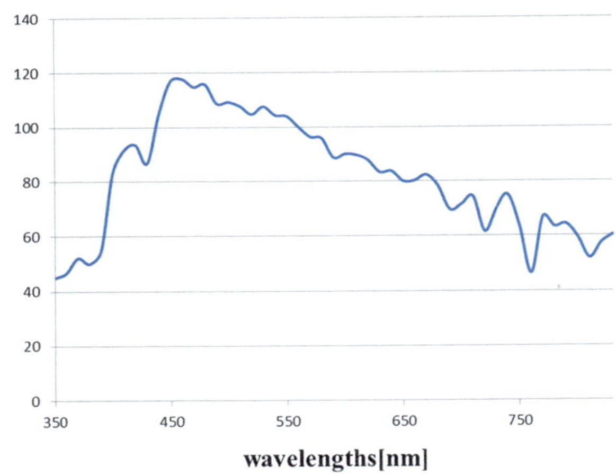

Color correction is performed with reference to the illumination spectrum of D65. If the illumination spectrum is different from D65, the chromaticity of the same object in the same scene will be perceived different in colors. Thus the chromaticity for the current light spectrum has to be corrected to be perceived similar to that of D65, since the chromaticity under D65 is most natural to average people. The attribute of light source is numerically characterized by the color temperature. Thus, the current color temperature should be changed to match D65. This process is called AWB (Auto-White Balance) and it is to compensate for the color distortion caused by the light spectrum different from D65. The key technology of AWB is to measure the color temperature of the current light source. To do this, achromatic-colored regions in the scene are used to estimate the color temperature because the color there reflects the color temperature of the light source. Gray or white regions are typical achromatic-colored regions. Achromatic color region is where the ratios between R, G, and B components are identical. Thus, the AWB process is modeled by Eq. (6), where average values of R, G, and B components in achromatic-colored region are denoted as $\bar{R}, \bar{G}, \bar{B}$ respectively. It is desirable for AWB to be performed before color correction.

$$\begin{bmatrix} R' \\ G' \\ B' \end{bmatrix} = \begin{bmatrix} \bar{G}/\bar{R} & 0 & 0 \\ 0 & 1 & 0 \\ 0 & 0 & \bar{G}/\bar{B} \end{bmatrix} \begin{bmatrix} R \\ G \\ B \end{bmatrix} \tag{6}$$

However, locating achromatic-colored region is almost impossible in practice; even human eyes sometimes cannot identify it from the natural scene. However, it is possible to measure the color variations of achromatic-colored region when the color temperature of the ambient light is changed. Such color variation is confined in a small area of the color gamut. Such area can be identified experimentally in the C_B-C_R plane by plotting the C_B-C_R components of achromatic-colored regions for all allowable color temperatures, as shown in Fig. 16. Then we can find connected regions where at least one achromatic-colored pixel exists. Among them, we can assume that there exists one region which reflects the current ambient color temperature. In this way, we can estimate the ambient color temperature. There are many heuristic ways to decide which area gives a good estimate for the ambient

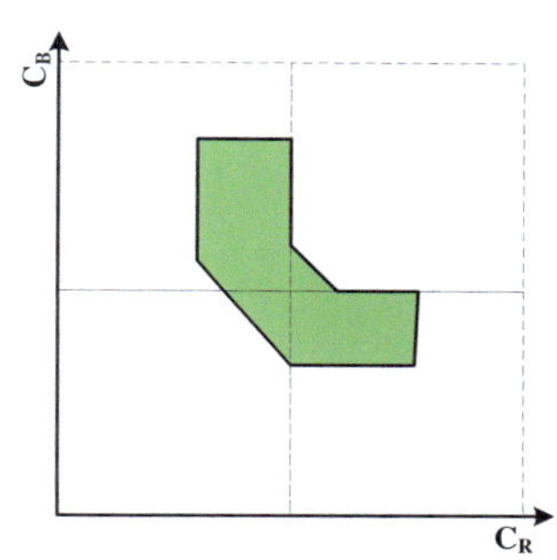

Fig. 16 Determination of the chrominance variation of achromatic-colored regions in the C_B-C_R plane

color temperature. Besides, instead of using C_B/C_R components, other terms can be used such as G-R and G-B, G/B and G/R, and so on.

Chromaticity is an objective specification of a color regardless of its brightness, and is further represented by hue and saturation. The white point is a neutral reference, which is characterized by chromaticity. All other chromaticities are defined with respect to the white point using polar coordinates (an angle and the distance from the origin). Hue is "the degree to which a stimulus can be described as similar to or different from stimuli that are described as red, green, and blue" [16].

HSL (Hue-Saturation-Lightness) and HSV (Hue-Saturation-Value) are the two most common cylindrical-coordinate representations of points in an RGB color space [17]. Figure 17 shows the hue from 0° to 360°. In case of emphasizing or deemphasizing particular color, we first find the hue corresponding to that color and then emphasize or deemphasize all R, G, and B values that have the same hue. In case of adjusting color of the whole image consistently, it is done by rotating the hue of each RGB data around the white point as much as needed.

Calculating the hue from R, G, and B data requires very complicated operation. So the Y-C_R-C_B color space can be regarded as a cost-effective substitute for hue. Mapping all colors in the C_R-C_B plane is shown in Fig. 18. Hue control in the C_R-C_B plane can be performed by Eq. (7a). The constant 128 indicates that the values of Y-C_R-C_B are in 8-bit, which is replaced by 512 when using 10-bit data. Saturation control is the same as amplifying the C_B and C_R components according to Eq. (7b). It is performed by Eq. (7c) when both hue and saturation controls are conducted simultaneously.

$$\begin{bmatrix} C'_B - 128 \\ C'_R - 128 \end{bmatrix} = \begin{bmatrix} \cos\theta & -\sin\theta \\ \sin\theta & \cos\theta \end{bmatrix} \begin{bmatrix} C_B - 128 \\ C_R - 128 \end{bmatrix} \tag{7a}$$

$$\begin{bmatrix} C'_B - 128 \\ C'_R - 128 \end{bmatrix} = \begin{bmatrix} S_b & 0 \\ 0 & S_r \end{bmatrix} \begin{bmatrix} C_B - 128 \\ C_R - 128 \end{bmatrix} \tag{7b}$$

$$\begin{aligned} \begin{bmatrix} C'_B - 128 \\ C'_R - 128 \end{bmatrix} &= \begin{bmatrix} S_b & 0 \\ 0 & S_r \end{bmatrix} \begin{bmatrix} \cos\theta & -\sin\theta \\ \sin\theta & \cos\theta \end{bmatrix} \begin{bmatrix} C_B - 128 \\ C_R - 128 \end{bmatrix} \\ &= \begin{bmatrix} \cos\theta & -\sin\theta \\ \sin\theta & \cos\theta \end{bmatrix} \begin{bmatrix} S_b & 0 \\ 0 & S_r \end{bmatrix} \begin{bmatrix} C_B - 128 \\ C_R - 128 \end{bmatrix} \\ &= \begin{bmatrix} S_b\cos\theta & -S_b\sin\theta \\ S_r\sin\theta & S_r\cos\theta \end{bmatrix} \begin{bmatrix} C_B - 128 \\ C_R - 128 \end{bmatrix} \end{aligned} \tag{7c}$$

Various functions are included in an ISP for reproducing correct colors that human eyes perceive, and are performed in different color domains. AWB is

Fig. 17 Hue in the HSB/HSL encodings of RGB

Fig. 18 Color distribution in the C_R-C_B plane

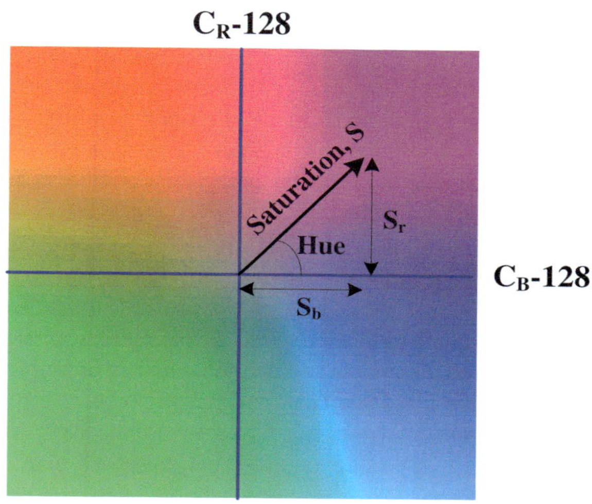

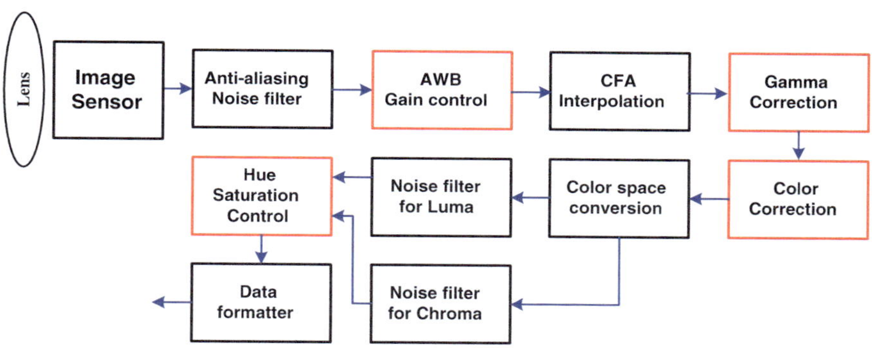

Fig. 19 ISP architecture for color reproduction

performed usually in the Bayer domain, both gamma correction and color correction are done in the RGB domain, and the hue/saturation control is conducted in the Y-C_R-C_B domain (See Fig. 19).

4 ISP Architecture with Pre-/Post-processing

Some additional pre-/post-processing functions are added to the baseline ISP pipeline described above. The purpose of pre-processing is to compensate for the sensor or camera distortions such that robust images can be acquired through a legacy ISP pipeline. The role of post-processing is to give a better visual quality from the standpoint of human visual system.

An image sensor has permanent bright or dark pixels due to physical defects. They are called dead or defective pixels, and the function to remove them is called DPC (Deal Pixel Concealment). Dead pixels are far brighter and much darker than their neighbors and generate salt-and-pepper noise. They can be easily removed by using a median filter, which always leads to a blurred image. A special noise filter has been developed to effectively remove the salt-and-pepper noise. This filter detects the dead pixels in real time and corrects them by replacing them with neighbor pixel data. Another method is to correct predefined dead pixels whose locations are searched and stored in the memory in advance. This method is free from the risk of blurring by a legacy noise filter because it conceals only predetermined defect pixels. The more coordinates of defective pixels are stored, the more high-cost memory is consumed. So, an appropriate memory storage should be determined in terms of cost and performance.

Sensor response does not have perfect linearity. Each pixel of an image sensor is a capacitive photodiode, and the charge in each pixel is discharged according to the incident photons. The discharged charge is sampled in voltage, and is regarded as a pixel value. Naturally it is not possible to detect no-light condition because the photodiode is always discharged by the reverse bias current, even in no-light condition. To detect sensor response corresponding to no-light, any image sensor has the dedicated sensor region called optical black area. The optical black area has the same structure as that of normal pixels, but it is made intentionally not to be exposed to light by covering photo-diodes with metal. Thus, it is possible to estimate the sensor response at no-light condition. Because there are R, G, and B pixels in the optical black area, it is possible to have sensor responses to '0' in no-light condition if their averaged values in the optical black area are subtracted from the sensor output appropriately. The function to do this is called BLC (Black Level Compensation) and is implemented by using Eq. (8), where OB_R, OB_G, and OB_B are the average values of red, green, and blue pixels in the optical black area, respectively. BLC should be the function to operate at the earliest stage in an ISP pipeline because only this function can make the sensor response linear.

$$\begin{bmatrix} R' \\ G' \\ B' \end{bmatrix} = \begin{bmatrix} R \\ G \\ B \end{bmatrix} - \begin{bmatrix} OB_R \\ OB_G \\ OB_B \end{bmatrix} \tag{8}$$

The magnitude and brightness of each pixel will have linearity after BLC. However, the linear slope of each pixel is not constant but varies randomly

Fig. 20 Flat-field image without lens-shading correction

according to its spatial position. The image is brightest in the center of optical axis and becomes monotonically darker as one goes to the edge of the field-of-view. The shading might be caused by nonuniform illumination or nonuniform camera sensitivity. In general this shading effect is mainly due to a lens system, and is called lens-shading distortion. Figure 20 shows a lens-shading image which is an originally flat-field image having a constant value all over the plane.

LSC (Lens-Shading Correction) is the process to compensate for the disparity of linear gain of each pixel due to lens shading, such that all pixels can have the same light-to-voltage gain regardless of their locations in the sensor array. The simplest and robust solution for LSC is to compensate for shading by the correction gain, which was estimated for each pixel in advance and then stored in the memory. This method is called FFC (Flat Field Compensation) [20]. FFC consists of two numbers for each pixel, the pixel's gain and its dark current. The corrected image $C(x, y)$ at the pixel location (x, y) is obtained by Eq. (9).

$$C(x,y) = \frac{R(x,y) - D(x,y)}{F(x,y) - D(x,y)} \cdot m \tag{9}$$

where, $D(x, y)$ is a dark frame, $R(x, y)$ is a raw image, $F(x, y)$ is a flat-field image, and m is the average value of $F(x, y)–D(x, y)$. The dark frame and the flat field are captured experimentally by taking the flat-field scenes in a very dark lighting condition and in a marginally unsaturated lighting condition. FFC is not appropriate in a baseline ISP because it requires sufficiently large memory to store the entire image. Instead, an appropriate mathematical model for the LSC gain map is used.

Noise reduction is performed after consistent linearity is obtained for the whole pixels. Noise sources in an image are various. The need for noise reduction is increasing as the resolution of image sensor is increased with the pixel size being drastically reduced. Noise reduction is considered as a key component to determine the performance of camera systems, and consumes the most computational power of legacy ISPs.

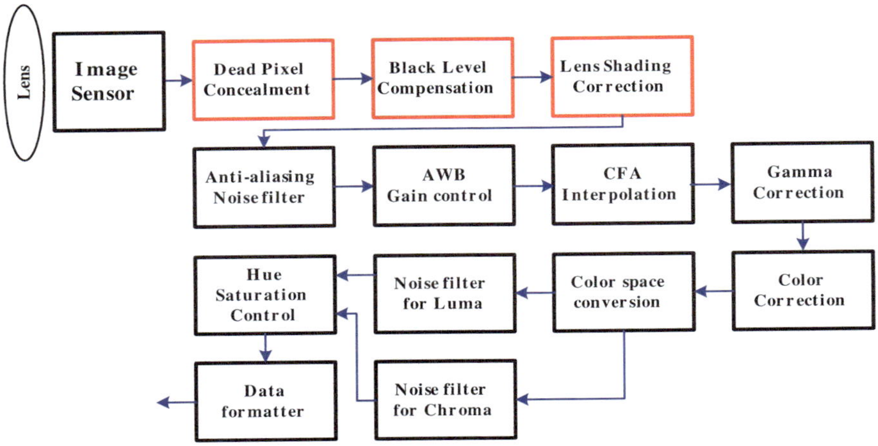

Fig. 21 ISP architecture for handling sensor derating factors

The enumerated methods so far are the functions to let the sensor response to be linear and to compensate for derating factors of an image sensor or a camera module. These are not mandatory functions in a baseline ISP, but need to be considered whether to implement or not, since they try to compensate for the imperfection of a camera system. Figure 21 shows an ISP pipeline with such compensation functions. LSC should be located after BLC, but there is no strict restriction on DPC location if and only if the DPC is located before color interpolation in a baseline ISP chain. Thus it is often desirable to embed DPC function in anti-aliasing noise filter.

Let's examine typical methods to enhance subjective visual quality. Mach bands [18] are an optical illusion, which can be seen in an image patch where there are two wide bands, one light and the other dark, separated by a narrow strip with a light-to-dark gradient. Human eyes perceive two narrow bands of different brightness at either side of the gradient that are not present in the original image (See Fig. 22).

Fig. 22 Mach bands as optical illusion

Edge enhancement is a digital processing technique to improve the sharpness of an image by intentionally emphasizing Mach band effect. The creation of bright and dark highlights on either side of any line makes the line look contrasted from a distance. It only increases the perceptual sharpness. Some artifacts are raised by edge enhancement. The enhancement is not completely reversible, and some detail in the image can be lost as a result of enhancement. Repeated sharpening operations on the resulting image compound the loss of detail, and lead to artifacts known as ringing. Most sharpening filters are based on the first and the second-order derivatives. Among them, Laplacian filter has been the most popular tool. Equation (10) describes one of the Laplacian filters for the pixel value $I(x, y)$, where x and y are horizontal and vertical coordinates in an image.

$$\begin{aligned} L(x,y) = \nabla^2(x,y) &= \frac{\partial^2 I(x,y)}{\partial x^2} + \frac{\partial^2 I(x,y)}{\partial y^2} \\ &= I(x-1,y) + I(x+1,y) + I(x,y-1) + I(x,y+1) - 4I(x,y) \end{aligned} \tag{10}$$

Contrast is the difference in color and light that makes an object distinguishable from others and the background. The human visual system is more sensitive to contrast than absolute luminance. The contrast-controlled value is acquired by Eq. (11), where K_c, K_r and K_b are contrast control gain, reference luminance, brightness control offset, respectively. The contrast gain K_c is a fractional number ranging from 0 to 1. The reference luminance K_r is defined as 2^{B-1} if B-bit codes are used for luminance representation. The brightness offset K_b is used to increase or decrease the average brightness level.

$$Y' = K_c(Y - K_r) + K_b + K_r \tag{11}$$

Figure 23 shows the proposed baseline ISP pipeline considering all related standards. The proposed ISP chain will be the minimum configuration for designing a baseline ISP.

5 Further Works on ISP

The ISP itself is a pipelined chain of functional units, whose inputs are fed from the previous unit and the processed outputs are transferred to the next unit. In each functional unit, every pixel of an image is processed sequentially. When a pixel is processed, only its adjacent pixels are utilized and a small window is defined around the pixel such that some lines of the incoming image have to be stored in the memory. For defining an $N \times N$ window, at least $N - 1$ lines are to be saved into the memory. In other words, it is often said that $N - 1$ line memories are required. In this way, an ISP only utilizes the spatially localized information. One of the hot functions in a legacy ISP is mainly focused on the true color reproduction. As the

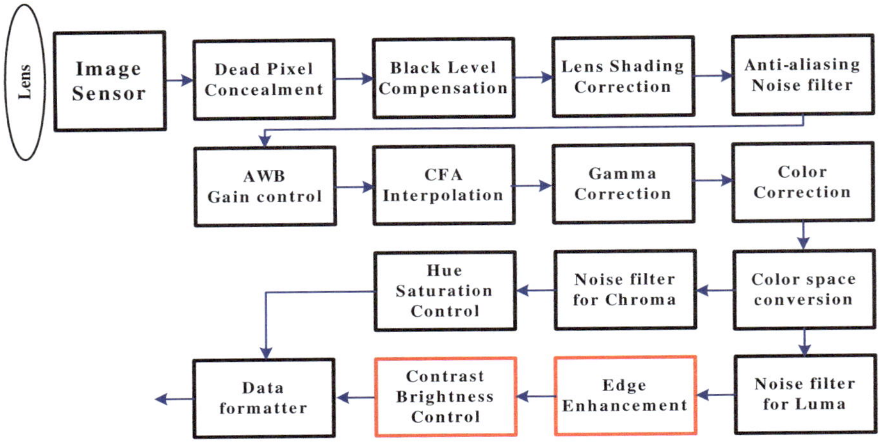

Fig. 23 Proposed baseline ISP pipeline

ambient color temperature changes, the color-related functions begin to degrade the subjective color quality. Realizing the robust color quality over the ambient color temperature is becoming a critical requirement for high-quality ISP implementation. It is because the drastic change of chrominance is annoying to human eyes while the drastic change of luminance is perceived as natural and often can be ignored without annoying our eyes.

When global information is necessary, an entire image has to be saved in the memory. In this case, the amount of memory requirement is so huge such that the frame memory is realized by using an external SDRAM. When the frame memory is available, a more sophisticated function can be conducted in software by using a powerful CPU and/or a GPGPU (General Purpose Graphic Processing Unit). Nowadays, many computer vision applications have been implemented in an intelligent camera. In a legacy ISP, however, those functions requiring the frame memory are not considered since they cannot be embedded inside an image sensor. Thus, they are not regarded as further works for ISP. They are highly related to the intelligent camera. Recently many researches have been done to expand the dynamic range of the image sensor. WDR (Wide Dynamic Range) or HDR (High Dynamic Range) implies such a technique to expand the dynamic range by utilizing two or more frames respectively and is not considered in a legacy ISP pipeline since it requires the frame memory.

There remain a few functions to be implemented in an ISP. Nevertheless, color interpolation and noise reduction are always key functions that need more improvement. Besides, the false color suppression or pseudo-color removal is also becoming a major function since the false color critically distorts the human eyes.

Acknowledgments This work is supported by the Center for Integrated Smart Sensors funded by the Ministry of Science, ICT & Future Planning as the Global Frontier Project.

References

1. Recommendation ITU-R BT.601-7, Studio encoding parameters of digital television for standard 4:3 and wide-screen 16:9 aspect ratios, ITU, March 2011
2. Recommendation ITU-R BT.656-4, Interface for digital component video signals in 525-line and 625-line television systems operating at the 4:2:2 level of Recommendation ITU-R BT.601, ITU, Dec 2007
3. Recommendation ITU-R BT.709-5, Parameter values for the HDTV standards for production and international programme exchange, ITU, Feb 2004
4. Recommendation ITU-R BT.2020-1, Parameter values for ultra-high definition television systems for production and international programme exchange, ITU, July 2014
5. http://en.wikipedia.org/wiki/Foveon_X3_sensor
6. Bayer BE (1976) US Patent 3971065. Color imaging array. Accessed 20 July 1976
7. Gunturk Bahadir K, Glotzbach John, Altunbasak Yucel, Schafer Ronald W, Mersereau Russel M (2005) Demosaicking: color filter array interpolation. IEEE Signal Process Mag 22 (1):44–54
8. Buades A, Coll B, Morel JM (2005) A review of image denoising algorithms, with a new one. Multisc Model Simul 4(2):490–530
9. http://en.wikipedia.org/wiki/Retina
10. http://en.wikipedia.org/wiki/Color_space
11. http://en.wikipedia.org/wiki/RGB_color_space
12. http://en.wikipedia.org/wiki/SRGB
13. http://en.wikipedia.org/wiki/CIE_1931_color_space
14. http://en.wikipedia.org/wiki/Illuminant_D65
15. http://en.wikipedia.org/wiki/Standard_illuminant#Illuminant_E
16. http://en.wikipedia.org/wiki/Hue
17. http://en.wikipedia.org/wiki/HSL_and_HSV
18. http://en.wikipedia.org/wiki/Mach_bands
19. Laroche CA, Prescott MA (1994) Apparatus and method for adaptively interpolating a full color image utilizing chrominance gradients. US Patent 5,373,322
20. http://en.wikipedia.org/wiki/Flat-field_correction

An Ultra-Low-Power Image Signal Processor for Smart Camera Applications

Zhenhong Liu and Nam Sung Kim

Abstract Among thriving cyber physical systems (CPS), smart camera applications require to run both image sensors and image signal processors (ISPs) to capture images whenever necessary. Due to the nature of such applications (i.e., constantly capturing images and analyzing the images to detect any event of interest), the image sensor and ISP become the two most energy consuming components in smart camera applications. In this chapter, we start with our intuition that the perceptive quality of images is not strongly correlated with the accuracy of object detection algorithms and propose three techniques that require only minor modifications to the baseline ISP but dramatically reduce the ISP energy consumption in object detection mode for smart camera applications. When joining three proposed techniques, we demonstrate that our ISP consumes only 3 % of the baseline ISP energy while degrading face detection accuracy by 3–4 %.

Keywords Image signal processor · Smart camera · Face detection

1 Introduction

An image signal processor (ISP), which processes a raw image from a CMOS image sensor, is a specialized processor and it is an essential component in a digital cameras. Figure 1(left) shows a raw image from an image sensor and it is far from what the human eyes perceive. After the ISP processes the image, we see the image in Fig. 1(right), which is close to what the human eyes perceive. The ISP is comprised of many processing functions such as white balance, gamma correction, format processing, geometric correction and color filter array interpolation.

Z. Liu (✉) · N.S. Kim
University of Wisconsin-Madison, Madison, WI, USA
e-mail: zliu238@wisc.edu

N.S. Kim
e-mail: nskim3@wisc.edu

C. Kyung (ed.), *Theory and Applications of Smart Cameras*,
KAIST Research Series, DOI 10.1007/978-94-017-9987-4_3

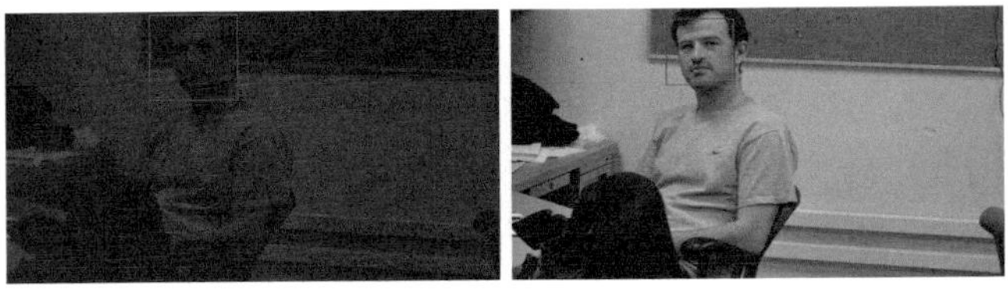

Fig. 1 Images before (*left*) and after (*right*) ISP processing

The image processing functions of the ISP are normally implemented in a pipelined fashion and a given image passes through pipeline stages implementing these functions operating in tandem. A more expensive ISP often integrates more sophisticated processing functions such as red-eye removal and image stabilization. In general, the image processed by the ISP is normally output to a display or a storage device.

The increasing capability of digital cameras and processors in mobile devices has given rise to emerging applications in which images are acquired for purposes other than picture-taking. For example, a smart camera may be used to recognize the user of a mobile device and to unlock features of the device, or the camera may be used as an input device to recognize gestures by the user. For these emerging applications, the ISP may be teamed up with a (general-purpose) processor running various recognition algorithms to extract characteristics of given objects or gestures, where the ISP is required to operate continuously. Even the energy consumption of a simple ISP is comparable to that of an image sensor [1] and a few times higher than a general-purpose processor running recognition algorithms in our smart camera environment; our baseline ISP consumes ~20 mW while an ARM Cortex-M0 processor consumes ~4–10 mW in 65 nm technology [2] Thus, the energy consumed by the ISP for such applications can tax the capacity of the batteries used in mobile devices.

In this chapter, we observe that the task of optimizing images for human perception may not align with the requirements of recognition algorithms (e.g., face detection in this chapter). Accordingly, we propose an ISP that may operate in at least two modes, one mode optimizing the image signal processing for human vision and the other mode optimizing the image signal processing for object detection and gesture recognition, where this latter mode may provide degraded perceptual image quality. Though the perceptual image quality is degraded, we show that such degraded image quality negligibly impacts on the accuracy of object detection while significantly reducing the ISP energy consumption. More specifically, we propose three techniques for an ISP that can receive images from an image sensor and output images optimized for either human vision or object detection.

- First, we hypothesize that many ISP stages are only needed for human perception and propose to skip some ISP stages after identifying which stages are critical for a face detection algorithm. Our experiment shows that only gamma correction and demosicking stages are critical for a face detection algorithm we adopt.

This can reduces the energy consumption of the ISP by 33 % while degrading the true positive and negative face detection accuracies by only 3 and 5 %, respectively.

- Second, we observe that the ISP often processes many pixels with similar values and propose to interpolate the output pixel values based on their neighboring output pixel values, if their neighboring pixels have similar values. This reduces the energy consumption of the ISP by 30 % while degrading the true positive and negative face detection accuracies by only 3 and 9 %, respectively.
- Third, we propose to reduce the number of pixels to process (i.e., scaling input images). For example, we process only one out of every four pixels. This reduces the energy consumption of the ISP by 93 % while degrading the true positive face detection accuracy by only 3 %. However, this degrades the true negative face detection accuracy by 14 %, which may not be acceptable and cannot be used alone.
- Finally, we propose to join three techniques after we observe that the negative impact of each technique on the detection accuracy is not cumulative. This reduces the energy consumption by 97 % while degrading the true positive and negative face detection accuracies by only 5 and 4 %, respectively.

The rest of the chapter is organized as follows. Section 2 describes background on our baseline ISP and a face detection algorithm. Section 3 presents our experimental methodology. Section 4 describes three proposed techniques and provides evaluations. Section 5 discusses the related work. Section 6 concludes this study.

2 Background

2.1 Baseline ISP Design

We design a baseline ISP for a video recording system. Our baseline ISP is comprised of 10 pipeline stages as depicted in Fig. 2 and it supports the image resolution of up to 1920×1080 pixels. The baseline ISP can produce one 8-bit output per cycle. We describe the functionality of key ISP stages below.

Black Level Compensation (BLC): Ideally, a "black" pixel from the image sensor should have a value of 0. However, this is not the case in reality due to the leakage current of the photodiode in CMOS sensor. Therefore, the BLC stage is required to subtract this bias from each pixel value.

Lens Shading Correction (LSC): Due to the deflection of the light on the lens surface, the uniform in-coming light will distribute non-uniformly across the image sensor. The light will be the brightest at the center and decrease gradually with the distance from the center. To correct such a lens shading effect, we multiply every pixel with a gain factor, which is a function of the distance between the pixel and the center of the image sensor in our ISP.

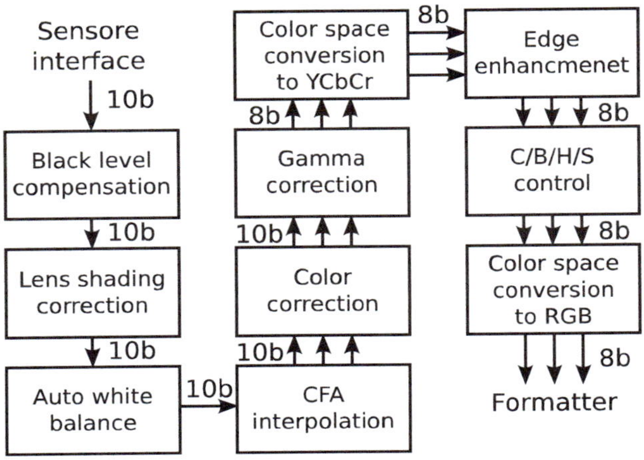

Fig. 2 Baseline ISP

Auto White Balance (AWB): An image sensor responses differently to the power spectrum of diverse light sources. Since the ambient light source is not a standard illuminant D65, its color values need to be re-scaled to be like those of D65 (i.e., AWB). In our ISP design we implement the gray-world algorithm that uses the average pixel values of each color channel to calculate the scaling parameters for each color for AWB [3].

Color Filter Array (CFA) interpolation (also known as demosaicing): To produce a color image with a *single* image sensor, a CFA is placed over the photodiodes [4] to separate RGB values for each pixel. Since the pixel values of the 3 color channels cannot fill the whole image plane, some interpolation is needed as illustrated in Fig. 3. We use a gradient-based interpolation algorithm for RGGB CFA pattern in our baseline ISP design.

Color Correction (CC): The difference between the spectral sensitive of photodiodes and the human visual system makes the color sensed by the image sensor inaccurate. If we directly use the RGB values from the image sensor, the color of the image will be significantly diverged from the color perceived by human eyes.

Fig. 3 CFA interpolation

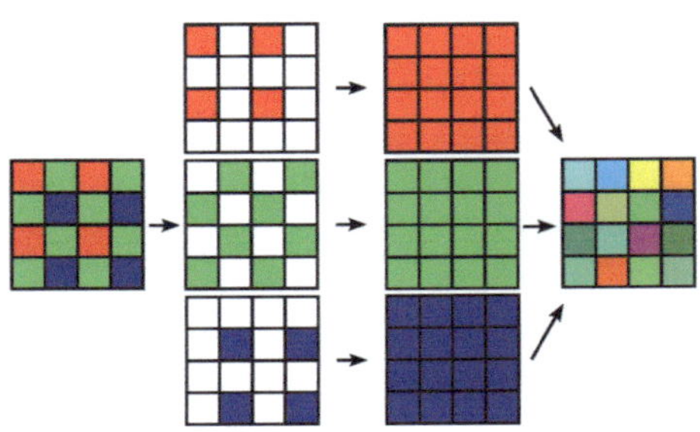

Thus, we correct the image sensor's RGB values through a color space conversion (i.e., matrix multiplications) in our ISP design.

Gamma Correction (GC): For human eyes perception, 8-bit color depth is sufficient while image sensors produce 10-bit color depth. The linear conversion of 10-bit color values to 8-bit ones will lead to perceptually very dark images. Thus, following the ITU-R. BT.709 standard, we non-linearly map each 10-bit value from the sensor to 8-bit one to show more details in the dark part of images.

RGB to YCbCr color space conversion (R2Y): We convert colors from RGB to YCbCr color space [5] in our ISP design because the following two stages can be implemented much more easily in YCbCr color space.

Edge Enhancement (EDGE): We enhance the contrast of edges in images (i.e., sharper images) by decreasing/increasing the Y component of the pixels on or near edges in our ISP design.

Contrast/brightness/hue/saturation control (CTR): We adjust the contrast, brightness, hue and saturation of an image by applying a linear transformation to YCbCr components in our ISP design.

YCbCr to RGB color space conversion (Y2R): We convert the colors back to the RGB color space.

2.2 *Face Detection Algorithm*

In this chapter, we use a face detection algorithm proposed by Viola and Jones [6]. This is one of the most popular and widely used face detection algorithms and well supported by OpenCV. It extracts Haar-like features from the gray-scale integral image and uses a series of cascaded weak classifiers in order to achieve real-time performance. Figure 4(left) shows some of the basic Haar-like feature masks. A mask is placed on a gray-scale image and a feature is calculated by subtracting the sum of the pixels under the black area from the sum of the pixels under the white area. Extracting the features from the original gray-scale images is very

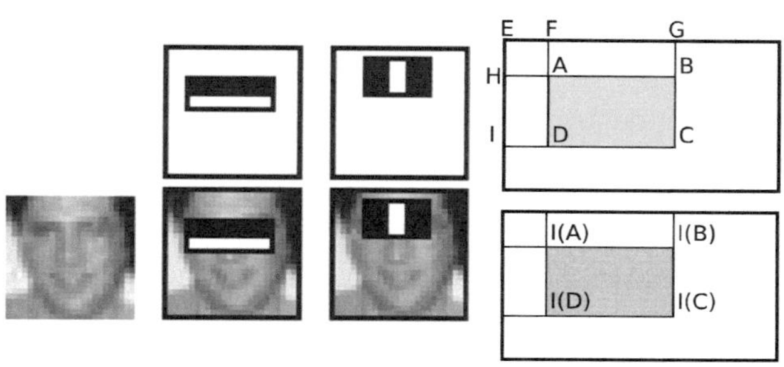

Fig. 4 Haar-like features [6] (*left*). Original image (*right-top*). Integral image (*right-bottom*)

compute-intensive since it needs to scan and accumulate the rectangular area under the feature mask every time. Viola and Jones used an integral image for efficiently extracting features. An integral image (Fig. 4(right-top)) is essentially a lookup table of the same dimension as the original image (Fig. 4(right-bottom)), but each pixel on the integral image is the sum of all the pixels in its up-left rectangular area, as shown in Fig. 4(right-bottom). With the integral image, the sum of pixels in rectangle ABCD can be efficiently calculated with a few lookups and additions/subtractions: $sum(\text{ABCD}) = I(\text{A}) + I(\text{C}) - I(\text{B}) - I(\text{D})$ where $I(\text{A}) = sum(\text{EFAH})$ and $I(\text{B}) = sum(\text{EGBH})$.

3 Evaluation Methodology

Raw image preparation: To evaluate the impact of our proposed techniques, we first need to obtain (raw) images from an image sensor or images that are not yet processed by an ISP. While a typical digital camera only offers images processed by its embedded ISP, a Canon 650D digital single-lens reflex camera allows us to store unprocessed 14-bit pixel values from its image sensor in a lossless JPEG format (ITU-T81) embedded in.CR2 format [7]. We use this camera to take images for this study. After decoding a lossless JPEG in a CR2 file, we right-shift each 14-bit pixel value by 4 bits to produce a 10-bit unsigned integer value and make it compatible with our baseline ISP. The most important characteristics of the raw images, such as noises and distribution of the pixel values, are unchanged after mapping them to 10-bit values. Since the baseline ISP supports the image resolution of up to 1920 × 1080 pixels, the taken images are cropped accordingly. The Bayer pattern of the camera sensor is RGGB and the way in which the images are cropped makes this pattern unchanged. However, the raw images obtained using this method have negligible lens shading effect because the original images are much larger and cropped around the center part. Therefore, a pseudo lens shading effect is applied to these images by scaling the pixel values according to their coordinates on the image. The pseudo lens shading effect uses parameters from the lens shading effect of a small camera.

Face detection algorithm: A face detection program is implemented using OpenCV v2.4.7 APIs [8]. It takes an image as an input and outputs the same image with detected faces marked with rectangles. Training a classifier requires positive and negative image sets. A positive image contains one or more faces while a negative one does not contain any face. We use about 3000 negative images from [9] and about 3000 positive images generated from 800 faces taken by our Canon 650D camera. Finally, the training uses OpenCV built-in programs: *opencv_createsamples* to create sample from images with one or more faces and *opencv_traincascade* to train the classifier. The final output is an *xml* file, which is used to initialize the classifiers in the face detection application.

ISP power estimation: We synthesize the baseline ISP with an IBM 65 nm standard cell library targeting 250 MHz, which is fast enough for processing

Table 1 Percentage power breakdown of the baseline ISP

Stage	Power (%)	Stage	Power (%)
BLC	0.9	LSC	17.3
AWB	3.1	CFA	30.6
CC	6.2	GC	2.9
R2Y	3.4	EDGE	1.2
CTR	2.8	Y2R	3.0
SRAM	24.0	Misc.	4.1

1920 × 1080 video frames at 60 frames per second, and validate the design after performing post-synthesis simulations. We use actual raw images to generate input trace files to estimate the average power consumption of various ISP configurations. The power consumption of the SRAM in CFA interpolation stage is calculated separately using IBM 65 nm memory compiler. The baseline ISP consumes 21.2 mW total and its power breakdown is given in Table 1.

Evaluation metric: We use 100 images with one or more faces and 100 images without any face as an evaluation image set. The images are taken under various light conditions and scenes. In the positive evaluation images, about 85 % of the images have at least one face, 10 % have two faces and the rest have three to five faces. To describe how accurate the face detection is, we use per-image based metrics (whether or not we detect at least one face in a given image), which is more appropriate for our target application that is intended to start to record and transmit images upon detection of one or more faces. We define four detection outcomes: (i) true positive (TP), (ii) true negative (TN), (iii) false positive (FP) and (iv) false negative (FN). Consider that a given image has one or more faces and the detector recognizes at least one face. This is a TP event that wakes up the smart camera system for recording and transmitting images. Similarly, suppose that a given image has no face and the detector recognize no face. This is a TN event in which the system stays in a sleep mode. The rate of TP, TN, FP, and FN is defined as follows:

$$\mathrm{TP} = \frac{\#\text{ of correctly recognized images w/faces}}{\#\text{ of images w/faces}} \tag{1}$$

$$\mathrm{TN} = \frac{\#\text{ of correctly recognized images w/no face}}{\#\text{ of images w/no face}} \tag{2}$$

$$\mathrm{FP} = 1 - \mathrm{TN} \tag{3}$$

$$\mathrm{FN} = 1 - \mathrm{TP} \tag{4}$$

Note that achieving a high TN (i.e., low FP) rate is critical because an FP event may unnecessarily wake up the system, wasting the energy.

4 Optimizing ISP for Face Detection

In this section, we start with our hypothesis that the perceptual image quality is not directly correlated with face detection accuracy. Founded on this hypothesis, we propose three techniques that can reduce the ISP energy consumption while maintaining high TP and TN detection accuracies. Then we join these three techniques and demonstrate that the joined techniques have a synergistic effect on TP and TN detection accuracies while dramatically reducing the ISP energy consumption.

4.1 Skipping Non-critical ISP Stages

Based on our earlier hypothesis, we speculate that some ISP stages may not be critical for face detection algorithms or they marginally improve face detection accuracy. For example, In the Viola-Jones face detection algorithm, a given color image is first converted to a gray-scale image and only the gray-scale image is used for detection. Therefore, all stages related to modifying or enhancing the color of images can be unnecessary for high detection accuracy.

Implementation details: Skipping most of the ISP stages is straightforward. We only need multiplexers at the output port of each ISP stage to select between the input and output of the stage. If the input of the stage is selected, the stage is bypassed and all flip-flops in this stage are disabled. However, skipping CFA and GC stages requires some additional logic. As shown in Fig. 5, when the CFA interpolation stage is skipped, the pixel value for only the color channel will be duplicated so the other two color channels have the same value. Since the RGB values in this pixel are the same, the pixel represents a gray color in the output image, which looks like a normal gray-scale image with a transparent checkerboard on it. When the GC stage is skipped, the pixels in all three color channels are right shifted by two bits to replace the non-linear mapping, since we still need to map the

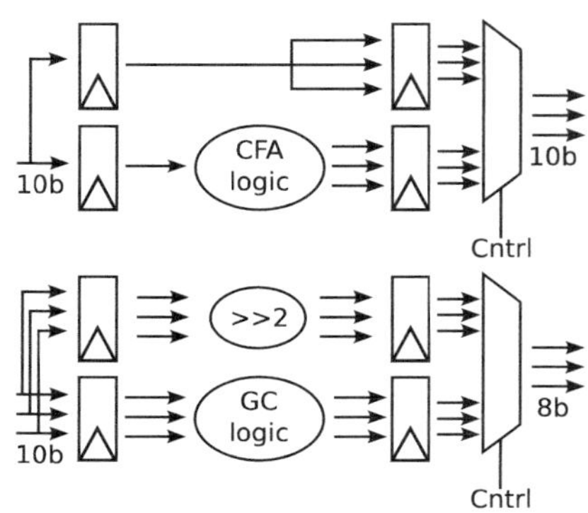

Fig. 5 Skipping CFA and GC

10-bit color to 8-bit and this is the simplest method; note that later our evaluation indicates that we must keep these two CFA and GC stages but this modification is useful for the second technique that will be described in Sect. 4.2. Finally, we do not consider skipping the two color space conversion stages: R2Y and Y2R, because they are required to pre-process and post-process the image for CTR and EDGE. Therefore, as long as any one of CTR and EDGE is kept, the color space conversion must not be skipped. On the other hand, R2Y and Y2R stages, which actually do not modify the images, are automatically skipped when both CTR and EDGE are skipped.

Impact on TP, TN, SSIM, and energy consumption: In Fig. 6 we show the TP (left) and TN (right) detection accuracies of for various ISP configurations in which we skip one or more ISP stages. We also show a structural similarity index (SSIM) value [10] to correlate human perception image quality with detection accuracy for each explored ISP configuration. Since we have a large number of combinations of which stages to skip, we do not evaluate all of them exhaustively. Instead, we first evaluate the detection accuracy of skipping a single stage at a time. Analyzing "single stages" in Fig. 6 (left), we see that skipping either the CFA or GC stage can significantly degrade the TP detection accuracy. The ISP configurations skipping the GC and CFA stages lead to only 61 and 85 % TP detection accuracy while the baseline ISP and other ISP configurations offer 98 and 92–97 % TP detection accuracies, respectively. *This suggests that the CFA and GC stages are the most critical ones for face detection purpose;* all the other stages only modify the image to enhance the perceptual quality while CFA and GMA reconstruct images from 10-bit Beyer pattern images.

Observing "single stages" in Fig. 6(left), *we also see that there is a very weak correlation between SSIM and TP detection accuracy*. For example, skipping the CFA stage exhibits notably lower TP accuracy (i.e., 85 vs. 94 %) but higher SSIM than skipping the AWB stage (i.e., 0.88 vs. 0.71). In contrast, skipping the GC stage shows a significant negative impact on both SSIM and detection accuracy (i.e., 61 % and 0.46). After studying "single stages" in Fig. 6(left), we skip two ISP stages but always keep the CFA and GC stages. We see that even skipping two stages have a minor negative impact on the TP accuracy. Thus, we attempt to skip all the ISP stages except for the CFA and GC stages (cf. "B" in Fig. 6(left)). This ISP configuration offers 3 % lower TP accuracy (95 %) than the baseline ISP

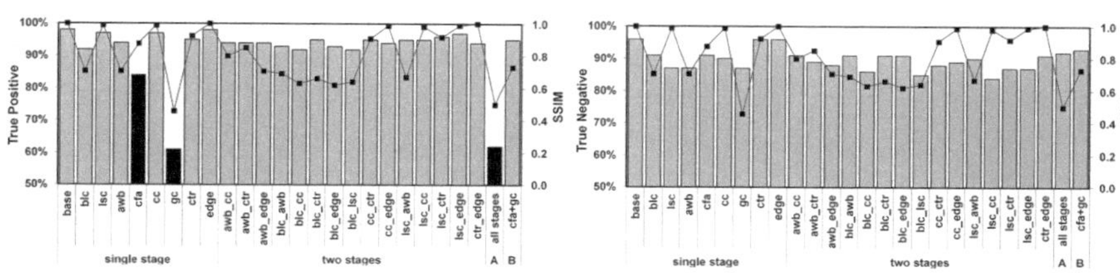

Fig. 6 Detection accuracy (shown in bars) and SSIM after skipping ISP stages: (*left*) true positive and (*right*) true negative

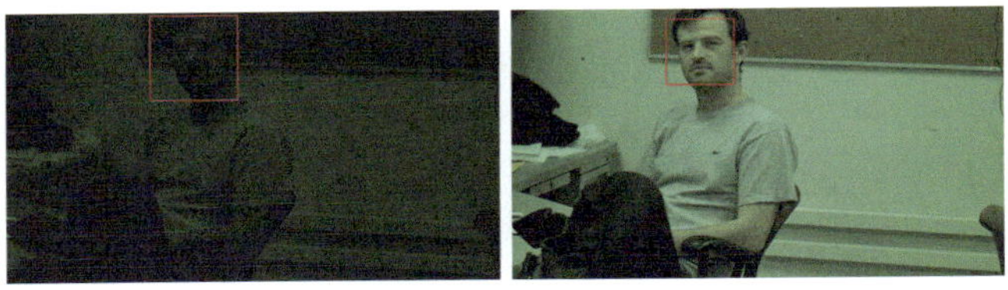

Fig. 7 A raw image (*left*). The image after CG and CFA processing (*right*)

(98 %), while reducing the energy consumption by 33 %; Fig. 7(right) shows the image after only the CFA and CG are applied to the raw image in Fig. 7(left).

Although only two out of ten stages are used, the energy reduction is limited to 33 % because the CFA stage and the logic associated with the CFA consume more than 50 % of the total energy consumption of the baseline ISP. Finally, we also attempt to skip all the ISP stages including the CGA and GC stages (i.e., feeding raw images from the image sensor directly to the classifier). As expected, we observe huge TP accuracy degradation due to missing the CFA and GC stages (cf. "A" in Fig. 6(left)). *Finally, we observe that the negative effect of skipping ISP stages is not cumulative because some ISP stages and the classifier are not linear systems.* Similar non-linear characteristics are observed in our other techniques (in particular when we join three proposed techniques).

Skipping ISP stages have a much smaller negative impact on TN detection accuracy. The lowest TN detection accuracy among all the ISP configurations is 87 %. This does not mean the perceptive image quality has less impact on TN detection accuracy. Note that TN detection accuracy is also high for skipping CFA and GC. This is because the TP detection accuracy for skipping these two stages is so low, simply rejecting many images with faces as ones that do not have any face. We see that the TN detection accuracy is 93 % when skipping all stages except CFA and GC, which is only 3 % lower than the baseline ISP.

4.2 Interpolating Pixel Values

Most pixel values in a natural image change gradually while it is the image's edges and corners that contribute to the accuracy of face detection algorithms. Exploiting this observation, we propose to skip processing pixels if their neighboring pixels have similar values. For the skipped pixels we replace their value with the processed pixel values at the ISP output stage. By doing so, we can still preserve the most obvious edges and corners while reducing the number of pixels to process. Some details in the image can be lost, negatively impacting the perceptual image quality. However, we demonstrate that such a technique negligibly impacts the detection accuracy.

Implementation details: Since our baseline ISP takes and processes one 10-bit pixel value per cycle, we augment a stage that buffers input pixels from the image sensor and compare their values in that stage. Figure 8(top) shows a row of R and G pixels. When our adaptive pixel interpolation feature is enabled, R_{even} and G_{even} will always be processed by the ISP (e.g., (R_0, G_0) and (R_2, G_2) in Fig. 8(top)). As an example for processing R pixels, refer to Fig. 8(bottom) where we check whether R_1 will be processed or not after comparing its value with the average value of R_0 and R_2. If the difference between R_1 and $(R_0 + R_2)/2$ is smaller than a given threshold value (Th), processing R_1 is skipped and the output pixel value of R_1 at each stage is replaced with the average value of the output R_0 and R_2 pixel values at each stage. Whether or not a certain pixel is skipped is determined by the raw pixel values from the image sensor at the ISP input stage and the decision for each pixel is propagated through the ISP pipeline, and each ISP pipeline stage either processes the input value or does nothing at a given cycle to reduce dynamic energy consumption. The threshold value controls the quality of the interpolation; as the threshold value is smaller, fewer pixels will be interpolated. Skipping some pixel processing leverages the hardware modification illustrated in Sect. 4.1 since each pixel is processed sequentially.

Note that we cannot simply skip all the ISP stages and interpolate the values of pixels chosen for interpolations based on the final output values of the neighboring pixels at the last ISP stage. For example, to process one pixel (that we do not interpolate), the CFA stage needs the (intermediate) values of the neighboring 48 pixels in a 7 × 7 pixel matrix; the intermediate values denote the values after the pixels are processed by the proceeding ISP stages. These pixels stored in the SRAM may include the pixel values that are supposed to be skipped and interpolated later. Thus, we need to interpolate the intermediate values of the skipped pixels based on those of the neighboring pixels at the output of the AWB stage that feeds the intermediate values of pixels to the CFA stage (cf. Fig. 2). Nonetheless, the number of pixels processed by the CFA stage (i.e., the energy consumption of the CFA stage) is also reduced leveraging the feature described in Fig. 5 of Sect. 4.1.

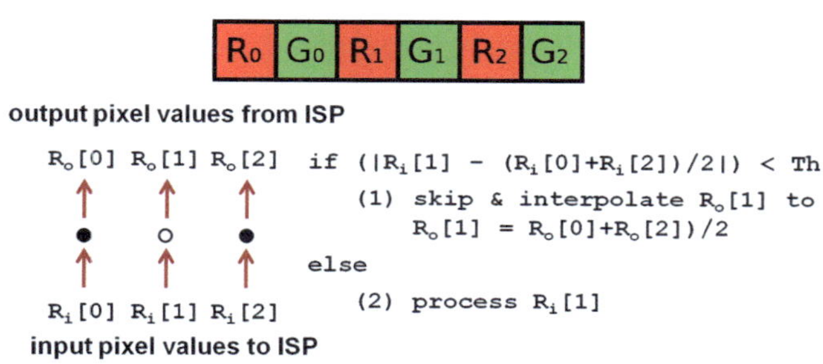

Fig. 8 A row of pixels (*top*). Adaptive pixel value interpolation (*bottom*)

Impact on TP, TN, and energy consumption: In Fig. 9 we show the TP (left) and TN (right) detection rates of for various pixel interpolation threshold values. We also show how many pixels are interpolated for each threshold value. Analyzing Fig. 9, we see that our adaptive pixel interpolation technique shows a very minor impact on TP detection accuracy and a modest impact on TN detection accuracy. As we increase the threshold values, more pixels can be interpolated but the TP detection accuracy asymptotically decreases. When the threshold value is set to 20, 44 % of the pixels in (raw) images can be skipped, while the TP detection accuracy is still as high as 95 % (i.e., 3 % lower than the baseline ISP); Fig. 10 shows The TN detection accuracy simply stays at 89 % for all four threshold values. This is because most part of the background is much smoother than the faces and further increasing the effect of interpolation on the background has little effect on it. This adaptive pixel interpolation technique can reduce the energy consumption of all ISP stages except for the SRAM in the CFA stage and its controller. Note that the logic to support the proposed pixel interpolation technique is very simple (i.e., a few comparators and multiplexers). When the threshold value is set to 20, we can reduce the energy consumption of the ISP by 30 %.

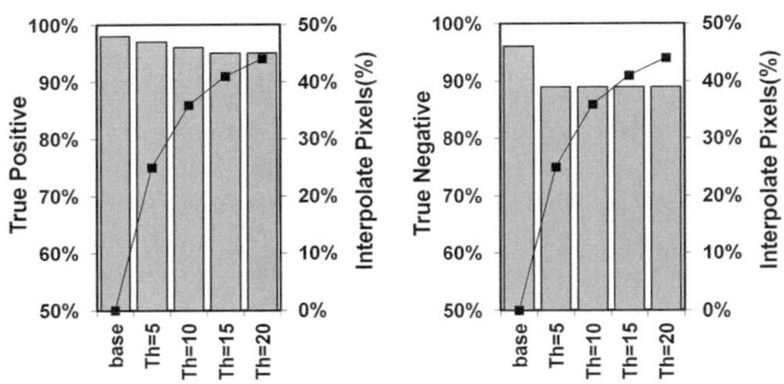

Fig. 9 Detection accuracy of interpolating pixel values adaptively: true positive (*left*) and true negative (*right*)

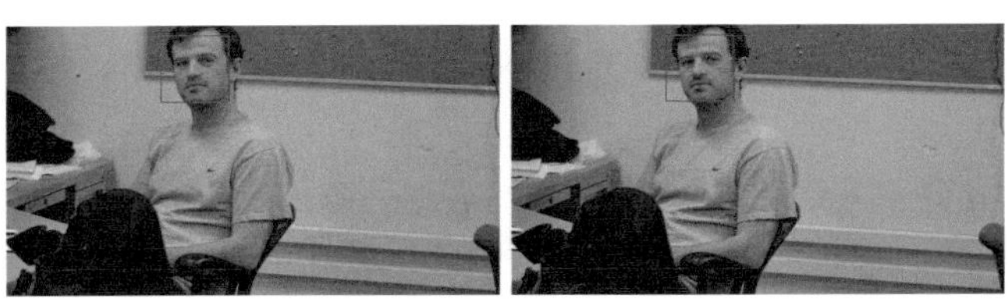

Fig. 10 Fully-processed image before (*left*) and after (*right*) the interpolation (Th = 20)

4.3 Scaling Images

A practical implementation of a face detection algorithm must be able to detect faces of different sizes. Usually, there is a minimal size of face that the algorithm can detect. As long as the size of a face in the input image is larger than the minimal size, the face can be detected. Therefore, it is not always necessary to maintain the original dimension of the image. Instead, we can scale the images down to greatly reduce the total number of pixels to be processed by the ISP. A smaller input image can also benefit the face detection application and reduce energy spent on the detection.

Implementation details: We can either scale the input pixels at the interface between the image sensor and the ISP, or we can use a simple scaling algorithm to efficiently decrease a given image size and replace the CFA interpolation. Figure 11 shows a part of the color filter array on an image sensor. For 2 × 2 scaling, we merge four sub pixels in a 2 × 2 RGGB square to a single pixel with all 3 RGB components. The RGB value for the merged pixel is $\{R, G, B\} = \{R_{00}, (G_{01} + G_{10})/2, B_{11}\}$.

This scaling algorithm also "interpolates" the CFA and outputs a smaller color image since we get all RGB values for a pixel in the scaled image. Thus, when this scaling algorithm is applied, the CFA stage in the ISP can be bypassed. Scaling image by 2 × 2 can reduce the total number of processed pixels to only 25 % of the original image. Similarly, to scale the image by 4 × 4, we merge the sub pixels in a 4 × 4 square to a single pixel. The merged pixel has a value of $\{R, G, B\} = \{(R_{00} + R_{02} + R_{20} + R_{22})/4, (G_{01} + G_{03} + G_{10} + G_{12} + G_{21} + G_{23} + G_{30} + G_{32})/8, (B_{11} + B_{13} + B_{31} + B_{33})/4\}$.

Impact on TP and TN: Three scaling algorithms are evaluated in our experiment: 2 × 2, 4 × 4 and 8 × 8 in Fig. 12 shows the detection accuracy of the scaled images. The TP detection accuracy along with SSIM shows a simple decreasing trend with the scaling factor increasing from 2 × 2 to 8 × 8. When the images are scaled by 2 × 2 or 4 × 4, the TP detection accuracy shows only a modest decrease to about 95 %. However, when the scaling factor goes up to 8 × 8, the TP detection accuracy decreases to about 80 %. The TN detection accuracy has a quite different behavior. When the images are scaled by 2 × 2, the TN detection accuracy drops to 82 %. Then, with the scaling factor increasing, the TN detection accuracy gradually increases to 90 % at 8 × 8. Since both TP and TN detection accuracy is important,

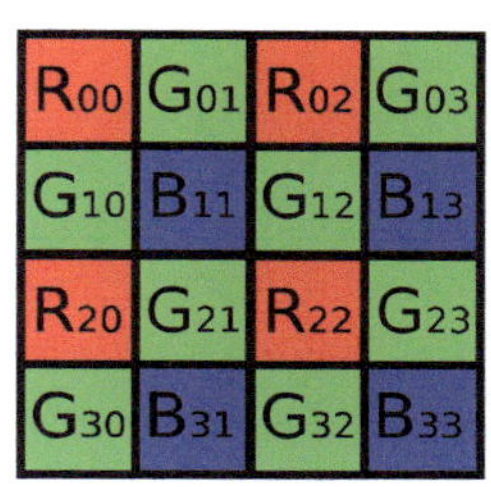

Fig. 11 Part of the color filter array

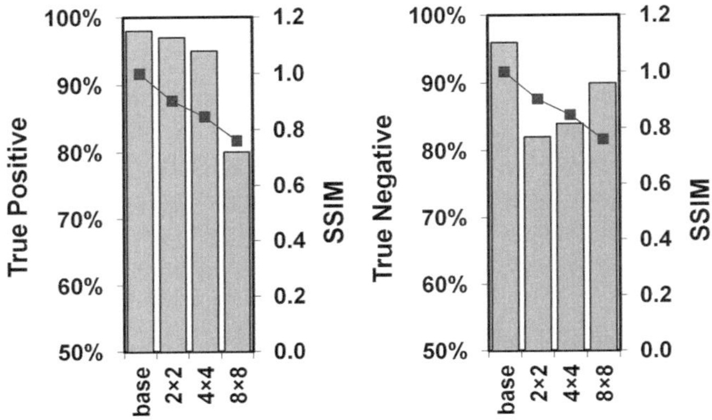

Fig. 12 Detection accuracy of scaling images: true positive (*left*) and true negative (*right*)

we select the 4 × 4 scaling algorithm as the best one. It reduces the total number of pixels to process to 6.25 % of the original image thus reduces the ISP energy consumption (almost) proportionally. The energy consumption can be even lower if we use the algorithm described in section—to replace the CFA interpolation stage with the much simpler logic for scaling.

4.4 *Putting It Together*

The three proposed techniques are exploiting somewhat orthogonal characteristics of the face detection algorithm and the ISP. Therefore, they can be joined together to further reduce ISP energy consumption. To evaluate how the three techniques perform when they are joined, we enable only the CFA and GC stages in the ISP and vary the interpolation threshold value and scaling factor. Figure 13 shows the results of joining all the three proposed techniques. The configuration is denoted by "threshold value + scaling factor" in Fig. 13. For example, "5 + 2 × 2" denotes that the threshold value is 5 and the scaling factor is 2 × 2. Our evaluation shows that when all techniques are joined, the TN detection accuracy for some configurations improves over simple image scaling due to the non-linear effect of various techniques on the face detection algorithm, as discussed in Sect. 4.1. Furthermore, the high frequency artifacts introduced by scaling is filtered when our adaptive interpolation is applied. Considering both TP and TN detection accuracies and ISP energy consumption, we see that the "20 + 4 × 4" configuration becomes the most desirable one; it reduces ISP energy consumption by 97 % while degrading TP and TN detection accuracies by only 5 and 4 %, respectively.

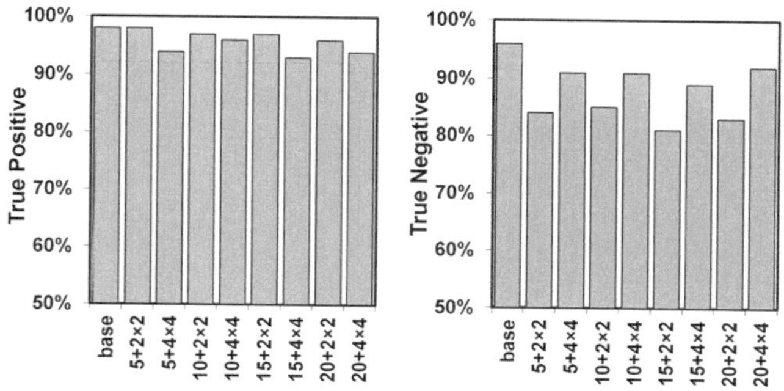

Fig. 13 Detection accuracy of three combined techniques: true positive (*left*) and true negative (*right*)

5 Related Work

The prior work closest to the objective of our study is [11] where LiKamWa aims to improve power efficiency of the ISP especially for computer vision applications. In this "extended abstract," it proposes to slow down the frame rate and apply other low-power techniques such as putting the ISP into a sleep state between frames, but no detailed energy analysis and/or impact on the detection accuracy was provided. Also, there are other studies targeting to reduce the power consumption of ISP, such as [12, 13]. However, those studies only optimize the ISP for high perceptual quality and do not consider the optimization of object detection applications. In [14], a mixed-signal processor for feature extraction is proposed to reduce the system power consumption. However, it focuses on reducing power consumed of the I/O between the CMOS sensor and the processor.

6 Conclusion and Future Work

In this chapter, we propose three techniques to reduce the energy consumption of the ISP after observing that the image quality for human perception is not strongly correlated with the image quality for object detection algorithms. We demonstrate that our three techniques, which require minor modification on the baseline ISP, can reduce the energy consumption of our ISP by 97 % while degrading true positive and negative detection accuracies by only 5 % and 4 %, respectively. Furthermore, the proposed ISP design can be easily re-configured to be running at either low-energy object detection or normal mode. Consequently, it can provide either images of high perceptual quality or images processed only for detection to reduce the energy consumption.

Acknowledgments This work is supported by the Center for Integrated Smart Sensors funded by the Ministry of Science, ICT & Future Planning as the Global Frontier Project and an NSF grant (CCF-0953603).

References

1. BYD Microelectronics Co., Ltd., QVGA CMOS Image Sensor BF3901 Datasheet
2. DDC PowerShrink™ Platform. http://www.suvolta.com/technology/ddc/
3. Buchsbaum G (1980) A spatial processor model for object colour perception. J Franklin Inst 310(1)
4. Bayer BE (1976) Color imaging array. USA Patent US3971065 (A), 20 July 1976
5. Parameter values for the HDTV standards for production and international programme exchange. http://www.itu.int/rec/R-REC-BT.709/en
6. Viola P, Jones M (2001) Rapid object detection using a boosted cascade of simple features. In: EEE computer society conference on computer vision and pattern recognition
7. Inside the Canon RAW Format Version 2. http://lclevy.free.fr/cr2
8. OpenCV Library. http://opencv.org/
9. Tutorial on OpenCV HaarTraining. https://code.google.com/p/tutorial-haartraining/
10. Wang Z, Bovik A, Sheikh H, Simoncelli E (2004) Image quality assessment: from error visibility to structural aimilarity. IEEE Trans Image Proc 13(4)
11. LiKamWa R (2014) Extended abstract: efficient image processing for continuous mobile vision. ACM PhD Forum
12. Kim K, Park I-C (2006) Combined image signal processing for CMOS image sensors. In: IEEE international symposium on circuits and system (ISCAS)
13. van Dalen E, Pestana S, van Wel A (2006) An integrated, low-power processor for image signal processing. In: IEEE international symposium on multimedia (ISM)
14. Bong K et al (2014) An 1.61 mW Mixed-signal column processor for BRISK feature extraction in CMOS image sensor. IEEE international symposium on circuits and systems (ISCAS)

Foundations and Applications of 3D Imaging

Min H. Kim

Abstract Two-dimensional imaging through digital photography has been a main application of mobile computing devices, such as smart phones, during the last decade. Expanding the dimensions of digital imaging, the recent advances in 3D imaging technology are about to be combined with such smart devices, resulting in broadened applications of 3D imaging. This chapter presents the foundations of 3D imaging, that is, the relationship between disparity and depth in a stereo camera system, and it surveys a general workflow to build a 3D model from sensor data. In addition, recent advanced 3D imaging applications are introduced: hyperspectral 3D imaging, multispectral photometric stereo and stereo fusion of refractive and binocular stereo.

Keywords Stereo imaging · Hyperspectral 3D imaging

1 Foundations of 3D Imaging

The history of measuring a 3D shape started with pantography, which measures a 3D shape by directly contacting surface points with a mechanical linkage. The distance measurements over the surface with a stylus allow for duplicating the 3D shape of an object on paper [25]. Contact-based measurement systems have been used to transfer engravings or 3D shapes. However, such contact-based approaches tend to damage fragile surfaces while they are being measured. It is also time-consuming to capture the entire 3D shape of an object by contact. Alternatively, non-contact-based methods are more commonly practiced using modulated light patterns over the object's surface nowadays.

M.H. Kim (✉)
Computer Science Department, KAIST, 291 Daehak-ro, Yuseong-gu, Daejeon, Korea
e-mail: minhkim@kaist.ac.kr

C. Kyung (ed.), *Theory and Applications of Smart Cameras*,
KAIST Research Series, DOI 10.1007/978-94-017-9987-4_4

1.1 Passive 3D Imaging

Passive 3D imaging approaches reconstruct 3D shapes from camera signals only, while active approaches capture shapes by projecting structured light or modulated illumination over an object's surface. In general, passive methods utilize an optical phenomenon such as parallax disparity. Stereo imaging is one of the most popular passive methods, allowing for point-wise depth of a scene. Stereo imaging can be divided into binocular and multi-view approaches.

1.1.1 Binocular Stereo

Binocular stereo utilizes two cameras with a specific displacement between the cameras. Two images captured by the cameras contain a pair of corresponding pixels projected from a surface. Binocular disparity describes pixel-wise displacement of parallax between the corresponding points on a pair of stereo images.

Figure 1 shows a schematic diagram of a classic binocular system. Rays from a point of an object on a surface are projected into two imaging planes, passing through the optical centers of the cameras. Since there is a distance between the two imaging planes, so-called baseline *b*, these rays reach different pixel positions on

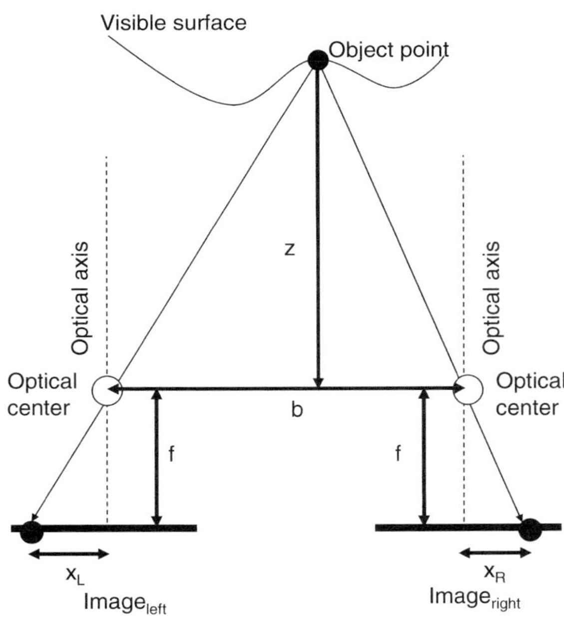

Fig. 1 Suppose there is a point on an object, at which two cameras point. The point is projected at different pixel positions on two image planes. The difference between these two positions changes depending on the depth distance of the point from the camera plane, the focal length of the cameras and the baseline (the distance between the two cameras). Image courtesy of © 2015 Elsevier Computer Vision and Image Understanding [2]

these two planes. The pixel-wise displacement between these corresponding pixel positions is called disparity d of the object point:

$$d = |X_L - X_R|, \tag{1}$$

where X_L and X_R are the distances between the projected point and the center on each image plane. This parallax disparity d of these corresponding points, in Eq. (1), is inversely proportional to the depth distance z from the camera plane to the object surface. Now we can apply trigonometry to the four distances: the depth distance of the point d, the focal length of the cameras f, the baseline b, and the disparity d, allowing us to recover the depth z from given image measurements. Supposing we know the focal length and the baseline, the depth information can be calculated as:

$$z = \frac{fb}{d}. \tag{2}$$

Since disparity is the pixel displacement of two different rays projected from the same point on the object, per-pixel disparity can be obtained by searching the corresponding pixel points along the axis of the aligned cameras. Suppose the two cameras' image planes are aligned perfectly on a line, the corresponding points exist in different columns in the images along the line, the so-called epipolar line. Assuming there is a single epipolar line for this corresponding pair, we can narrow down the search range of correspondence in the image within a single line of a certain width.

Computing a fine disparity map from stereo is a four-step process: finding matching cost of corresponding pixels along the epipolar line, aggregating the matching costs to near pixels, building an initial disparity map, and refining the disparity map.

In the process of searching for matching costs, we assume that the scene consists of Lambertian surfaces, where the color of the surface is assumed to be identical in any directions. Often non-Lambertian surfaces, such as specular plastic or metal surfaces, cannot be scanned properly using a stereo system.

However, the initially obtained matching costs contain severe noise and errors. Aggregating the matching costs is equivalent to filtering noisy data and propagating sparse disparity information. There are two popular approaches to cost aggregation: local and non-local methods. The critical difference between these two approaches is the search window size, where all pixels are tested for the similarity of matching costs, i.e., non-local cost aggregation enforces the cost similarity to all pixels in the test image, while local aggregation only accounts for the matching costs of pixels within the local window.

Next, the initially computed disparity candidates need to be selected by minimization of the sum of the matching costs within a boundary condition. There are local or global approaches to this optimization process. Local methods incur less computational cost than the global methods, while global methods yield a more elaborate depth map than that of local methods.

Lastly, we could refine the disparity estimates using a relevant filtering method, such as a median filter or a box filter, yielding a refined depth map. Once we have a disparity map from the last stage, we can calculate the depth information from the computed disparity, the baseline of stereo, and the focal lengths of the cameras using Eq. (2).

1.1.2 Multi-view Stereo

Binocular stereo is fundamental in stereo imaging, but the performance of this stereo is affected by many parameters. In particular, the baseline between the cameras is critical to the performance of a stereo system. However, the baseline must be adapted to the scene configuration for optimal performance. There is no universal configuration of the baseline for real-world conditions.

Wide-baseline stereo reserves more pixels for disparity than narrow-baseline stereo does. Therefore, wide-baseline systems can discriminate depth with a higher resolution. On the other hand, the search range of correspondences increases, and in turn, it increases the chances of false matching. The estimated disparity map is plausible in terms of depth, but it includes many small regions without depth as spatial artifacts (of holes) on the depth map. This missing information is caused by occlusion and false matching in featureless or pattern-repeated regions, where the corresponding point search fails.

Narrow-baseline stereo has a relatively short search range of correspondence. The search range of matching costs is shorter than that of the wide-baseline stereo. There are fewer chances for false matching so that accuracy and efficiency in cost computation can be enhanced. In addition, the level of spatial noise in the disparity map is low because the occluded area is small. However, narrow-baseline stereo reserves a small number of pixels for depth discrimination. The depth-discriminative power decreases accordingly, whereas the spatial artifacts in the disparity map are reduced. It trades off the discriminative power for the reduced spatial artifacts in the disparity map.

This fundamental limitation of the baseline in binocular stereo has been addressed by the use of more than two cameras, so-called multi-baseline or multi-view stereo. Okutomi and Kanade [37] proposed a multi-baseline stereo method, which is a variant of multi-view stereo. The proposed system consists of multiple cameras on a rail. They presented the matching cost design for the multi-baseline setup. Instead of computing the color difference of a pixel on the reference view and the corresponding point on the other view, the color differences of all views are summed up. This multi-baseline stereo gives more accurate depth estimates than binocular stereo does.

Furukawa and Ponce [14] presented a hybrid patch-based multi-view stereo algorithm that is applicable to objects, scenes, and crowded scene data. Their method produces a set of small patches from matched features, which allows for filling in the gaps between neighboring feature points to be filled in, yielding a fine mesh model. Gallup et al. [15] estimated the depth of a scene by adjusting the baseline and resolutions of images from multiple cameras so that depth estimation becomes computationally efficient. This system exploits the advantages of multi-baseline stereo while requiring the mechanical support of the moving cameras. Nakabo et al. [32] presented a variable-baseline stereo system on a linear slider. They controlled the baseline of the stereo system depending on the target scene to estimate the accurate depth map.

Zilly et al. [47] introduced a multi-baseline stereo system with various baselines. Four cameras are configured in multiple baselines on a rail. The two inner cameras establish a narrow-baseline stereo pair while two outer cameras form a wide-baseline stereo pair. They then merge depth maps from two different baselines. The camera viewpoints in the multi-baseline systems are secured mechanically at fixed locations in general. This design restricts the spatial resolution along the camera array while the depth map is being reconstructed. Refer to [38] for the in-depth investigation of other multi-view methods.

1.1.3 Refractive Stereo

Refractive stereo estimates depth using the refraction of light via a transparent medium. Suppose a 3D point p in a target scene is projected to p_d on an image plane through the optical center of an objective lens C directly without any transparent medium (Fig. 2a). Inserting a transparent medium in the light path changes the transport of the incident beam from p, and it reaches at p_r on the image plane with a

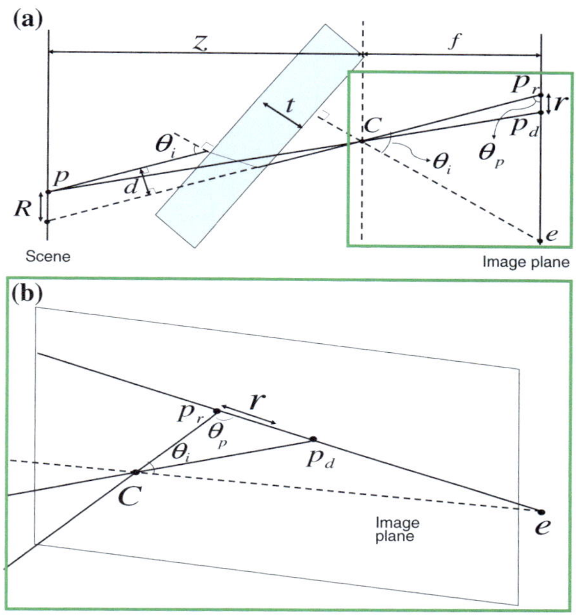

Fig. 2 **a** A cross-section view of the light path in refractive stereo. **b** A close-up view of refractive light transport in 3D. Image courtesy of © 2015 Elsevier Computer Vision and Image Understanding [2]

lateral displacement d (between with and without the medium). The displacement between p_d and p_r on the image plane is called *refractive disparity*.

Now we formulate the depth z of p using simple trigonometry as follows [16, 17]:

$$z = f\frac{R}{r}, \tag{3}$$

where r is a refractive disparity completed by searching a pair of corresponding points, f is the focal length, and R is the ratio of lateral displacement d to $\sin(\theta_p)$:

$$R = \frac{d}{\sin(\theta_p)}, \tag{4}$$

Here θ_p is the angle between $\overrightarrow{p_rC}$ and the image plane. To obtain the value of R, we first compute $\cos(\theta_p)$ as

$$\cos(\theta_p) = \frac{\overrightarrow{p_re} \cdot \overrightarrow{p_rC}}{\left|\overrightarrow{p_re}\right|\left|\overrightarrow{p_rC}\right|}. \tag{5}$$

Then, we simply assign $\sin(\theta_p)$ into Eq. (4) after computing $\sin(\theta_p)$ with a simple equation:

$$\sin^2(\theta_p) + \cos^2(\theta_p) = 1. \tag{6}$$

Lateral displacement d, the parallel-shifted length of the light passing through the medium, is determined as [20]

$$d = \left(1 - \sqrt{\frac{1 - \sin^2(\theta_i)}{n^2 - \sin^2(\theta_i)}}\right) t \sin(\theta_i), \tag{7}$$

where t is the thickness of the medium, n is the refractive index of the medium, and θ_i is the incident angle of the light. Here, $\sin(\theta_i)$ can be obtained in a similar manner as the case of $\sin(\theta_p)$ using the following equation:

$$\cos(\theta_i) = \frac{\overrightarrow{p_rC} \cdot \overrightarrow{eC}}{\left|\overrightarrow{p_rC}\right|\left|\overrightarrow{eC}\right|}. \tag{8}$$

The refracted point p_r lies on a line, the so-called *essential line*, passing through an *essential point* e (an intersecting point of the normal vector of the transparent medium to the image plane) and p_d (Fig. 2b). This property can be utilized to narrow down the search range of correspondences onto the essential line, allowing us to compute matching costs efficiently. It is worth noting that disparity in refractive stereo depends on not only the depth z of p but also the projection position p_d of light and the position of the essential point e, whereas the disparity in traditional stereo depends on only the depth z of the point p. Prior to estimating a depth, we calibrate these optical properties in refractive stereo in advance.

Nishimoto and Shirai [36] first introduced a refractive camera system by placing a refractive medium in front of a camera. Rather than computing depth from refraction, their method estimates depth using a pair of a direct image and a refracted one, assuming that the refracted image is equivalent to one of the binocular stereo images. Lee and Kweon [26] presented a single camera system that captures a stereo pair with a bi-prism. The bi-prism is installed in front of the objective lens to separate the input image into a stereo pair with refractive shift. The captured image includes a stereo image pair with a baseline. Depth estimation is analog to the traditional methods. Gao and Ahuja [16, 17] proposed a seminal refractive stereo method that captures multiple refractive images with a glass medium tilted at different angles. This method requires optical calibration of every pose of the medium. It was extended by placing a glass medium on a rotary stage in [17]. The rotation axis of the titled medium is mechanically aligned to the optical axis of the camera. Although the mechanical alignment is cumbersome, this method achieves more accurate depth than the previous one does.

Shimizu and Okutomi [39, 40] introduced a mixed approach that combines the refraction and the reflection phenomena. This method superposes a pair of reflection and refraction images via the surface of a transparent medium. These overlapping images are utilized as a pair of stereo images. Chen et al. [11, 12] proposed a calibration method for refractive stereo. This method finds the pairs of matching points on refractive images with the SIFT algorithm [30] to estimate the pose of a transparent medium. They then search corresponding features using the SIFT flow [29]. By estimating the rough scene depth, they recover the refractive index of a transparent medium.

1.1.4 Other Approaches

In addition to these binocular, multi-view and refraction-based approaches, Levin et al. [27] introduced a coded aperture-based approach, in which they insert a coded aperture blade inside a camera lens instead of a conventional aperture. It allows to estimate depth to be estimated by the evaluation of blur kernels of the coded aperture.

Bando et al. [3] presented a color-filtered aperture in a commodity camera, where the sub-apertures of red, green and blue colors are windowed at different positions. This optical design enables the camera to form three color channels with geometric shift at different positions to yield depth. They extract depth from the shifted channels, analogous to traditional depth from defocus.

Recently, Baek and Kim [1, 2] introduced a hybrid approach, so-called stereo fusion that combines binocular and refractive stereo, using a refractive medium on a binocular base. The performance of depth reconstruction in binocular stereo relies on how adequate the predefined baseline for a target scene is. Wide-baseline stereo is capable of discriminating depth better than the narrow one, but it often suffers from spatial artifacts. Narrow-baseline stereo can provide a more elaborate depth map with fewer artifacts, while its depth resolution tends to be biased or coarse due to the short disparity. Therefore, Baek and Kim [1, 2] proposed an optical design of heterogeneous stereo fusion on a binocular imaging system with a refractive medium, where the binocular stereo part operates as wide-baseline stereo; the refractive stereo module works as narrow-baseline stereo. They then introduced a stereo fusion workflow that combines the refractive and binocular stereo algorithms to estimate fine depth information through this fusion design. Their stereo fusion system outperforms homogeneous stereo approaches in measuring depth.

1.2 Active 3D Imaging

As mentioned in section "Passive 3D Imaging", it is necessary to search corresponding points in estimating a depth per pixel. The most ambiguous part of passive 3D imaging is determining how to search corresponding points. If we could identify corresponding point more confidently, it would enable a higher accuracy in measuring the 3D shape of an object or a scene. This active 3D imaging approach is often called 3D scanning. The most common methods are 3D scanning with swept-planes, structured lighting and photometric stereo. Assuming that a target object is static, these methods require the time to sweep the shadow plane across the object. Recently, some time-of-flight techniques were introduced to overcome this limitation of static objects; however, the spatial resolution of these methods is relatively low. In this section, we briefly survey the foundations of 3D scanning with swept-planes, structured lighting and photometric stereo.

1.2.1 3D Scanning with Swept-Planes

Bouguet and Perona [7] introduced a seminal work on 3D scanning. It is a simple and inexpensive solution for extracting the 3D shape of static objects. This method requires a desk-lamp, a stick and a checker board. The camera captures the object illuminated by the desk-lamp. This method is based on a simple idea. The user is supposed to move the stick in front of the light source, casting a moving shadow over the surface of the object. Then the 3D shape can be obtained from the spatial and temporal location of the captured shadow.

1.2.2 3D Scanning with Structured Lighting

Bouguet and Perona's [7] idea has been developed with a projector. A camera and a projector are coupled to establish correspondences through calibration in a 3D scanning system with structured lighting. Once the correspondences are established, a 3D point cloud is reconstructed using ray-plane triangulation. Instead of a simple swept-plane sequence, spatially-encoded planes, temporally-encoded planes, or a combination of both spatial and temporal encodings in a projector-based system allows us to reconstruct still objects or even dynamic scenes.

1.2.3 Photometric Stereo

Photometric stereo, also known as *shape-from-shading*, estimates surface gradients using images taken under multiple light directions, assuming that the surface reflection satisfies the Lambertian constraints. For deeper understanding of photometric stereo, it is necessary to understand the foundations of light transport, formulated in the beginning of computer graphics.

Light Transport. Kajiya [23] models recursive light transport w.r.t. the incident Θ' and the exitant Θ light directions in the hemispherical domain as a rendering equation:

$$L(x \rightarrow \Theta) = L_e(x \rightarrow \Theta) + \int_{\Omega_x} \rho(x, \Theta' \rightarrow \Theta) L(x \leftarrow \Theta') \cos(N_x, \Theta') d\omega_{\Theta'},$$

where $\rho()$ is a reflectance function, and N_x is a surface normal at the point x.

In a diffuse environment, self-emitted radiance $L_e()$ and reflectance $\rho()$ do not depend on the incident and exitant light directions. Although the incident radiance, say $L(x \leftarrow \Theta')$, still depends on the incident direction, the light transport on purely diffuse surfaces can be simplified:

$$L(x) = L_e(x) + \int_{\Omega_x} \rho(x) L(x \leftarrow \Theta') \cos(N_x, \Theta') d\omega_{\Theta'}.$$

The spherical integral over the hemisphere Ω_x can be transformed into an integral over all surfaces S in the scene. Hence no directions appear anymore in the rendering equation:

$$L(x) = L_e(x) + \rho(x) \int_S K(x, y) L(y) dA_y,$$

where $K(x, y)$ is the product of a binary visibility $V(x, y)$ and the geometrical relationship $G(x, y)$ between the illuminating surface y and the reflected surface x at a distance r_{xy}:

$$G(x, y) = \cos(N_x, \Theta') \cos(N_y, -\Theta') / r_{xy}^2. \tag{9}$$

We now rewrite the above in a discrete matrix-vector form:

$$L = L_e + (\rho K) L, \tag{10}$$

where L is the radiance vector of each infinitesimal patch dA, L_e is the self-emitted radiance vector of each patch, $K = V \cdot G$, and ρK is the exitant diffuse illumination vector (so-called radiosity). Assuming a state of equilibrium for light transfer, we can rewrite Eq. (10) as $L = (I - \rho K)^{-1} L_e$. Again we can expand this into a Neumann series:

$$L = L_e + (\rho K)L_e + (\rho K)^2 L_e + \cdots + (\rho K)^n L_e. \tag{11}$$

In this form, we can easily model how much light energy is contributed from n-bounded light; the nth order of the polynomial is equivalent to the effect of n-bounded light. This allows us to remove indirection illumination from reflection.

Surface Normals. A shading illuminated by a point light, I can be calculated as a dot product between the incident radiance L and surface normal N: $I = L \cdot N$.

Supposing we obtain i images under a different light in photometric illumination, we can obtain the following linear system for each point of the surface:

$$\begin{bmatrix} I_1 \\ I_2 \\ \vdots \\ I_i \end{bmatrix} = \begin{bmatrix} L_{1,x} & L_{1,y} & L_{1,z} \\ L_{2,x} & L_{2,y} & L_{2,z} \\ & \vdots & \\ L_{i,x} & L_{i,y} & L_{i,z} \end{bmatrix} \begin{bmatrix} N_x \\ N_y \\ N_z \end{bmatrix}.$$

To solve the linear system above, the row-rank of the matrix L should be at least three. With more than three light sources used, it becomes an over-constrained linear system, and it can be solved using least-squares to obtain a normal estimation N.

Shape From Normals. Once we obtain a surface normal per pixel, each point on the image now has a normal N of $\{N_x, N_y, N_z\}$; therefore, we solve for the height field z at (x, y) by minimizing an objective function [5]:

$$\Gamma(z) = \sum_{x,y} \left(N_z \frac{\partial z(x,y)}{\partial x} + N_x \right)^2 + \left(N_z \frac{\partial z(x,y)}{\partial y} + N_y \right)^2.$$

where we approximate the ratios of the partial derivatives of z to x and y assuming orthographic projection:

$$\frac{\partial z(x,y)}{\partial x} = z(x+1, y) - z(x, y);$$
$$\frac{\partial z(x,y)}{\partial y} = z(x, y+1) - z(x, y).$$

Once we obtain a set of point clouds (a set of $\{x, y, z\}$), we recover the 3D shape of the surface by indexing neighboring points.

Interreflection in Photometric Stereo. Photometric stereo estimates surface gradients using images taken under multiple light directions, assuming that the surface reflection observes the Lambertian constraints. However, interreflection breaks this assumption, and it causes critical problems in photometric stereo. A few works have addressed these problems by removing interreflection in photometric stereo.

Nayar et al. [34] presented an iterative method to estimate non-biased surface normals. They first estimated a pseudo shape, a shape that contains interreflection in its shape, and reflectance, and they iteratively corrected the pseudo-shape so that it would converge to the real shape. They also showed the convergence property of their iterative algorithm. Nayar et al. [35] introduced a method that removes interreflection using structured light patterns. They used high frequency illumination patterns in order to separate the direct and the indirect illuminations of a scene. While two illumination patterns are enough to separate the indirect illumination theoretically, three illumination patterns were used in practice. For photometric stereo, they used the high frequency illumination patterns for each light source; thus at least triple-number of images were required. Liao et al. [28] presented an active method to remove n-bounded light from photometric stereo using colored multiplex lighting. The proposed algorithm theoretically assumes that there are at least two images of an object with the same illumination but varying surface albedos. They modeled and solved an interreflection problem based on monochromatic surface albedo. Gupta and Nayar [18] presented a micro-phase shifting technique, i.e., sinusoidal illumination patterns with high frequency. They also reduced the number of illumination patterns needed for shape recovery using micro phase shift. By using only high frequency patterns, the indirect illumination effects become the same for every image under each illumination pattern so that indirect illumination effects can be removed. Most of the prior interreflection-removal methods have employed active approaches with the specially designed illumination patterns or the spectrum of the modified light sources. Recently, Nam and Kim [33] proposed an interreflection removal technique that uses multispectral imaging and does not rely on any structure of the light source. They presented a novel multispectral photometric stereo method that allows us to remove interreflection on diffuse materials to be removed by the use of multispectral reflectance information. Their proposed method can be easily integrated into an existing photometric stereo system by simply substituting the current camera with a multispectral camera, as their method does not rely on additional structured or colored lights. They demonstrated several benefits of their multispectral photometric stereo method such as interreflection removal and the reconstruction of the 3D shapes of objects with high accuracy.

2 Applications of Advanced 3D Imaging

Reconstructing a 3D object model from multiple overlapping geometry scans has been an active area in the past decade in computer graphics [6]. In general, there are two different types of 3D scanning systems: typically a triangulation-based system for small objects and a time-of-flight system for large scale objects such as a building. Such 3D scanning systems have been coupled to a color imager and lights to capture surface color information as texture. Camera calibration and registration between the camera system and the three-dimensional scanner are necessary to investigate the interrelationship between the spectral and geometric information.

Bernardini and Rushmeier [6] surveyed and summarized the general 3D scanning pipeline that has been employed by many research projects and commercial systems. Recently, Holroyd et al. [22] presented a two-way 3D imaging system that allows us to extract both three-dimensional shapes and reflectance functions from the same set of image data. The system achieves high accuracy in registration of the shape and reflectance and also explores the directional changes of material appearance. However, the spectral resolution of the system is limited to the trichromatic RGB channels.

Recently, a straightforward approach to hyperspectral 3D imaging has been introduced, which swaps out the standard RGB camera used in current 3D scanning systems and replacing it with a two-dimensional hyperspectral imager. As one of the seminal approaches, Mansouri et al. [31] attempted to integrate a two-dimensional multispectral imager into a three-dimensional range scanning system. Similar to the method of Brauers et al. [9], a set of seven bandpass filters is employed and accompanied by an LCD projector. This projector illuminates the surface of the target object to measure the topology of the 3D surface. This system captures a hyperspectral image and maps it to a scanned surface as a texture map. This seminal system is limited to capturing a flat surface only.

2.1 3D Imaging Spectroscopy

Kim et al. [24] introduced a 3D imaging spectroscopy (3DIS) system the integrates 2D imaging spectroscopy and 3D scanning, which is the first complete hyperspectral 3D imaging system to yield complete 3D scanning models. This enables the measurement of physically-meaningful 3D hyperspectral patterns of arbitrarily-shaped solid objects with high accuracy. In particular, they proposed a modification of a dispersion-based hyperspectral imaging design [45] to achieve high enough spatial and spectral resolution to build a 3D hyperspectral pattern from the captured 2D hyperspectral images. Figure 3 shows an overview of the hyperspectral 3D imaging system.

2.1.1 The Dispersion-Based Hyperspectral Imager

The design of the hyperspectral imager [24] is based on the snapshot-based design. They coupled dispersive prism optics and a coded aperture mask to resolve spatio-spectral information for radiometric sampling. They then addressed the under-determined problem by solving sparsity-constrained optimization problems.

Unlike bandpass filter-based systems, their imaging system measures continuous hyperspectral patterns from NUV-A (359 nm) to NIR (1 μm). To increase the efficiency of the UV spectrum, this system was built with specialized optical materials, such as fused silica (FS) and calcium fluoride (CaF_2). These optical components enable this system to exceed the spectral range of traditional imaging

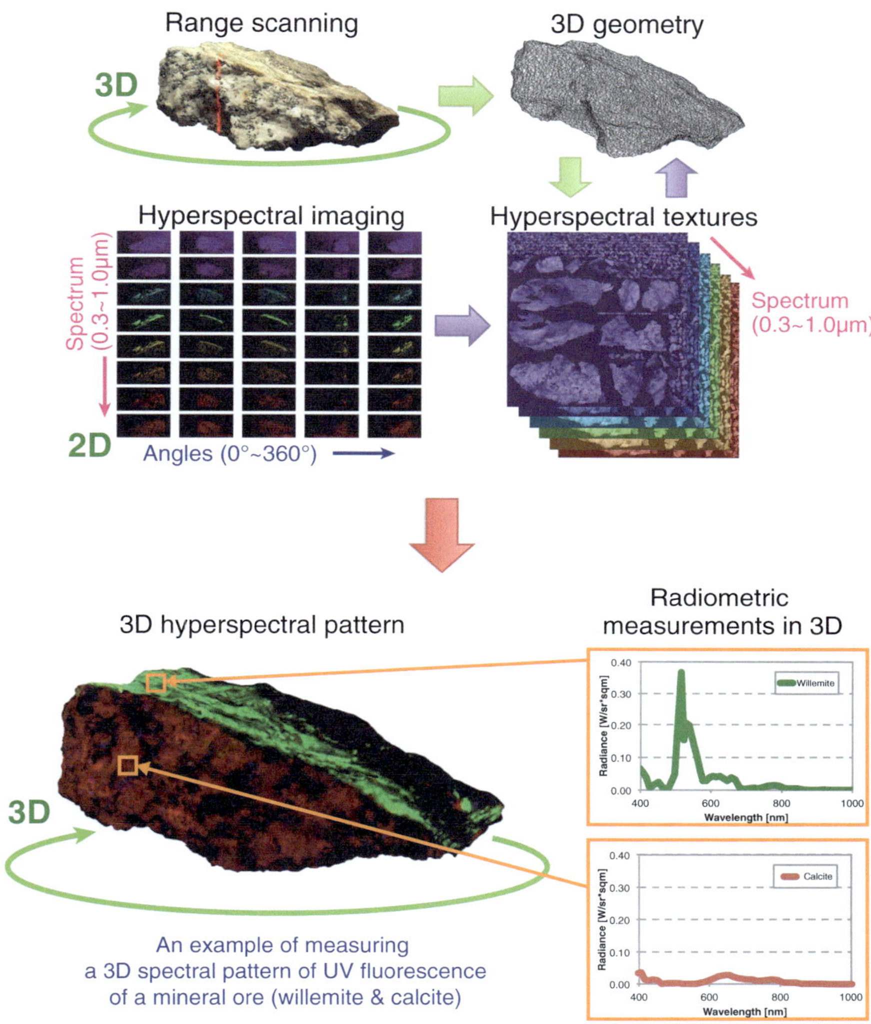

Fig. 3 An overview of the hyperspectral 3D imaging system built by Kim et al. [24] for measuring 3D hyperspectral patterns on 3D solid objects. This 3D imaging spectroscopy system, so-called 3DIS, measures 3D geometry and hyperspectral radiance simultaneously. Piecewise geometries and radiances are reconstructed into a 3D hyperspectral pattern. This 3DIS system is used to acquire physically-meaningful 3D hyperspectral patterns of various wavelengths for scientific research. Image courtesy of © 2012 ACM Transactions on Graphics [24]

systems, where UV transmittance below 400 nm decreases rapidly due to absorption effects. In contrast to the IR band, the UV band is challenging for imaging due to the inherent transmittance characteristics of the optical substrate of glass components.

In the hyperspectral imager, a random-pattern coded aperture is lithographically etched on a quartz substrate. A piezoelectric translation stage modulates the aperture, and the aperture code is then directly relayed onto the monochromatic imaging sensor. The system includes relay lenses and a double Amici prism to disperse the incoming rays. The light sources used in this system are a Xenon light bulb and a UV fluorescence light for measuring UV fluorescence.

2.1.2 3D System Integration

The 2D hyperspectral imager is integrated into a 3D imaging pipeline in order to measure not only the 3D shape but also the reflectance of 3D solid objects. Their 3D scanning pipeline is based on the classic 3D imaging workflow described in [13]. The hyperspectral imager, a laser range scanner and a Xenon light source are mounted together on a standard optical table located in a darkroom. The positions of the imager system, the light source and turntable axis are calibrated in terms of the laser scanner coordinate system, using the standard calibration targets.

Figure 4 presents the design of the hyperspectral 3D imaging system. A laser projector shines a thin sheet of light onto the object. On each scan line, the laser sensor detects the reflected laser light to produce a depth map. A Xenon light source (or a UV fluorescent light) illuminates the surface with a broad spectrum. The hyperspectral imager measures reflected radiance to compute a reflectance map with respect to the reference white.

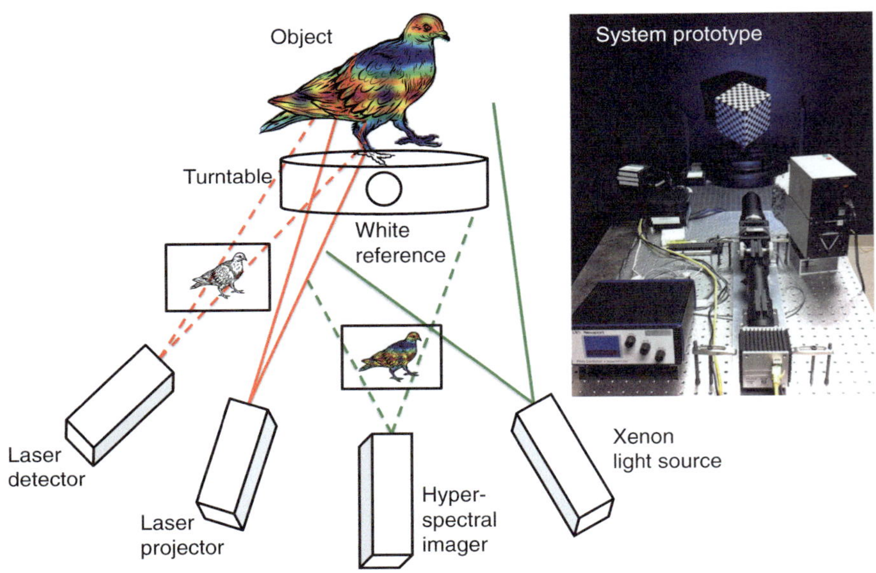

Fig. 4 Principal design of the 3DIS system [24]. *Inset* a photograph of the prototype system. Image courtesy of © 2012 ACM Transactions on Graphics [24]

2.1.3 Measuring 3D Hyperspectral Patterns

The 3DIS system is demonstrated with various practical and scientific applications for measuring hyperspectral patterns on 3D solid objects. The imaging system was used for the non-destructive measurement of the 3D reflectance and fluorescence patterns of biological organisms, minerals and an archaeological artifact in collaboration with Yale Peabody Museum of Natural History. Figure 5 shows an example of a scanned 3D object.

2.2 *Hyperspectral Photometric Stereo*

2.2.1 Measuring Surface Normals

Photometric stereo is a 3D imaging technique that has been commonly performed to capture the shape of 3D solid objects in computer vision for more than three

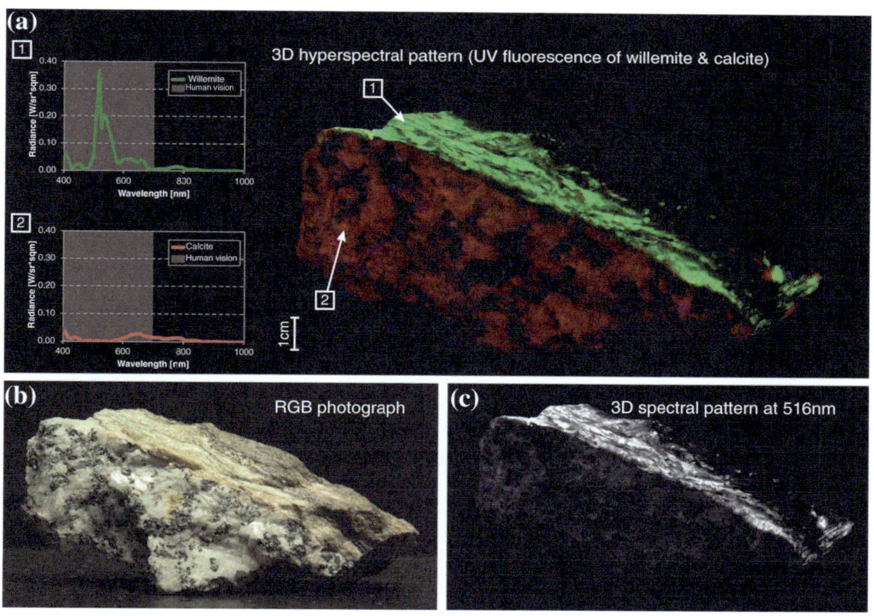

Fig. 5 An example of scanning an ore sample that includes willemite, calcite and magnetite, captured by the 3DIS system [24]. **a** Is a physically-based rendering result of the 3D hyperspectral pattern of UV fluorescence with associated spectral readings. *Inset* (*top*) NUV-induced fluorescent radiance of willemite. *Inset* (*bottom*) fluorescent radiance of calcite. An ultra-violet spectrum (260–390 nm) illuminates the object, and the emitted fluorescent radiance (excluding reflected UV) is measured and rendered in 3D. **b** Shows the appearance of the ore under white light captured in a photograph. **c** Is a rendering of the 3D spectral pattern at 516 nm, where the willemite presents a spectral peak of fluorescence. Image courtesy of © 2012 ACM Transactions on Graphics [24]

decades. Photometric stereo estimates surface normal vectors over the surface of the 3D solid objects, yielding a normal map, the surface topology description of the 3D shape orientation. Photometric stereo captures the shading information over the surface by varying the position of point light sources. This allows us to estimate the normal vectors from shading, measured as pixel intensities by a monochromatic camera [5]. In contrast, the classical binocular stereo estimates the depth information from the parallax disparity, caused by placing the two cameras at a distance away from each other along a base line. One of the virtues of photometric stereo is that it produces a high-resolution normal map from a relatively simple setup, which includes a camera and multiple light sources. However, photometric stereo and hyperspectral imaging have been rarely combined and practiced together in the computer vision and hyperspectral imaging.

2.2.2 Combining Hyperspectral Imaging with Photometric Stereo

In photometric stereo, many optical phenomena occur as obstacles, such as indirect illumination, specular reflection and self shadows, degrading the accuracy of the shape measured by photometric stereo. Much research has focused on reconstructing surface normals from Lambertian and non-Lambertian reflections by removing self shadows and specular reflections from photometric stereo [4, 5, 10, 21, 41, 43, 46]. However, removing the indirect illumination effect from photometric stereo [28, 34] has received less attention.

Interreflection is an optical phenomenon that occurs over a concave surface. When one of the sides is illuminated, the reflected light illuminates the neighboring side in the concave shape, where two points over the surface face each other. Most photometric stereo methods are designed with the general assumption that the surface has Lambertian reflection and the geometric proxy of the object is a convex shape, which does not suffer from interreflection. However, objects in the real world comprise a mixture of convex and concave shapes. Removing interreflection in photometric stereo is a traditional chicken-and-egg problem as we need to account for interreflection without knowing geometry. This is a typical problem when we capture a 3D surface geometry with concave shape using photometric stereo. See Fig. 6c for examples.

Nam and Kim [33] proposed a hyperspectral (a.k.a. multispectral, which means acquiring visible spectral information in hyperspectral imaging) photometric stereo method to remove interreflection from diffuse materials of a concave surface while capturing a 3D shape with photometric stereo. They estimate the amount of interreflection on a monochromatic diffuse surface using the reflectance information of the visible spectrum. The method is integrated into a typical photometric stereo system by simply substituting a multispectral imager for the RGB camera as the method does not rely on additional structured or colored lights. Figure 6 shows a schematic overview of their hyperspectral photometric stereo method.

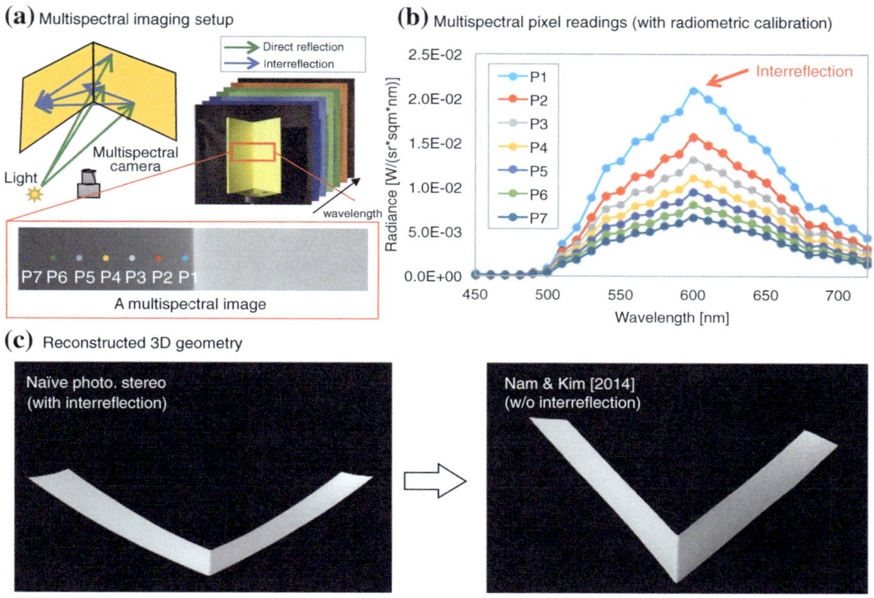

Fig. 6 Schematic diagram of the multispectral photometric stereo method proposed by Nam and Kim [33]. This method allows the removal of interreflection over a monochromatic diffuse surface. **a** Presents the imaging setup and an example of a captured multispectral image of an L-shaped object in 90°. As shown in the closeup view, the inner faces of the object present interreflection along the direct reflection. **b** Shows the spectral power distribution measured by the multispectral imager over the seven points in the closeup view. **c** compares the surfaces reconstructed by ordinary photometric stereo (*upper*) and their proposed method (*lower*). The reconstructed geometry using the Nam and Kim's [33] method is much closer to the physical shape (L-shaped in 90°) of the captured object, as compared to the geometry obtained by a naïve photometric stereo method. Image courtesy of © 2014 IEEE Computer Graphics and Applications [33]

2.2.3 Removing Interreflection

Removing interreflection is challenging as the effect is integrated in a light path, where a ray of light is emitted from a light source, travels through a medium such as air, reflects on the surface of an object, and enters the camera. The multiple-bounded light is affected by the surface albedo of the reflecting surface. The multiple-bounded light becomes a new light source, so-called radiosity, and the reflection is added at each point. Most of energy that enters the camera is either directly from the light source or one bounded light from the object surface, which is the product of the light and surface reflectance. The portion of the second and higher bounded light varies from scene to scene. Some objects or scenes are more susceptible to indirect illumination. In this case, the radiance measured by a camera is affected not only by the illumination and the surface albedo, but also by the reflectance of the surrounding surfaces.

The key insight of this hyperspectral photometric stereo method is that a hyperspectral camera can capture spectrum-dependent albedo with many channels. Therefore, multiple bounces of wavelength-dependent interreflection are modelled as a polynomial function, and the interreflection effect is optimized through hyperspectral reflectance analysis. This allows us to separate interreflection over diffuse surfaces from measured radiance. Their hyperspectral photometric stereo does not rely on multiplexing spectral lights as the method in [42, 44] do; thus the Nam and Kim's [33] method is capable of acquiring any arbitrary shape and illumination without the help of structured light and colored light.

2.2.4 Measuring a Shape with Hyperspectral Imaging

Nam and Kim [33] demonstrated several benefits of the hyperspectral (a.k.a. multispectral) photometric stereo method, such as the removal of interreflection and the reconstruction of the 3D shapes of objects with high accuracy.

Figure 7 compares the performance variation of their method according to the number of input spectral channels. They scanned a concave-shaped soap, which would have high interreflection. As the soap had specular reflection, polarizing filters were attached in front of the light sources in order to prevent the specular reflection coming into the sensor directly. This figure compares three different 3D photometric stereo results to the ground truth obtained by a 3D laser scanner (NextEngine). Figure 7b shows the reconstruction results of the normals and 3D geometry with the naïve photometric stereo approach (without the removal of interreflection). The reconstructed geometry is flattened compared to the ground truth. Figure 7c, d present the normals and 3D models using the proposed method with two different cameras: an RGB camera and a hyperspectral imager. In Fig. 7c, although the reconstructed geometry is still somewhat flattened, there is an improvement in terms of sharpness at the edges. Figure 7d shows the results of the hyperspectral photometric stereo system using 29 channels. The proposed method yields a geometry that is virtually identical to the ground truth. Using a sufficient number of channels, high-frequency details of the object surface can be obtained, yielding high-fidelity normals and 3D shapes.

2.3 Stereo Fusion of Refractive and Binocular Stereo

The performance of depth reconstruction in binocular stereo relies on how adequate the predefined baseline for a target scene is. Wide-baseline stereo is capable of discriminating depth better than the narrow-baseline stereo, but it often suffers from spatial artifacts. Narrow-baseline stereo can provide a more elaborate depth map with fewer artifacts, while its depth resolution tends to be biased or coarse due to the short disparity. Baek and Kim [1, 2] proposed a novel optical design of

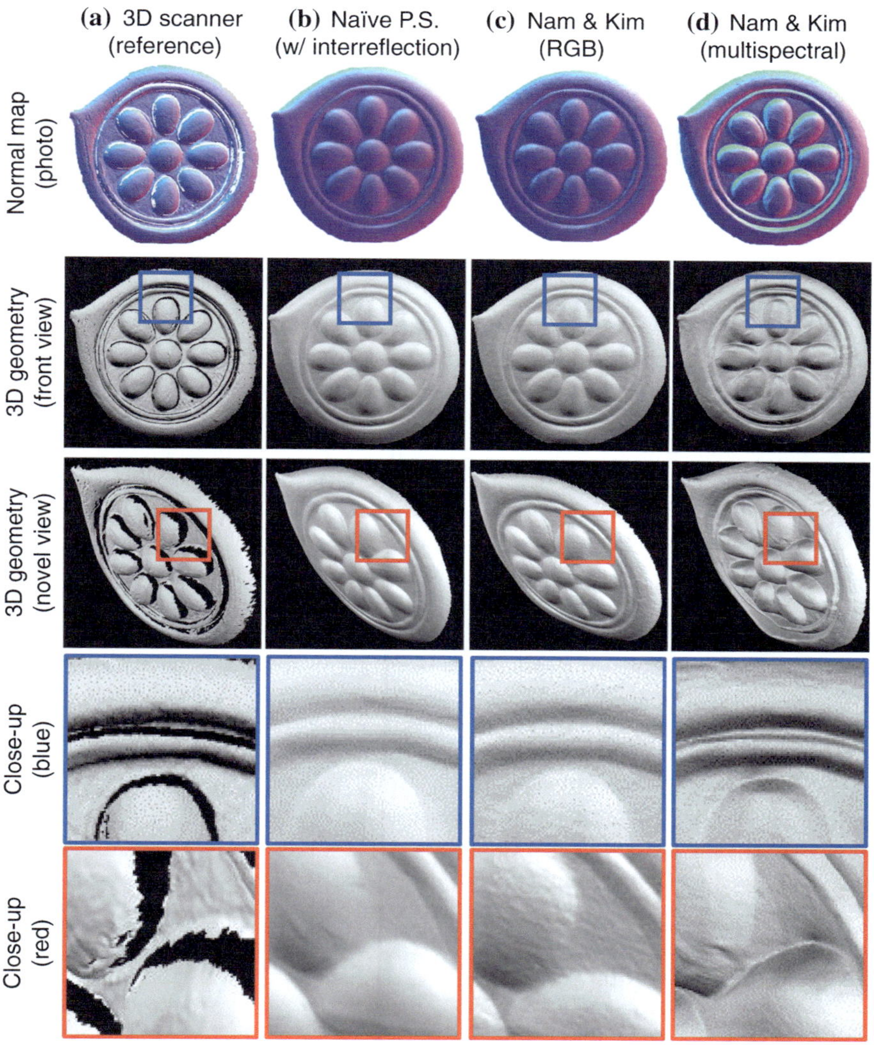

Fig. 7 Nam and Kim [33] compared the reconstruction 3D models depending on the number of input spectral channels. **a** Shows the ground truth obtained by a 3D laser scanner (NextEngine). **b** Is the result of naïve photometric stereo without the removal of interreflection. **c** Is the result of application of their hyperspectral method applying it to an RGB camera using three spectral channels. **d** Is the result of the proposed hyperspectral photometric stereo using 29 visible channels. The reconstructed shape is virtually identical to the ground truth. Image courtesy of © 2014 IEEE Computer Graphics and Applications [33]

heterogeneous stereo fusion on a binocular imaging system with a refractive medium, where the binocular stereo part operates as wide-baseline stereo, and the refractive stereo module works as narrow-baseline stereo. They then introduced a stereo fusion workflow that combines the refractive and binocular stereo algorithms to estimate fine depth information through this fusion design.

Hardware Design. Their stereo fusion system consists of two cameras and a transparent medium on a mechanical support structure. The both camera lenses have the same focal length of 8 mm. The cameras are placed on a rail in parallel with a baseline of 10 cm to configure binocular stereo. A transparent medium is placed on a rotary stage for refractive stereo in front of one of the binocular stereo cameras. See Fig. 8 for the hardware design. Note that this refractive stereo module is equivalent to narrow-baseline stereo while the binocular stereo structure is equivalent to wide-baseline stereo in their system.

Stereo Fusion. Their stereo fusion workflow consists of two main steps. They first estimate an intermediate depth map from a set of refractive stereo images (from the camera with the medium) and reconstruct a virtual direct image. Then, this virtual image and a direct image (from the other camera without the medium in a baseline) are used to estimate the final depth map referring to the intermediate depth map from refractive stereo. Figure 9 overviews the workflow of their stereo fusion method.

They compared their method with a renowned graphcut-based algorithm [8] with an image of the same resolution. In general, global stereo methods allow for an accurate depth map to be obtained, but they incur high computational cost. It is not surprising that this global method was about eight times slower than their method (see Fig. 10a). They also compared their method with a local binocular method [19], which computes the matching cost as the norm of intensity difference and aggregates the cost using the weight of the guided filter [19]. Its computing time was ~212 s with the same scene (see Fig. 10b). This local method struggles with typical false matching artifacts. A refractive method using SIFT flow [12] is compared to their method (Fig. 10c, d). The same number of six refractive images were employed for both methods. While the refractive method suffers from wavy

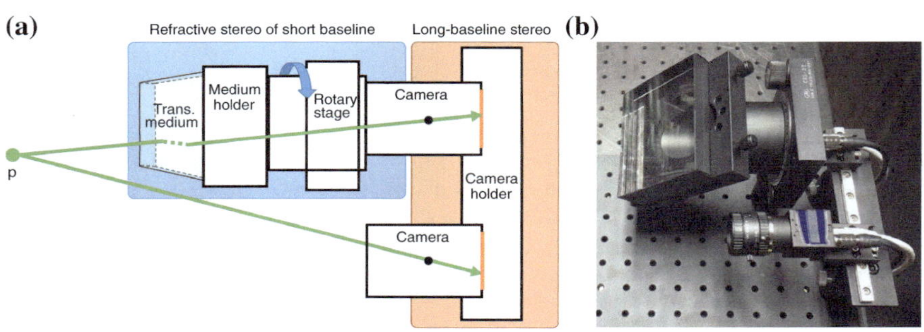

Fig. 8 **a** The schematic diagram of the Baek and Kim's [1] stereo fusion system. A point p is captured by both the refractive stereo and the binocular stereo module. **b** The Baek and Kim's [1, 2] system prototype. Image courtesy of © 2015 Elsevier Computer Vision and Image Understanding [1, 2]

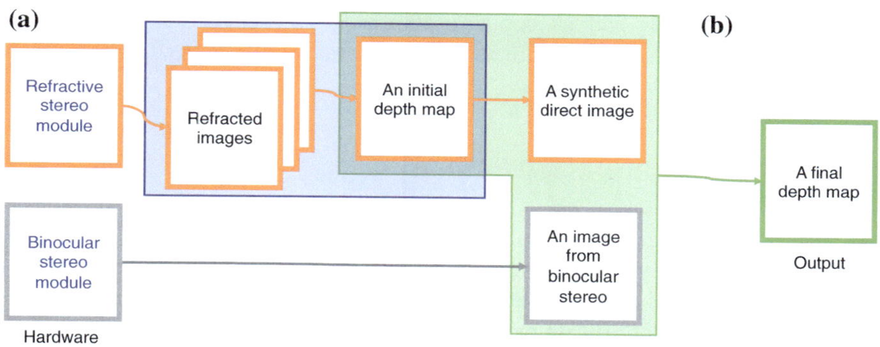

Fig. 9 Schematic diagram of the Baek and Kim's [1] stereo fusion method. **a** Their refractive stereo method estimates an intermediate depth map from refractive stereo. **b** Their stereo fusion method reconstructs a final depth map from a pair of an image, one from binocular stereo and a synthetic direct image obtained using the intermediate depth map. Image courtesy of © 2015 Elsevier Computer Vision and Image Understanding [1, 2]

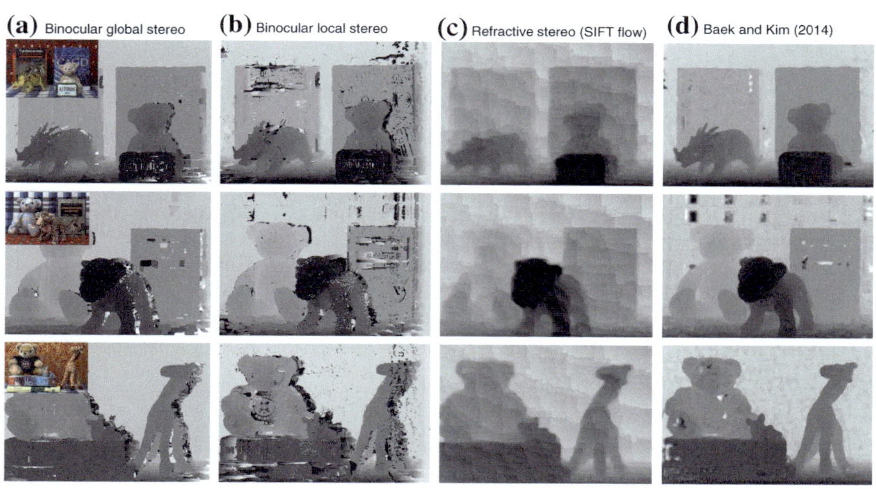

Fig. 10 Depth maps of three different scenes in each row were computed by four different methods. The *first two columns* (**a**) and (**b**) show results obtained with global [8] and local binocular stereo [19] methods. The *third column* (**c**) presents the results obtained by the refractive stereo method [12]. The Baek and Kim's [1] method (**d**) estimates depth accurately without suffering from severe artifacts. Image courtesy of © 2015 Elsevier Computer Vision and Image Understanding [1, 2]

artifacts of SIFT flow and its depth resolution is very coarse, as is typical of refractive stereo, their method estimates depth accurately with fewer spatial artifacts in all test scenes.

3 Conclusions

This chapter briefly surveyed the foundations of 3D imaging: the relationship between disparity and depth in stereo imaging and popular a 3D imaging method that enables the building of 3D models. In addition, related applications related to advanced 3D imaging have been introduced: hyperspectral 3D imaging, multi-spectral photometric stereo, and stereo fusion of refractive and binocular stereo. Expanding the dimensions of digital imaging, the recent advances in 3D imaging technology are about to be combined with smart devices, resulting in broadened applications of 3D imaging.

Acknowledgements This work was supported by a Korea NRF grant (2013R1A1A1010165) and the Center for Integrated Smart Sensors, funded by the Ministry of Science, ICT & Future Planning, as the Global Frontier Project.

References

1. Baek SH, Kim MH (2014) Stereo fusion using a refractive medium on a binocular base. In: Proceedings Asian conference on computer vision (ACCV 2014). Springer, LNCS, Singapore, pp 1–16
2. Baek SH, Kim MH (2015) Stereo fusion: combining refractive and binocular disparity. In: Computer vision and image understanding (CVIU), pp 1–42
3. Bando Y, Chen BY, Nishita T (2008) Extracting depth and matte using a color-filtered aperture. ACM Trans Graph 27(5):134:1–134:9
4. Barsky S, Petrou M (2003) The 4-source photometric stereo technique for three-dimensional surfaces in the presence of highlights and shadows. IEEE Trans Pattern Anal Mach Intell 25 (10):1239–1252
5. Basri R, Jacobs DW, Kemelmacher I (2007) Photometric stereo with general, unknown lighting. Int J Comput Vision 72(3):239–257
6. Bernardini F, Rushmeier H (2002) The 3D model acquisition pipeline. Comput Graph Forum 21(2):149
7. Bouguet JY, Perona P (1998) 3D photography on your desk. In: ICCV, pp 43–52
8. Boykov Y, Veksler O, Zabih R (2001) Fast approximate energy minimization via graph cuts. IEEE Trans Pattern Anal Mach Intell 23(11):1222–1239
9. Brauers J, Schulte N, Aach T (2007) Modeling and compensation of geometric distortions of multispectral cameras with optical bandpass filter wheels. In: 15th European signal processing conference, vol 15, pp 1902–1906
10. Chandraker M, Agarwal S, Kriegman D (2007) Shadowcuts: photometric stereo with shadows. In: IEEE conference on computer vision and pattern recognition, CVPR'07. IEEE, pp 1–8
11. Chen Z, Wong K, Matsushita Y, Zhu X, Liu M (2011) Self-calibrating depth from refraction. In: Proceedings international conference on computer vision (ICCV), pp 635–642

12. Chen Z, Wong KYK, Matsushita Y, Zhu X (2013) Depth from refraction using a transparent medium with unknown pose and refractive index. Int J Comput Vision 8:1–15
13. Farouk M, Rifai IE, Tayar SE, Shishiny HE, Hosny M, Rayes ME, Gomes J, Giordano F, Rushmeier HE, Bernardini F, Magerlein K (2003) Scanning and processing 3D objects for web display. In: Proceedings international conference on 3D digital imaging and modeling (3DIM), pp 310–317
14. Furukawa Y, Ponce J (2010) Accurate, dense, and robust multiview stereopsis. IEEE Trans Pattern Anal Mach Intell 32(8):1362–1376
15. Gallup D, Frahm JM, Mordohai P, Pollefeys M (2008) Variable baseline/resolution stereo. In: Proceedings on computer vision and pattern recognition (CVPR), pp 1–8
16. Gao C, Ahuja N (2004) Single camera stereo using planar parallel plate. In: Proceedings international conference on pattern recognition (ICPR), vol 4, pp 108–111
17. Gao C, Ahuja N (2006) A refractive camera for acquiring stereo and super-resolution images. In: Proceedings on computer vision and pattern recognition (CVPR), pp 2316–2323
18. Gupta M, Nayar SK (2012) Micro phase shifting. In: Proceedings on computer vision and pattern recognition (CVPR), pp 813–820
19. He K, Sun J, Tang X (2010) Guided image filtering. In: Proceedings on European conference on computer vision (ECCV). Springer, pp 1–14
20. Hecht E (1987) Optics. Addison-Wesley, Reading
21. Hernández C, Vogiatzis G, Cipolla R (2008) Shadows in three-source photometric stereo. In: Computer vision–ECCV 2008. Springer, pp 290–303
22. Holroyd M, Lawrence J, Zickler T (2010) A coaxial optical scanner for synchronous acquisition of 3D geometry and surface reflectance. ACM Trans Graph (Proc SIGGRAPH 2010) 29(3):99:1–99:12
23. Kajiya JT (1986) The rendering equation. In: Proc. ACM SIGGRAPH Computer Graphics '86, vol 20, pp 143–150
24. Kim MH, Harvey TA, Kittle DS, Rushmeier H, Dorsey J, Prum RO, Brady DJ (2012) 3D imaging spectroscopy for measuring hyperspectral patterns on solid objects. ACM Trans Graph (Proc SIGGRAPH 2014) 31(4):38:1–38:11
25. Lanman D, Taubin G (2009) Build your own 3D scanner. ACM SIGGRAPH 2009 Courses on—SIGGRAPH 2009. ACM Press, New York, pp 1–94
26. Lee D, Kweon I (2000) A novel stereo camera system by a biprism. IEEE Trans Rob Autom 16(5):528–541
27. Levin A, Fergus R, Durand F, Freeman WT (2007) Image and depth from a conventional camera with a coded aperture. ACM Trans Graphics 26(3):70:1–70:9
28. Liao M, Huang X, Yang R (2011) Interreflection removal for photometric stereo by using spectrum-dependent albedo. In: Proceedings on computer vision and pattern recognition (CVPR), pp 689–696
29. Liu C, Yuen J, Torralba A (2011) Sift flow: dense correspondence across scenes and its applications. IEEE Trans Pattern Anal Mach Intell 33(5):978–994
30. Lowe DG (2004) Distinctive image features from scale-invariant keypoints. Int J Comput Vision 60(2):91–110
31. Mansouri A, Lathuiliere A, Marzani FS, Voisin Y, Gouton P (2007) Toward a 3d multispectral scanner: an application to multimedia. IEEE MultiMedia 14(1):40–47
32. Nakabo Y, Mukai T, Hattori Y, Takeuchi Y, Ohnishi N (2005) Variable baseline stereo tracking vision system using high-speed linear slider. In: Proceedings international conference on robotics and automation (ICRA), pp 1567–1572
33. Nam G, Kim MH (2014) Multispectral photometric stereo for acquiring high-fidelity surface normals. IEEE Comput Graphics Appl 34(6):57–68
34. Nayar SK, Ikeuchi K, Kanade T (1991) Shape from interreflections. Int J Comput Vision 6(3):173–195
35. Nayar SK, Krishnan G, Grossberg MD, Raskar R (2006) Fast separation of direct and global components of a scene using high frequency illumination. ACM Trans Graph 25(3):935–944

36. Nishimoto Y, Shirai Y (1987) A feature-based stereo model using small disparities. In: Proceedings on computer vision and pattern recognition (CVPR), pp 192–196
37. Okutomi M, Kanade T (1993) A multiple-baseline stereo. IEEE Trans Pattern Anal Mach Intell 15(4):353–363
38. Seitz SM, Curless B, Diebel J, Scharstein D, Szeliski R (2006) A comparison and evaluation of multi-view stereo reconstruction algorithms. In: Proceedings on computer vision and pattern recognition (CVPR), pp 519–528
39. Shimizu M, Okutomi M (2006) Reflection stereo-novel monocular stereo using a transparent plate. In: Proceedings Canadian conference on computer and robot vision (CRV). IEEE, pp 14–14
40. Shimizu M, Okutomi M (2007) Monocular range estimation through a double-sided half-mirror plate. In: Proceedings Canadian conference on computer and robot vision (CRV). IEEE, pp 347–354
41. Sun J, Smith M, Smith L, Midha S, Bamber J (2007) Object surface recovery using a multi-light photometric stereo technique for non-lambertian surfaces subject to shadows and specularities. Image Vis Comput 25(7):1050–1057
42. Takatani T, Matsushita Y, Lin S, Mukaigawa Y, Yagi Y (2013) Enhanced photometric stereo with multispectral images. In: International conference on machine vision applications (MVA). IAPR. pp 1–4
43. Verbiest F, Van Gool L (2008) Photometric stereo with coherent outlier handling and confidence estimation. In: Proceedings on computer vision and pattern recognition (CVPR), pp 1–8
44. Vogiatzis G, Hernández C (2012) Self-calibrated, multi-spectral photometric stereo for 3d face capture. Int J Comput Vision 97(1):91–103
45. Wagadarikar AA, Pitsianis NP, Sun X, Brady DJ (2009) Video rate spectral imaging using a coded aperture snapshot spectral imager. Opt Express 17(8):6368–6388
46. Wu TP, Tang KL, Tang CK, Wong TT (2006) Dense photometric stereo: a markov random field approach. IEEE Trans Pattern Anal Mach Intell 28(11):1830–1846
47. Zilly F, Riechert C, Mller M, Eisert P, Sikora T, Kauff P (2013) Real-time generation of multi-view video plus depth content using mixed narrow and wide baseline. J Visual Commun Image Represent 25(4):632–648

E-R-D Optimization in Video Compression

Hyuk-Jae Lee, Hyun Kim and Chae-Eun Rhee

Abstract In mobile multimedia devices with video compression capability, a reduction of the power consumption in H.264/AVC compression is important to increase battery lifetime. This chapter presents a power-aware design to determine the best combination of operation conditions for multiple power-scaling schemes. To derive the best combination of existing power-scaling schemes, the power saving and rate-distortion (R-D) performances of individual schemes are presented. The combined effects of these schemes on power saving and R-D loss are modeled and the best operation combination is derived. The largest power saving can be achieved with the smallest R-D degradation by selecting an optimized combination from among all possible combinations. The optimized combinations are defined as a power level table comprising ten levels. Depending on the size and motion speed of a video, four different power level tables are designed to achieve performance improvements. For application of such tables to the encoder, the usable power for each period is calculated and the power level suitable for the calculated power budget is selected. This application method uses the given power budget as much as possible and shows a better performance. The presented power level tables are suitable for power control in real-time applications because the tables are developed in advance. The presented power-aware design is tested with four popular power-saving schemes and simulations with these four schemes show that a power saving of about 30 % is achieved for slow-motion videos, whereas these amounts are about 20 % for fast-motion videos at the sacrifice of less than 0.1 dB Bjontegaard Delta PSNR degradation.

H.-J. Lee (✉) · H. Kim
Department of Electrical and Computer Engineering, Inter-University Semiconductor Research Center, Seoul National University, Seoul, Korea
e-mail: hjlee@capp.snu.ac.kr

H. Kim
e-mail: snusbkh0@capp.snu.ac.kr

C.-E. Rhee
School of Information and Communication Engineering, Inha University, Incheon, Korea
e-mail: chae.rhee@inha.ac.kr

C. Kyung (ed.), *Theory and Applications of Smart Cameras*,
KAIST Research Series, DOI 10.1007/978-94-017-9987-4_5

Keywords H.264/AVC • Power-aware design • Power estimation model • Power reduction • Power-scaling scheme • Real-time application

1 Power-Aware Design for Hardware-Based Video Compression

The H.264/AVC video compression standard proposed by the Joint Video Team (JVT) is widely used for multimedia devices due to its high compression efficiency. Reduction of power consumption during H.264/AVC compression is important given the increasing use of mobile multimedia devices in which battery capacity is limited. Extensive research has been undertaken to reduce power consumption by the H.264/AVC encoder. Particular efforts have been made to reduce power consumption of motion estimation (ME), which accounts for a major portion of the power consumption in an H.264/AVC encoder [1–6]. The low-power algorithms for the H.264/AVC encoder significantly reduce the power consumption, but they fail to consider the rate-distortion (R-D) degradation resulting from that power reduction. In the approaches in [7–9], power consumption is controlled by adjusting the encoder's operation condition adaptively depending on battery status, user preferences, and operating environment. In [7], the concept of a power-aware design is introduced and various power-scaling schemes for hardware accelerators in H.264/AVC encoders are presented. In the approach in [8], the power consumption target is defined and the power consumption in the encoder is controlled to meet that target. To control power consumption in the encoder, various operation conditions for integer motion estimation (IME), fractional motion estimation (FME), and intra prediction (IP) are determined. In [9], the operation conditions of the encoder are controlled based on video complexity and the remaining battery capacity. In [10–12], power-rate-distortion (P-R-D) models are proposed in which the relationships between power, rate, and distortion are derived. A large number of simulations are needed when the video sequence is changed or when a new scheme is added because all combinations of schemes should be simulated in each video sequence. Therefore, it is necessary to develop a method to speed up the simulation time in order for the P-R-D models to be practically used to determine the best operation condition for a power consumption target.

This chapter presents a power-aware design for an H.264/AVC encoder; a design that optimizes R-D performance. The proposed power-aware design changes the operation condition in the encoder by using various power-scaling schemes. For an effective combination of various power-scaling schemes, this chapter formulates the model to estimate the cross-effects on power consumption among the scaling schemes. By using the proposed modeling approach, the number of simulations required to determine the best combination of power-scaling schemes is markedly reduced. Consequently, the set of power-scaling schemes is easily composed. Based on the estimated power saving and the simulated R-D performance, a power level table is defined. Each power level in that table has corresponding operation conditions that minimize the R-D loss for a given power consumption target.

To achieve better R-D performance, four different power level tables are defined independently, and dependent on video size and motion speed, in order to characterize power consumption behaviors that are appropriate for the video content. By using these pre-defined power level tables and utilizing the operation conditions derived by power-scaling schemes from pre-defined levels, optimized power control can be easily applied at run time. Decisions related to selection of the proper power level are made periodically. The available power budget for the current period is calculated by considering both the power consumed in previous periods and the power that is to be used in future periods. Subsequently, the power level for the current period is chosen based on the available power budget. In order to evaluate the proposed approach, four popular power-scaling schemes are selected and then the proposed power-aware design is simulated with these four schemes which are prediction mode reduction [13], search range control, early SKIP mode decision, and intra-frame period control [14]. Simulation results show that the average bitrate is increased by 1.34 and 3.42 % for CIF (352 × 288) and HD (1280 × 720) resolution videos, respectively, in slow-motion videos while achieving power savings of 25 %. In fast-motion videos, the increase in the average bitrate is 7.91 and 6.9 % for CIF and HD resolutions, respectively, achieving power savings of 25 %.

2 Implementation of Power-Aware Design

The power-aware design aims to control power consumption to meet a power budget target by combining various individual power-saving algorithms that are widely used for H.264/AVC encoders. The input for the power-aware design includes multiple power-scaling algorithms, each of which has various operating conditions that have a trade-off between R-D performance and power consumption. After applying these power-scaling algorithms, a power level table is generated to define a set of operating conditions incorporating various power-saving algorithms that can achieve the best R-D performance for the target power budget. The power-aware design uses a power level table and selects the best operating condition in that table to achieve the power consumption target.

2.1 Power Consumption of Mobile Devices

This section discusses the impact of video compression on the power consumption of mobile devices. There are a number of previous studies that investigate the power consumption of mobile devices. The multimedia operation consumes about 25 % of the total power consumption and H.264 encoder contributes most of the multimedia power consumption [15]. In [16], the power consumption of four operation phases is shown. In Fig. 7 of [16], Phase J2 represents the video capture without encoding and Phase J3 represents the video capture with H.264 encoding. Therefore, the difference between J2 and J3 corresponds to the power consumption of H.264 encoding. In this case, about 40 % power consumption is made by encoding operation. In summary, the power consumption of H.264 encoding operation contributes about 25–40 % of the total power consumption of mobile devices.

In the case when 35 % of power consumption of H.264 encoding is saved by the proposed algorithm, then the amount of the total power consumption of the mobile devices is between 8.75 and 14 %. In the case of 30 % power saving by the proposed algorithm, the total power saving is between 7.5 and 12 %. In summary, the proposed algorithm may save about 7.5–14 % of the power consumption of a mobile device.

2.2 Generation of a Power Level Table

The power level table defines a list of operation conditions for various power-saving algorithms along with their power consumptions (i.e., the power consumptions that are required by video compression when the power-saving algorithm is performed under the listed operation condition). The operation conditions are chosen to achieve the best video quality for the given power consumption level.

Generation of a power level table includes six steps. In the first step, the power saving from application of an individual power reduction algorithm is estimated by simulation. For each algorithm, simulation is performed under various operation conditions and the power consumption for each of the conditions is estimated. Such simulations are performed for every power-saving algorithm that is to be used for power control. Note that the power consumption estimation for one power-scaling algorithm is performed independently from that for the other algorithms. In addition, for accurate power estimation, a post-layout simulation for each algorithm is desirable.

In the second step, the power savings obtained by applying various combinations of power-scaling algorithms are estimated. Estimation of power consumptions for all possible combinations of operation conditions is time consuming because post-layout simulations are necessary to estimate accurate power consumption for each of the combinations. To save time, the estimation is not performed by using post-layout simulation. Instead, a power consumption model is used to derive the power consumptions associated with the various power-scaling algorithm combinations. That model combines the power simulation results from the first step and enables estimation of power consumption. Details of the power consumption model used for this estimation are presented in Sect. 3.

In the third step, corresponding R-D loss is measured for the various algorithm combinations estimated in the second step. This measurement is obtained by software simulation, which takes far less time than hardware simulation, making it possible to obtain R-D losses for all possible combinations. From the power saving and R-D loss obtained in the second and third steps, respectively, the relationship between power saving and corresponding R-D loss is obtained for all possible combinations of power-saving algorithms. That relationship is derived in the fourth step. The fifth step selects a fixed number of power-saving targets. In the example presented in Sect. 4, ten power levels are defined. The number of the power levels affects the granularity of the power control. The final step chooses the operating conditions that minimize R-D loss. This set of operation conditions combines to constitute the power level table. By using examples, each of the above six steps is explained in Sect. 4.

2.3 *Effect of Video Characteristics*

The effectiveness of a power-scaling algorithm may be influenced by the characteristics of the video. To avoid influences related to video characteristics and to achieve effective power saving, the proposed power-aware design classifies the input video into four categories based on video size and motion characteristics. A power table (see Sect. 2.2) is generated separately for four categories: fast-large, fast-small, slow-large, and slow-small videos. A large video sequence has a video sequence width greater than 1000 pixels. Otherwise, the sequence is considered small. A video is further categorized by the average of its motion vectors (MVs). A video sequence is classified as slow-motion or fast-motion if the average magnitude of the MVs in the previous 30 frames is either smaller or larger, respectively, than a pre-defined threshold. In this study, the pre-defined threshold is one pixel, obtained by experiments with various test sequences. For the initial 30 frames of a video, the sequence is always categorized as slow-motion because no previous motion information is available.

2.4 *Dynamic Selection of the Power Budget Target*

The power level table defines the estimated power consumption of the encoder when power-scaling algorithms operate under the conditions defined in that table. Note that the power consumption given in the power level table is an estimated value; thus, the real power consumption may be different, depending on the characteristics of the input video (see Sect. 2.3). To meet the power budget target, the actual power consumption is compared with that estimated from the power level table. If the actual and expected power consumption levels are different, the encoder needs to compensate for that difference by controlling the power consumption in the remaining periods of the video sequence. To this end, the encoding period is divided into small intervals and the power consumption is controlled independently for each period.

Suppose that a total power budget (P_{TOTAL}) is given for an encoder to operate for a certain period of time. Then, let P_{CUR} denote the available power budget for the current period, P_{PAST} denote the power consumption used in the preceding period, and P_{FUR} denote the estimated power consumption to occur in the future. The current power budget P_{CUR} is then calculated by using:

$$P_{CUR} = P_{TOTAL} - (P_{PAST} + P_{FUR}) \tag{1}$$

For each period, P_{CUR} is derived from (1). Subsequently, the appropriate operation condition in the power level table is chosen for the desired power budget target. Note that the exact value of P_{FUR} cannot be evaluated because the characteristics of future periods are not known until all periods are encoded completely. Thus, an average value, which is calculated by dividing P_{TOTAL} by the expected number of periods, is allocated to each future period.

2.5 *Selection of a Power-Scaling Algorithm*

By combining the power level table and the power budget target, selection of a power-scaling algorithm that uses a power level appropriate for the current power budget can be developed. The flow of this algorithm through repeated periods of a video sequence is shown in Fig. 1. The first step in that flow is to calculate the power budget target for P_{CUR} by using (11). The second step selects the appropriate power level table among the four size/motion-based power level tables for the input video. The third step selects the appropriate power level from the table selected in step two that will meet the P_{CUR} target. After encoding the current period with the chosen level, the fourth step identifies the video characteristics of the current period and determines the amount of power consumed. Those four steps are repeated until the encoding of all periods is completed. In every period, the appropriate power

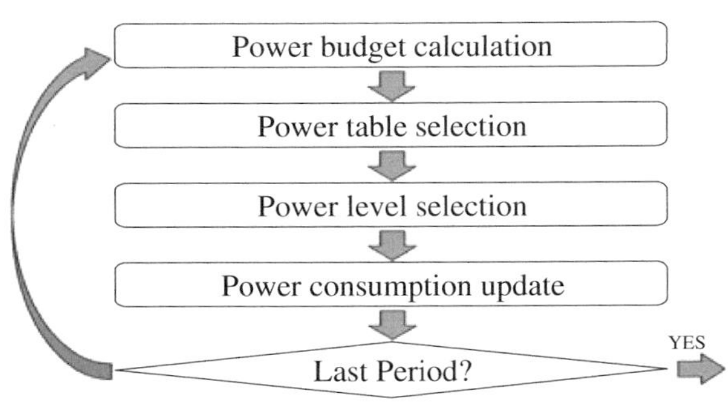

Fig. 1 Flow of a power-scaling algorithm

level for the current period is adjusted flexibly according to the power consumed in the past periods as well as that to be consumed in the future periods. In this way, the operating condition of an encoder can be adjusted to meet the power budget target. In this study, the selected power level update period is one minute. The update period and frame rate can be adjusted according to the requirement by an application and/or available power budget. Note that the computational complexity of this algorithm is very small because the power level tables are predefined; thus, the time needed to determine the optimal operation condition is negligible.

3 Power Estimation Model

There are a number of power-saving schemes that offer effective trade-offs between R-D loss and power reduction. This section proposes a model that determines power consumption when various algorithms are performed together.

3.1 *Formulation of a Power Estimation Model*

Power-scaling algorithms are classified into two types: one applicable for inter-frame prediction and the other suitable for intra-frame prediction. By applying those two types, total power saving (PS_{TOTAL}) is obtained by using:

$$PS_{TOTAL} = PS_{INTER} \times (1 - 1/IP) + PS_{INTRA} \times 1/IP \tag{2}$$

where PS_{INTER} and PS_{INTRA} denote the power savings achieved via inter-frame and intra-frame predictions, respectively, and where IP represents the period of the frames encoded by intra-frame prediction.

Of the two power saving terms in (2), PS_{INTER} is analyzed first. There are five main hardware-based operations within inter-frame prediction: IME, FME, IP, adaptive deblocking filter (ADF), and variable length coding (VLC). The total power saving is a summation of the power savings from these five hardware modules. A number of power-scaling algorithms are proposed for IME, FME, and IP. In contrast, the use of ADF and VLC for scaling power consumption is not reported extensively. Therefore, power scaling by ADF or VLC is not considered hereafter in this chapter. As a result, PS_{INTER} is a summation of PS_{IME}, PS_{FME}, and PS_{IP}, which denote the power savings achieved by applying IME, FME, and IP, respectively, as determined by using:

$$PS_{INTER} = PS_{IME} + PS_{FME} + PS_{IP} \tag{3}$$

Power-scaling algorithms for IME, FME, and IP can be classified into two categories. In the first category, the computational complexity of the operation is controlled and, consequently, the power consumption is also scaled. A search algorithm for IME is an example of that category. This type of algorithm controls the computational complexity of IME operations by adjusting the IME search range and, consequently, also scales the power consumption. In the second category, the execution frequency of the operation is controlled. For example, an early SKIP mode detection can result in elimination of ME operations. Depending on the decision made by this SKIP detection, the frequency of the ME operations is controlled. As a result, the power consumed by an ME module is also adjusted.

The two power control algorithm categories are reflected in a power-saving equation as discussed next. Let $PS_{IME,RC}$ denote the power saving achieved by the computational complexity of IME operations (representative of the first category) and $PS_{IME,RF}$ denote the power saving by reducing the frequency of IME execution (representative of the second category). $PS_{IME,RF}$ can be obtained by multiplication of $F_{IME,RF}$ which denotes the frequency of the IME operation that is not executed by the algorithm to achieve $PS_{IME,RF}$ and PC_{IME} which denotes the power consumption of conventional IME module. Then the power saving obtained by incorporating IME is determined by using:

$$PS_{IME} = [(1 - F_{IME,RF}) \times PS_{IME,RC}] + F_{IME,RF} \times PC_{IME} \tag{4}$$

Note that the term $(1 - F_{IME,RF})$ is multiplied to the first term of the right side of (4) because the frequency of the IME operation is reduced. Therefore, the power saving obtained by reducing the IME complexity is also reduced in proportion to the reduced frequency of the IME operation. Similarly, the power saving associated with FME and IP operations is determined by using:

$$PS_{FME} = [(1 - F_{FME,RF}) \times PS_{FME,RC}] + F_{FME,RF} \times PC_{FME} \tag{5}$$

$$PS_{IP} = [(1 - F_{IP,RF}) \times PS_{IP,RC}] + F_{IP,RF} \times PC_{IP} \tag{6}$$

where $F_{FME,RF}$ and $F_{IP,RF}$ denote the frequencies of the reduced FME and IP operations that are not executed by the algorithms used to achieve $PS_{IME,RF}$ and $PS_{IME,RF}$, respectively. PC_{FME} and PC_{IP} denote the power consumptions of conventional FME and IP modules, respectively. Then the total power saving for inter-frame prediction is obtained by summing (4)–(6) as in:

$$\begin{aligned} PS_{INTER} =& [(1 - F_{IME,RF}) \times PS_{IME,RC}] + F_{IME,RF} \times PC_{IME} \\ &+ [(1 - F_{FME,RF}) \times PS_{FME,RC}] + F_{FME,RF} \times PC_{FME} \\ &+ [(1 - F_{IP,RF}) \times PS_{IP,RC}] + F_{IP,RF} \times PC_{IP} \end{aligned} \tag{7}$$

There is an algorithm that reduces concurrently the execution frequencies of IME, FME, and IP. In this case, the algorithm affects the terms $F_{IME,RC}$, $PS_{IME,RF}$, $F_{FME,RC}$, $PS_{FME,RF}$, $F_{IP,RC}$, and $PS_{IP,RF}$. For the evaluation of (7), the frequencies $F_{IME,RC}$, $F_{IME,RC}$, and $F_{IME,RC}$ need to be obtained independently. In general, estimation of the frequencies is not difficult because it can be obtained from software simulation. In contrast, the terms $PS_{IME,RF}$, $PS_{FME,RF}$, and $PS_{IP,RF}$ do not have to be derived independently. Instead, only the summation of all three terms is needed. Therefore, only one hardware simulation is necessary to obtain the summation of all three terms. For algorithms that reduce only two of IME, FME, and IP, their impact on power saving can be obtained similarly.

To derive PS_{INTRA}, the second term in (2), the power saving in the intra-frame prediction is analyzed. Note that intra-frame prediction does not use the IME and FME modules and, consequently, requires less power consumption than that used by inter-frame prediction. Thus, intra-frame prediction inherently achieves power saving ($PS_{INTRA,INH}$) compared to that from inter-frame prediction. Additional power saving can be achieved by reducing the computational complexity of intra-frame prediction, which is achieved similarly to the savings described in (6). Thus, PS_{INTRA} is derived by using:

$$PS_{INTRA} = PS_{INTRA,INH} + [(1 - F_{IP,RC}) \times PS_{IP,RC}] \times PS_{IP,RF} \tag{8}$$

The power savings estimated from (2) through (8) are compared with results obtained by simulation in Sect. 5. The comparison indicates that the model estimates are similar to the simulation results.

3.2 *Impact of ADF and VLC on Power Consumption*

Previous study indicates that the computation complexity of ADF or VLC is much smaller than that of IME, FME or IP. In [17], the costs of hardware modules are presented in Table 2. In terms of gate count, ADF (DB in [17]) is 6.6, 5.01, and 16.65 % of IME, FME, and IP, respectively, whereas VLC (EC in [17]) is 9.61, 7.3, and 24.24 % of IME, FME, and IP, respectively. In terms of memory size, ADF is 6.64, 6.58, and 18.16 % of IME, FME, and IP, respectively, whereas VLC is 9.26, 9.19, and 25.35 % of IME, FME, and IP, respectively. Thus, the power consumption of ADF and VLC may not make a big impact on the total power consumption.

A comparison of the computational complexity of deblocking filter with ME and intra prediction is presented in [18] by utilizing the number of processor cycles to execute various functions for compression. Inter Prediction, Intra Prediction, and Deblocking Filter are compared and the complexity of Deblocking Filter is much less than both Inter Prediction and Intra Prediction. Thus, the amount of power consumption by Deblocking Filter may be much smaller than ME or Intra Prediction.

4 Power-Scaling Algorithms

In this section, four power-scaling algorithms are used to generate a power level table. The procedures used in the example are applicable to other power-scaling algorithms.

4.1 Four Power-Scaling Algorithms

This subsection briefly explains the four power-scaling algorithms chosen for this example. These algorithms are reported to be effective in the control of power consumption in trading off the complexity of an encoder. The selected four algorithms are briefly described in the following.

FME Prediction Mode Reduction The prediction mode reduction for FME decides the number of FME modes to be performed based on IME results [13]. A decreased FME complexity leads to the reduction of the power consumption. When the prediction mode reduction is not applied, FMEs are performed all prediction modes including 16 × 16, 16 × 8, 8 × 16 and four 8 × 8 partitions. This full FME operation is denoted by Mode 7. Mode 5 represents that two are selected from 16 × 16, 16 × 8 and 8 × 16 partitions and three are selected from four 8 × 8 partitions for FME operation. Mode 3 represents that one is from 16 × 16, 16 × 8, and 8 × 16 partitions and two are from four 8 × 8 partitions [13]. Mode 1 denotes that just one mode which is determined as the best mode in IME operation is chosen for FME operation.

IME Search Range Control The search range of ME significantly affects the power consumption, which increases in proportion to the data transferred from external memory and in proportion to the amount of ME computation. When the search range is small, ME is performed for a small number of positions in the reference frame. Therefore, power consumption for memory access is reduced because the data size as loaded from the external memory is decreased. Therefore, the power consumption can be controlled by adjusting the search range. To this end, the search range is reduced to 1/2, 1/4, 1/6 or 1/9 to increase the degree of power control. Table 1 shows the proposed scheme for the adjustment of the search range. The first column represents the ratio of the search range reduction. PMV_X and PMV_Y in the second column denote the horizontal and vertical components of the PMV, respectively. If the magnitude of PMV_X is larger than that of PMV_Y, an object in a video very likely moves in the horizontal direction. Thus, the width of the search range is set to a value that is equal to or greater than the height of the search range, which increases the probability that the best matched object is included in the search range. The adjusted width and height of the search range are presented in the third and fourth columns, respectively. Here, SR_X and SR_Y denote

Table 1 Search range control

Adjustment range	PMV condition	X size	Y size
1/2	$\lvert PMV_X\rvert >= \lvert PMV_Y\rvert$	SR_X	$SR_Y/2$
	$\lvert PMV_X\rvert < \lvert PMV_Y\rvert$	$SR_X/2$	SR_Y
1/4	–	$SR_X/2$	$SR_Y/2$
1/6	$\lvert PMV_X\rvert >= \lvert PMV_Y\rvert$	$SR_X/2$	$SR_Y/3$
	$\lvert PMV_X\rvert < \lvert PMV_Y\rvert$	$SR_X/3$	$SR_Y/2$
1/9	–	$SR_X/3$	$SR_Y/3$

the width and height of the original search range, respectively. The values of SR_X and SR_Y are obtained from the algorithm presented in [19], where the center position of the search range is the position at which the PMV points.

Early SKIP Mode Decision If the SKIP mode is determined early before ME operation, not only the computation time but also the power consumption is significantly reduced because both inter- and intra-frame predictions are skipped. Such a method is referred to as an early skip mode decision. However, the compression efficiency is degraded when the early skip mode decision is not correct. To this end, the early skip mode decision in this study is made after the IME. The three conditions for early skip decision are checked based on the result obtained from the IME. If the conditions for early skip decision are satisfied, the following FME and IP are skipped. Otherwise, the original skip mode conditions are tested again after all inter-frame predictions are finished.

For the proposed early skip mode decision, the best mode of the IME, denoted by $Mode_{IME}$, should be 16×16 (Condition 1). In Condition 2, the MV resulting from IME as denoted by MV_{IME} is compared with the PMV. As the skip mode decision is made before FME, only MV_{IME} is available for the decision. Condition 2 is divided into two cases depending on the motion characteristic of the video sequences. One is used for slow-motion videos and the other is for fast-motion videos. Let $Ratio_{SKIP}$ denote the ratio of the number of MBs encoded as the SKIP mode in the previous frame to the number of all MBs in the frame. If $Ratio_{SKIP}$ is greater than or equal to a threshold, the previous frames may have low image complexity and slow-motion characteristics. Assuming that successive frames have a high temporal correlation, the current frame may also have low image complexity and slow-motion characteristics. In this case, Condition 2 is satisfied when the difference between the MV_{IME} and the PMV is less than or equal to 1. In the case of slow-motion images with low complexity, the value of the MV is relatively small. Therefore, the image quality loss by an incorrect skip mode decision is insignificant. If $Ratio_{SKIP}$ is less than the threshold, the characteristic of the current frame is assumed to be complex with fast motion. In this case, the MV_{IME} should be equal to the PMV, where the condition is checked more strictly. TH_{LOW}, which is the threshold of $Ratio_{SKIP}$, is set to 30 % experimentally. Finally, Condition 3 is proposed to predict zero CBP using the sum of absolute differences (SAD). Let $SAD_{IME\_16\times16}$ denote the SAD of an MB from IME. Let $SAD_{IME\_4\times4}$ denote the SAD of a 4×4 block in the current MB. If $SAD_{IME\_16\times16}$ is less than the threshold, $TH_{16\times16}$, and sixteen $SAD_{IME_4\times4}$s in the current MB are less than the threshold, $TH_{4\times4}$, it is highly probable that the CBP of the MB is zero. If Conditions 1, 2 and 3 are all satisfied, the current MB is encoded as the SKIP mode and the computations for FME and IP are skipped.

Intra-frame Period Control Among the video frame types, the power consumption for I-frame compression is very small compared to that for P-frame or B-frame compression. This is because the loading of the reference data from external memory is not necessary. Furthermore, complex ME operations are not performed for an I-frame. Therefore, power consumption is reduced as the intra-frame period is decreased (i.e., as the intra-frame insertion frequency is increased).

4.2 Example Power Estimation Model

By applying the above four algorithms to (2), (7), and (8), an example power consumption model is derived. To that end, (2), (7), and (8) are re-formulated to consider the power savings of the four selected algorithms; that is, the FME prediction mode reduction (PS_{PMR}), IME search range control (PS_{SR}), early SKIP mode decision (PS_{ES}), and intra-frame period control ($PS_{INTRA,INH}$) algorithms. For the derivation in (4) (i.e., the power saving resulting from the IME operation), the following values in (9) are obtained for this example because the IME search range control is the only algorithm that affects the power saving associated with IME and the IME search range control is never skipped.

$$F_{IME,RC} = 0, \quad PS_{IME,RC} = PS_{SR}, \quad PS_{IME,RF} = 0 \tag{9}$$

For the derivation in (5) (i.e., the power saving resulting from the FME operation), both the FME mode reduction and the early SKIP schemes affect the amount of power saving. The FME prediction mode reduction decreases the power consumption of FME itself (i.e., $PS_{PME,RC} = PS_{PMR}$) whereas the early SKIP affects the operation frequency of FME. Let T_{NUM} denote the total number of MBs and ES_{NUM} denote the number of SKIP mode MBs as determined by the early SKIP mode decision algorithm. Then, the FME execution frequency is ES_{NUM}/T_{NUM} (i.e., $F_{FME,RC} = ES_{NUM}/T_{NUM}$).

$$F_{FME,RC} = ES_{NUM}/T_{NUM}, \quad PS_{FME,RC} = PS_{PMR} \tag{10}$$

In contrast, the early SKIP mode algorithm affects both FME and IP. Thus, it also contributes to the derivation in (6) (i.e., the power saving associated with intra-frame prediction). Thus, the power saving for the early SKIP mode decision algorithm is obtained by using:

$$PS_{ES} = PS_{FME,RF} + [(1 - F_{IP,RC}) \times PS_{IP,RC}] + PS_{IP,RF} \tag{11}$$

Thus, the total power saving associated with the inter-frame prediction is formulated as follows:

$$PS_{INTER} = PS_{SR} + (1 - ES_{NUM}/T_{NUM}) \times PS_{PMR} + PS_{ES} \tag{12}$$

For the derivation in (8) (i.e., the power saving resulting from the intra-frame prediction operation), no special power-scaling scheme is adopted. Therefore, only the inherent power saving from the intra-frame prediction ($PS_{INTRA,INH}$) contributes to the total power saving ($PS_{INTRA} = PS_{INTRA,INH}$). Therefore, the total power saving for both inter-frame and intra-frame prediction is determined by using:

$$\begin{aligned} PS_{TOTAL} = {} & (1 - 1/IP) \times [PS_{SR} \times (1 - ES_{NUM}/T_{NUM}) \\ & \times PS_{PMR} + PS_{ES}] + (PS_{INTRA,INH} \times 1/IP) \end{aligned} \tag{13}$$

4.3 *Power Simulation of Individual Algorithms*

The amount of power saving and the corresponding R-D loss associated with application of the four power-scaling algorithms are obtained independently by simulation. For the power simulation, the hardware-based H.264/AVC encoder [20] is synthesized by using a Synopsys Design Compiler with a 0.13 μm library and the power consumption is measured with post-layout simulation. The R-D performance is obtained by software simulation of the hardware reference model, which gives exactly the same result as that from the hardware-based encoder. For derivation of the R-D performance, twelve video sequences are used (Table 2): three slow-motion CIF videos (Container, News, and Sean), three fast-motion CIF videos (Table, Bus, and Stefan), three slow-motion HD videos (Aspen, Sunflower, and Intotree), and three fast-motion HD videos (Factory, Pedestrian area, and Tractor). The number of frames in each video is 100 while the quantization parameter (QP) values are 20, 24, 28, and 32. The encoding configuration uses a base profile and the group of pictures (GOP) structure is "IPPP…". The operating clock frequency is 50 MHz for CIF videos and 166 MHz for HD videos in order to obtain a frame rate of 30 frames per second (fps) for the hardware-based H.264 encoder [20].

4.3.1 FME Prediction Mode Reduction

Table 3 shows the power saving and R-D loss resulting from application of the FME mode reduction algorithm. The twelve test videos (Table 2) are grouped by size and motion characteristics in the first and second columns. The third column represents the number of prediction modes used during the FME operation. The fourth, fifth, and sixth columns show the average power savings achieved, the Bjontegaard Delta PSNR (BDPSNR) [21] change, and the Bjontegaard Delta Bitrate (BDBR) [21] change from the values obtained from application of Mode 7, respectively. The seventh column presents BDBR per power saving ratio, which reflects the effectiveness of the mode reduction scheme.

Among all FME mode reductions, the average BDPSNR drop is less than 0.1 dB while the average BDBR increase is 2.52 % (Table 3). When Mode 1 is used, the BDBR difference between the CIF and HD sizes is very large. The BDBR/power saving ratio in the seventh column also shows a similar trend. The differences in prediction error caused by application of the various modes are not as large in HD

Table 2 Test video sequences

Motion—size	CIF(352 × 288)	HD(1280 × 720)
Slow	Container	Aspen
	News	Sunflower
	Sean	Intotree
Fast	Table	Factory
	Bus	Pedestrian area
	Stefan	Tractor

Table 3 Performance of the prediction mode reduction

Size	Motion	Mode	Power saving (%)	BDPSNR (dB)	BDBR (%)	BDBR/PS
CIF	Slow	5	6.79	−0.02	0.46	0.067
		3	13.59	−0.1	2.56	0.188
		1	18.12	−0.31	8.17	0.451
	Fast	5	6.66	−0.02	0.45	0.068
		3	13.32	−0.08	1.84	0.138
		1	17.76	−0.31	7.19	0.405
HD	Slow	5	6.73	−0.01	0.26	0.039
		3	13.45	−0.06	1.9	0.141
		1	17.94	−0.08	2.71	0.151
	Fast	5	6.47	−0.01	0.41	0.063
		3	12.94	−0.06	1.85	0.143
		1	17.26	−0.08	2.46	0.143

videos as it is in CIF videos. This is due to the low spatial complexity in HD videos compared to that in CIF videos. Thus, in HD videos, the R-D loss from making an erroneous reduction mode decision is less critical than that in CIF videos.

4.3.2 IME Search Range Control

Table 4 shows the power saving and R-D performance obtained by applying the IME search range adjustment. The applied search range adjustment rates are 1/2,

Table 4 Performance of the search range control

Size	Motion	Adj Rate	Power saving (%)	BDPSNR (dB)	BDBR (%)	BDBR/PS
CIF	Slow	1/2	6.43	−0.01	0.22	0.034
		1/4	9.64	−0.02	0.4	0.041
		1/6	10.71	−0.03	0.73	0.068
		1/9	11.43	−0.14	3.66	0.321
	Fast	1/2	7.28	−0.05	1.22	0.168
		1/4	10.92	−0.44	9.9	0.907
		1/6	12.14	−0.52	11.99	0.988
		1/9	12.95	−1.08	24.35	1.88
HD	Slow	1/2	6.86	−0.01	0.42	0.061
		1/4	10.29	−0.03	0.9	0.087
		1/6	11.43	−0.06	1.97	0.172
		1/9	12.2	−0.2	5.98	0.49
	Fast	1/2	8.5	−0.16	4.59	0.541
		1/4	12.75	−0.31	9.34	0.733
		1/6	14.17	−0.51	14.55	1.027
		1/9	15.11	−0.77	22.09	1.462

1/4, 1/6, or 1/9 of the original range in both horizontal and vertical directions. Among the slow-motion video sequences, the average power saving is 9.87 %, whereas the average power saving in fast-motion sequences is 11.73 %. The average BDBR increases are 1.78 and 12.26 % for the slow-motion and 12.26 % for the slow-motion and fast-motion sequences, respectively. The BDBR/power saving ratio in the seventh column is smaller for the slow-motion sequences than that for the fast-motion sequences. This result is expected because the search range for IME is sensitive to motion characteristics; consequently, the presented IME search range control is particularly effective in a power-aware design for slow-motion sequences.

4.3.3 Early SKIP Mode Decision

Table 5 shows the power saving and R-D performance obtained by applying the early SKIP mode decision algorithm [14]. In the third column, the effect of the algorithm on power saving differs markedly between slow-motion and fast-motion CIF sequences. This is because there are many MBs determined as SKIP modes in slow sequences and it is easy to determine the SKIP mode by applying an early SKIP mode decision. However, fast sequences have a small number of MBs determined as SKIP modes. Thus, the results obtained by applying the early SKIP mode decision are somewhat inaccurate. The average BDPSNR drop is less than 0.06 dB and the average BDBR increase is 1.89 %. The CIF sequences show an average BDBR/power saving ratio of 0.025, whereas the average ratio for the HD sequences is 0.576, a marked difference. In addition, the slow-motion sequences show an average ratio of 0.199, whereas the fast-motion average is 0.402, markedly larger than that of the slow-motion sequences. Thus, application of an early SKIP mode decision scheme in a power-aware design is effective for CIF and slow-motion videos.

4.3.4 Intra-Frame Period Control

Table 6 shows the power saving and R-D performance obtained by applying the intra-frame period control scheme. The third column represents selected intra-frame period which varies among 10, 15, 30 and 60. The R-D performance as indicated by BDPSNR and BDBR decreases substantially as the number of intra-frame periods

Table 5 Performance of the early skip mode decision

Size	Motion	Power saving (%)	BDPSNR (dB)	BDBR (%)	BDBR/PS
CIF	Slow	14	−0.01	0.24	0.017
	Fast	5.15	−0.01	0.17	0.032
HD	Slow	8.67	−0.09	3.3	0.381
	Fast	5	−0.12	3.86	0.772

Table 6 Performance of the intra-period control scheme

Size	Motion	Period	Power saving (%)	BDPSNR (dB)	BDBR (%)	BDBR/PS
CIF	Slow	60	1.44	−0.27	6.79	4.713
		30	2.87	−0.87	21.62	7.533
		15	5.75	−1.85	44.47	7.735
		10	8.62	−2.84	67.62	7.845
	Fast	60	1.47	−0.06	1.4	0.955
		30	2.95	−0.2	4.56	1.546
		15	5.9	−0.41	9.52	1.614
		10	8.84	−0.62	14.11	1.596
HD	Slow	60	1.46	−0.06	2	1.372
		30	2.91	−0.18	5.52	1.896
		15	5.82	−0.39	11.45	1.967
		10	8.73	−0.66	18.86	2.16
	Fast	60	1.53	−0.04	1.34	0.874
		30	3.05	−0.12	3.62	1.188
		15	6.11	−0.25	7.81	1.278
		10	9.16	−0.4	11.97	1.307

decreases (i.e., as the intra-frame insertion frequency is increased). This result shows that the frequent insertion of I-frames has a significant effect on both power consumption and bitrate. The HD sequences have an average BDBR/power saving ratio of 1.505, whereas the CIF sequences have a markedly higher average BDBR/power saving ratio of 4.192. This is because the inter-frame prediction cost of CIF sequences is smaller than that of HD sequences due to the higher spatial complexity in CIF sequences. Moreover, fast-motion and slow-motion sequences show an average BDBR/power saving ratios of 1.295 and 4.402, respectively. This result demonstrates that the intra-frame period control scheme is useful in HD, fast-motion video sequences.

4.4 Estimation of the Combined Power Saving and Derivation of the Optimal Operating Conditions

This subsection describes the generation of a power level table based on the results of the four power-scaling algorithms described in Sect. 2.2. As a result of steps 2, 3, and 4 in Sect. 2.2, the relationships between BDBR and power saving obtained from various combinations of algorithm operation conditions are plotted in Fig. 2. The horizontal axis represents the power saving, whereas the vertical axis does the BDBR change. Each point in Fig. 2 represents the BDBR change and power saving derived from a given operation condition. The power savings are obtained from

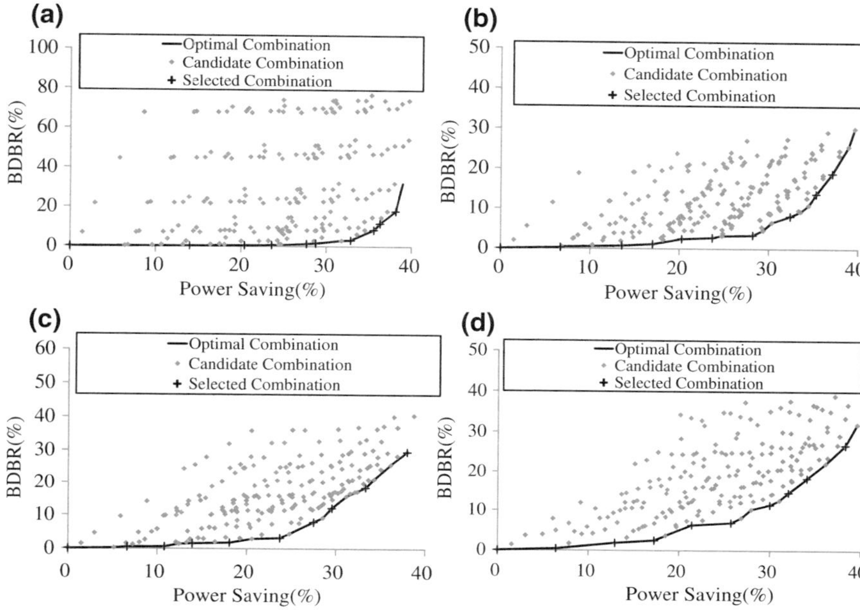

Fig. 2 BDBR change versus power consumption for various combinations of power-scaling schemes: **a** CIF slow-motion, **b** HD slow-motion, **c** CIF fast-motion, **d** HD fast-motion reproduced with permission from Kim et al.

(13) which is used to obtain the power simulation results of individual scaling algorithms. For estimation of BDBR change, all combinations of possible operation conditions are simulated by using reference software which gives exactly the same result as that from the hardware-based encoder. In Fig. 2, the points at the lower-right portion of the plots represent better power saving performance than those at the upper-left because the lower-right points have small BDBR increases at the same or similar power saving levels. Among the gray points in Fig. 2, the points that have the smallest BDBR change relative to the obtained power savings are connected with a segmented line. The points along the segmented line represent the operation conditions providing the least power consumption for a given BDBR change.

4.5 *Generation of a Power Table*

The final step in this example of the power-aware design is the generation of power levels that are associated with the optimal operating conditions of the four algorithms. To this end, a set of 10 operating conditions is chosen so as to have a regular interval of power saving. One advantage of a regular interval is that it can apply the most appropriate power level to the encoder based on the available power. In addition, changes in compression efficiency resulting from the use of different power levels are accomplished smoothly due to the regularity of the intervals. In this study, there are 10 power level entries in each power table, which result in power level intervals of approximately 5 % of power saving, similar to the interval used in [8].

The 10 operating conditions are selected from those represented in Fig. 2. Among the optimal operating conditions, those with regular power saving intervals are selected (marked with + in Fig. 2). From these operating conditions, power level tables are developed as shown in Table 7. In Table 7, the ten power levels are designated as level 0 to level 9. At level 0, none of the four power-scaling schemes is applied. Level 9 offers the largest amount of power saving among the ten power levels. The PMR, SR, ES, and IP columns represent the operation conditions of the prediction mode reduction, search range control, early SKIP mode decision and intra-frame period control, respectively. The PMR operation condition affects CIF and HD videos differently. For the CIF video sequences, prediction mode 1 (see Table 7) is used from level 7 to level 9 in slow-motion and from level 6 to level 9 in fast-motion, whereas for HD sequences prediction mode 1 is used from level 6 to

Table 7 Power level table reproduced with permission from Kim et al.

Level	CIF slow-motion				HD slow-motion			
	PMR	SR	ES	IP	PMR	SR	ES	IP
0	–	–	–	–	–	–	–	–
1	–	–	O	–	5	–	–	–
2	–	1/2	O	–	5	1/2	–	–
3	–	1/4	O	–	5	1/4	–	–
4	5	1/4	O	–	3	1/2	–	–
5	5	1/6	O	–	3	1/4	–	–
6	3	1/6	O	–	1	1/4	–	–
7	1	1/6	O	–	1	1/4	O	–
8	1	1/9	O	–	1	1/6	O	30
9	1	1/9	O	60	1	1/6	O	15
Level	CIF fast-motion				HD fast-motion			
	PMR	SR	ES	IP	PMR	SR	ES	IP
0	–	–	–	–	–	–	–	–
1	5	–	–	–	5	–	–	–
2	5	–	O	–	3	–	–	–
3	5	1/2	–	–	1	–	–	–
4	5	1/2	O	–	3	1/2	–	–
5	3	1/2	O	–	1	1/2	–	–
6	1	1/2	O	–	1	1/4	–	–
7	1	1/2	O	30	1	1/4	–	30
8	1	1/6	O	60	1	1/4	–	15
9	1	1/6	O	10	1	1/4	O	10

level 9 in slow-motion and from level 5 to level 9 in fast-motion sequences. In CIF videos, errors in the coefficients affect video quality more severely than errors in MVs because of the high spatial complexity of CIF videos. Thus, prediction mode reduction directly leads to degradation of R-D performance. In contrast, in HD videos, MV errors are more important than those in coefficients due to the low spatial complexity. As a result, R-D performance is not markedly degraded when the number of prediction modes is reduced. Therefore, the application of PMR is more useful for high resolution videos.

The ES algorithm is selected for use at lower power levels in CIF videos than those in HD videos as shown in the fourth, eighth, twelfth and sixteenth columns of Table 7. This is because the proportion of the 16 × 16 mode selected as the IME mode is higher in the low resolution videos than that in the high resolution videos. Thus, the ES algorithm is more useful for low resolution videos. The effects of SR and IP differ depending on the video's motion characteristics. The slow-motion video uses SR from level 2 to level 9 for both CIF and HD sequences and IP at level 9 in the CIF video and in levels 8 and 9 in the HD video. In contrast, fast-motion videos use SR from level 3 to level 9 in the CIF video and from level 4 to level 9 in the HD video and utilize IP for levels 7–9 for both CIF and HD videos. The maximum SR ratio is smaller in slow-motion (1/9) than in fast-motion (1/6) videos.

In slow-motion videos, a wide search range is not necessary because the MVs are relatively small. Thus, a degradation of R-D performance is not large when the search range is reduced. In contrast, the difference in the bit rate between intra-predicted and inter-predicted frames is very large due to the high temporal correlation between successive frames. Therefore, the SR algorithm has a great impact on power level selection, whereas the IP algorithm is not as effective due to its large R-D degradation. In contrast, most fast-motion videos require a wide search range compared to slow-motion videos due to their large MVs. Thus, a search range reduction directly leads to degradation of R-D performance. Also, the difference in bit rates between intra-predicted and inter-predicted frames is relatively small in fast-motion videos compared to that in slow-motion videos due to the low temporal correlation between successive frames. Therefore, the IP algorithm plays an important role in power saving for fast-motion videos, whereas the effect of SR in those videos is relatively small. In summary, the SR algorithm is more useful in slow-motion than fast-motion videos and the IP algorithm is more useful in fast-motion than slow-motion videos.

5 Performance of Power-Aware Design

In this section, the performance of the power-aware design is estimated and a comparison with the previous power-aware design is performed.

5.1 Performance Estimation of the Power-Aware Design

The performance of the power-aware design is assessed by performing simulations with 12 test video sequences. Each sequence consists of 100 frames and is encoded with four QP values (i.e., 20, 24, 28, and 32).

Table 8 summarizes the power savings and R-D performances for the 10 power levels presented in Table 7. From level 1 to level 9, the increases in power savings and the changes in BDPSNR and BDBR are presented in comparison to the power consumption and R-D performance at power level 0 (at which no power-saving algorithms are applied). For the slow-motion videos, the power savings for CIF and HD increase by 14 and 6.73 %, respectively, at level 1 while the corresponding R-D loss is insignificant. The largest power savings (38.137 and 37.138 % for CIF and HD videos, respectively) and the greatest increases in BDBRs (17.95 and 18.91 % for CIF and HD videos, respectively) are achieved at power level 9.

Table 8 Performance of the power level table

Level	CIF slow-motion			HD slow-motion		
	Power saving (%)	BDPSNR (dB)	BDBR (%)	Power saving (%)	BDPSNR (dB)	BDBR (%)
0	–	–	–	–	–	–
1	14	−0.01	0.24	6.7	−0.01	0.26
2	20.4	−0.02	0.46	13.6	−0.03	0.78
3	23.6	−0.03	0.71	17	−0.04	1.17
4	27.7	−0.05	1.34	20.3	−0.08	2.46
5	28.8	−0.06	1.62	23.7	−0.09	2.78
6	32.9	−0.12	3.19	28.2	−0.11	3.42
7	35.6	−0.32	8.39	32.4	−0.23	8.19
8	36.3	−0.44	11.68	35.3	−0.42	13.79
9	38.1	−0.7	17.95	37.1	−0.6	18.91
Level	CIF fast-motion			HD fast-motion		
	Power saving (%)	BDPSNR (dB)	BDBR (%)	Power saving (%)	BDPSNR (dB)	BDBR (%)
0	–	–	–	–	–	–
1	6.7	−0.02	0.45	6.5	−0.01	0.41
2	10.8	−0.03	0.61	12.9	−0.06	1.85
3	13.9	−0.06	1.55	17.3	−0.08	2.46
4	18.1	−0.07	1.8	21.4	−0.21	6.3
5	23.8	−0.13	3.17	25.8	−0.23	6.9
6	27.5	−0.33	7.91	30.3	−0.38	11.49
7	29.6	−0.51	12.16	32.1	−0.5	14.72
8	33.3	−0.82	18.55	34.1	−0.63	18.5
9	38	−1.31	29.69	38.3	−0.88	26.83

In fast-motion CIF videos, power level 1 saves 6.66 % of the power consumption with a concomitant 0.45 % increase in BDBR, whereas power level 9 saves 38 % of the power consumption along with a 29.69 % an increase in BDBR. For the HD videos, power level 9 results in a maximum 38.3 % power saving and a BDBR increase of 26.83 %. At a BDPSNR loss of less than 0.1 dB, about 28 and 23 % power savings are achieved for slow-motion CIF and HD videos, respectively, whereas about 18 and 17 % power savings are achieved for fast-motion CIF and HD videos, respectively. These results show that the R-D loss by power saving is larger in fast-motion videos than in slow-motion videos.

A further simulation evaluates the effectiveness of the four different power level tables according to video size and motion characteristics. For that comparison, a new power table (Table 9) is generated by using the procedure described in Sect. 2.2. Table 9 is obtained from the power saving and BDBR values averaged over all video sequences in Table 2, regardless of video size or motion characteristics. Figure 3 shows the relationships between power savings and BDBR values for the average power levels (i.e., those in Table 9) and the presented power levels (i.e., those in Table 7) for CIF slow-motion, CIF fast-motion, HD slow-motion, and HD fast-motion videos. For each of the video types in Fig. 3, the increase in BDBR from application of the optimized power levels is equal to or less than that obtained by using the average power levels. Compared to the optimized power level results, application of the average power levels increases the BDBR by more than 33 % when the power saving is 38 % as shown in Fig. 3a, whereas the BDBR is increased by more than 8.6 % when the power consumption is reduced by 34 % as shown in Fig. 3d.

Other simulations evaluate the accuracy of the total power saving model formulated by using (13). To evaluate the accuracy of the total power saving model, the amount of power saving obtained from (13) is compared to that obtained from measurement. Figure 4 compares the power levels and power savings for the four types of videos. The black curves labeled "Measured" show the average power savings obtained from measurements using the operation conditions in Table 7. The gray curves labeled "Model" show the power saving derived from (13). As shown

Table 9 Power level table without video classification

Level	PMR	SR	ES	IP
0	–	–	–	–
1	5	–	–	–
2	5	1/2	–	–
3	3	1/2	–	–
4	3	1/2	O	–
5	1	1/2	O	–
6	1	1/4	O	–
7	1	1/6	O	60
8	1	1/6	O	15
9	1	1/6	O	10

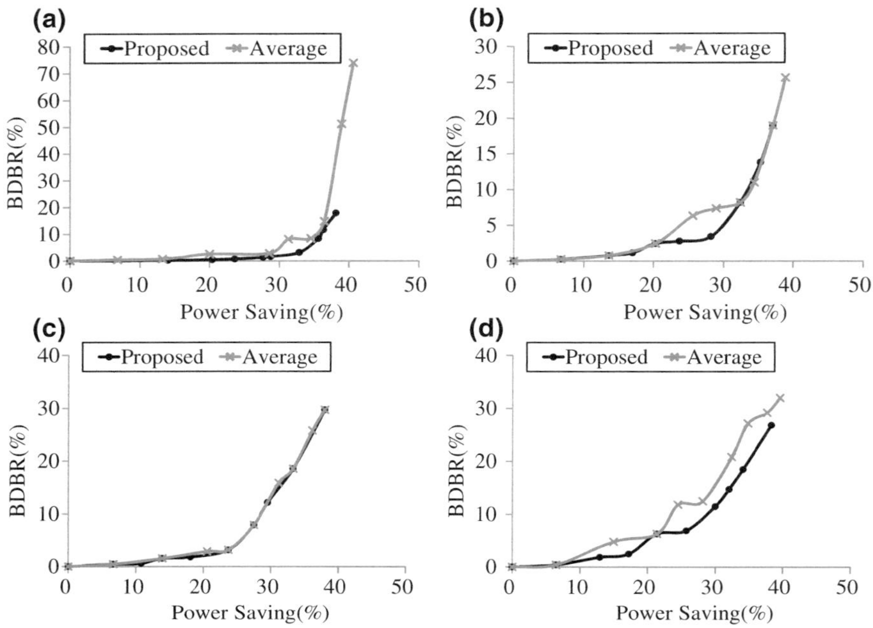

Fig. 3 Comparison of results from the proposed method and from the average power levels: **a** CIF slow-motion, **b** CIF fast-motion, **c** HD slow-motion, **d** HD fast-motion reproduced with permission from Kim et al.

in Fig. 4, the measured and modeled results for each of the four video types are nearly identical at most power levels. The maximum difference between the measured and optimized is 2.941 % at power level 3 for the CIF fast-motion video.

A further simulation evaluates the effectiveness of adaptively controlling the power level. In that simulation, power consumption is estimated at every 30 frames and the current power budget P_{CUR} in (1) is updated accordingly. Subsequently, the power level is adjusted adaptively according to the remaining power budget. The simulation results from the optimized adaptive approach are compared with the results from an approach in which a fixed power level is applied throughout the simulation from the beginning to the end. For both adaptive and fixed level approaches, the power saving target is 30 %. To achieve this goal, the fixed level control uses power levels 7 and 6 for CIF slow-motion and fast-motion videos, respectively. To achieve the 30 % goal from the adaptive approach, the total power P_{TOTAL} is chosen to achieve a 30 % power saving and P_{CUR} is chosen adaptively by using (1). A second, similar simulation with a power-saving target of 35 % is also performed. In this case, power levels 9 and 8 are chosen at the fixed power levels for the CIF slow-motion and fast-motion videos, respectively. For both simulations, three CIF-size fast-motion sequences and three CIF-size slow-motion sequences are used with each sequence comprising 1500 frames.

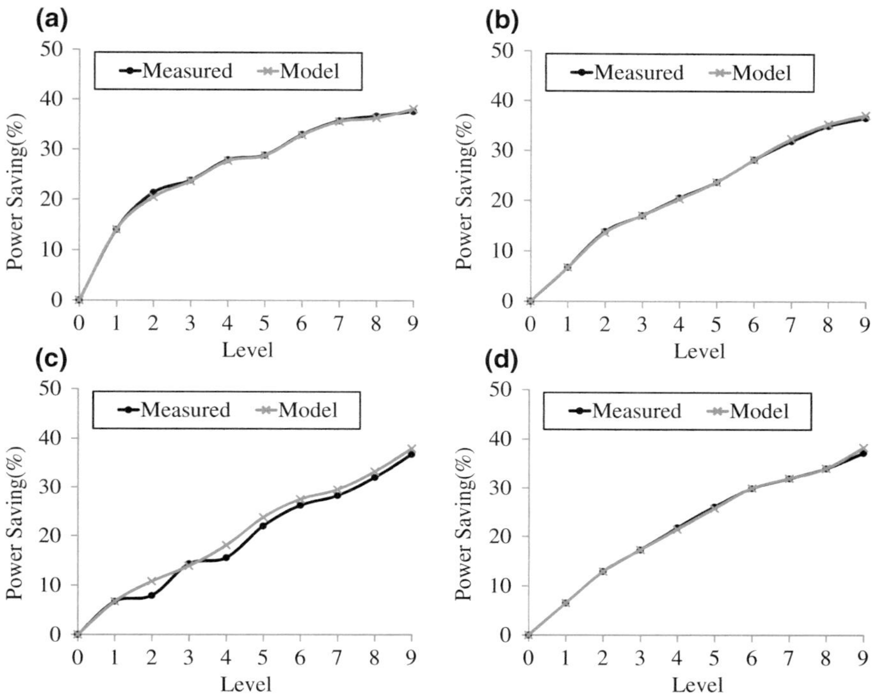

Fig. 4 Comparison of measured and modeled power level saving results: **a** CIF slow-motion, **b** CIF fast-motion, **c** HD slow-motion, **d** HD fast-motion reproduced with permission from Kim et al.

Table 10 presents the results of the adaptive power level approach compared with those from fixed power level approach. The test video sequences include FAST1 (a concatenated sequence of Foreman, Soccer, Crew, Ice, and Football sequences), FAST2 (a Bigbuckbunny sequence), FAST3 (an Elephants dream sequence), SLOW1 (a concatenated sequence of Akiyo, Coastguard, Mother_daughter, Silent, and Weather sequences), SLOW2 (a concatenated

Table 10 Increase of BDBR and BDPSNR by application of the adaptive level control

Sequence	30 % saving		35 % saving	
	BDBR (%)	BDPSNR (dB)	BDBR (%)	BDPSNR (dB)
FAST1	1.51	0.056	2.58	0.101
FAST2	9.98	0.504	16.59	0.951
FAST3	2.28	0.1	5.84	0.257
SLOW1	2.76	0.115	5.71	0.228
SLOW2	1.38	0.062	6.88	0.327
SLOW3	10.08	0.368	13.03	0.47

Table 11 Ratio of the consumed power to the available power

Sequence	30 % saving		35 % saving	
	Proposed method (%)	Fixed level (%)	Proposed method (%)	Fixed level (%)
FAST	99.93	95.38	99.9	95.33
SLOW	99.8	92.7	99.87	95.35

sequence of Hall monitor, Flower, Waterfall, Tempete, and Paris sequences), and SLOW3 (a Highway sequence). Overall, the results from the optimized adaptive control approach improve R-D performance compared to that from the fixed approach. The experimental results show that, in comparison with the fixed approach, BDBR decreases by an average of 6.46 % and BDPSNR increases by an average of 0.328 dB for fast sequences, whereas BDBR decreases by an average of 6.64 % and BDPSNR increases by an average of 0.262 dB for slow sequences when the adaptive approach is applied. Thus, the optimized adaptive approach produces better performance than the fixed approach because it consumes the available power as much as possible.

Table 11 presents the average percentage of the available power consumed after all video encoding is finished. For both the fast and slow sequences, the simulation results show that the optimized adaptive approach consumes more than 99.8 % of the power budget whereas a maximum of 95.4 % of the available power budget is consumed by the fixed power level approach. The remaining unused power budget results in R-D degradation, which is greater when the fixed power level approach is used.

5.2 Comparison with a Previous Power-Aware Design

In this subsection, the presented power-aware control system is compared to a previous power-aware design described in [8]. To compare the performance of the presented power-aware design and the one reported in [8], comparison of the R-D performance is made under the same power consumption targets.

Figure 5 presents the R-D performance of the original videos in comparison with that from the presented power-aware design and that reported in [8] when the power-saving target is a 35 % reduction. In this study and in [8], R-D curves for the slow-motion and fast-motion CIF-size Akiyo and Foreman sequences, respectively, are available. Thus, those videos are used for the comparison. In Fig. 5, the curve labeled “Original” represents the R-D performance with no power-scaling scheme applied, the curves labeled “Level 7” or “Level 8” represent the R-D results when the level 7 or level 8 operating conditions for CIF slow-motion or fast-motion, respectively, in Table 7 is applied, and the curve labeled “[8]” is the R-D curve from [8]. In Fig. 5a, the differences among the three curves are very small, as the Akiyo sequence is very static; moreover, the R-D performance does not change

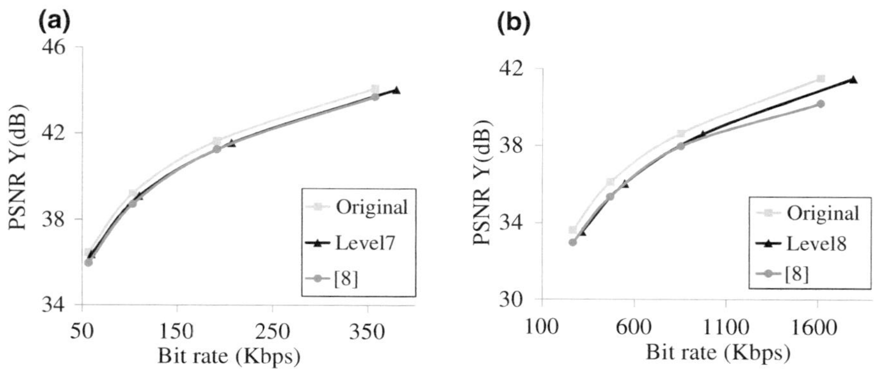

Fig. 5 Comparison of the R-D performance when PC_{TARGET} is 65 %: **a** Akiyo **b** Foreman

markedly even when the two power-scaling schemes are adopted. In Fig. 5b, the R-D curves show significantly higher performance when applying level 8 rather than the scheme in [8]. The difference in PSNR represented in Fig. 5b is about 0.5 dB at 1600 kbps.

In Fig. 6, the R-D performances in the original videos, as well as that obtained from the presented power-aware design and that in [8] are shown when the power-saving target is a 40 % reduction. Figures 6a, b show the results with Akiyo and Foreman sequences, respectively. For both sequences, level 9 in Table 7 is applied. Large differences between the results from the optimized power saving scheme and that in [8] are shown in Fig. 6. In the Akiyo sequence, the PSNR difference is larger than 2 dB at 200 kbps. In the Foreman sequence, the PSNR difference is larger than 2 dB at 1600 kbps. Note in Fig. 6 that the algorithm used in [8] results in a significant R-D performance degradation at a power saving target reduction of 40 %. In contrast, the optimized power-aware system incorporating

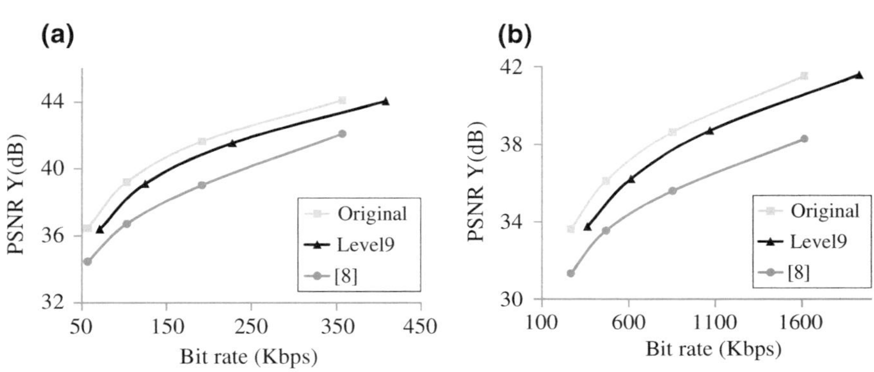

Fig. 6 Comparison of the R-D performance when PC_{TARGET} is 60 %: **a** Akiyo, **b** Foreman reproduced with permission from Kim et al.

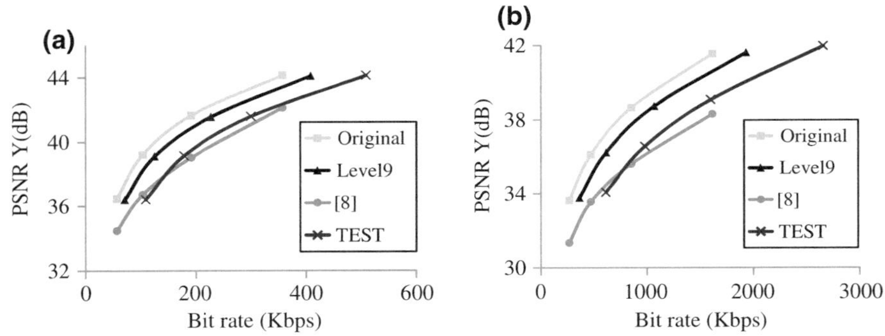

Fig. 7 Comparison of the R-D performance with the modification of the power-aware algorithm similar to [8]: **a** Akiyo, **b** Foreman

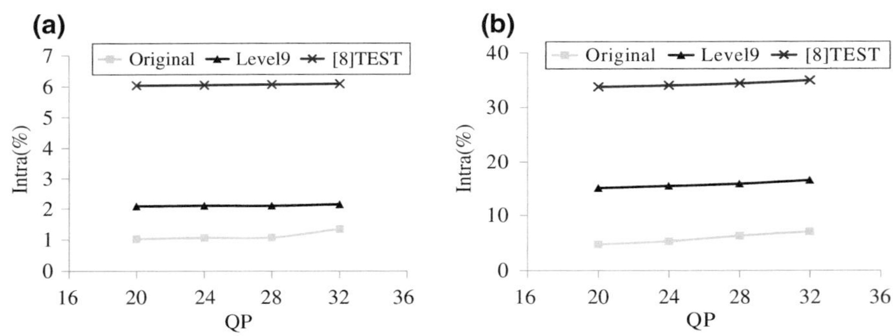

Fig. 8 Comparison of the proportion of intra prediction modes: **a** Akiyo, **b** Foreman

four types of algorithms can reduce power consumption by approximately 40 % without a marked R-D performance decrease.

The drop-off of the R-D performance for 40 % power saving in [8] may be caused by a significant increase of the number of blocks that are encoded as IP mode. In [8], the algorithm decides whether to perform IME or not based on the following condition. If IME is not performed, then the block is encoded as the intra prediction mode. For the decision, the following criterion is used

$$\begin{aligned}&\text{if}(\text{Power}_{\text{LefgAvg}} > \text{P}_{\text{IME}} + \text{P}_{\text{OTHERS}})\ \text{perform IME}\\ &\text{else perform Intra Prediction}\end{aligned} \tag{14}$$

where $\text{Power}_{\text{LeftAvg}}$ represents the remaining power budget which is calculated as follows.

$$\text{Power}_{\text{LefgAvg}} = \frac{\text{Power}_{\text{Budget}} - \sum_{i=1}^{k-1} \text{Power}^{i}_{\text{Usage}}}{n - (k - 1)} \tag{15}$$

P_{IME} represents the power consumption by IME and P_{OTHERS} represents the power consumption of HW modules other than IME, FME, and IP modules. In [8], it seems that the if-condition in (14) is satisfied for most blocks when the target power saving is 35 %. However, the power saving of 40 % causes the if-condition not to be satisfied for a substantial number of blocks. As a result, IME is not performed for these blocks and then these blocks are encoded as intra prediction mode which is the only possible option in [8] when IME is not performed. The increase of IP blocks consequently increases the bitrate significantly.

The main difference of the proposed design from [8] is that the presented power-aware design offers a much larger number of possible options than the algorithm in [8] does. This large number of options makes it possible to avoid a significant drop of the R-D performance for a small increase of target power saving. In the optimized power-aware design presented in this chapter, the power consumption model of (7) and (8) speeds up the estimation of power consumption for various operating conditions. Without this power consumption model, it may take too much time to estimate the power consumption of all these various options.

In order to justify this reasoning, the presented algorithm is modified to be very similar to that in [8]. To this end, the modified algorithm uses the same Eq. as (14) and selects whether to perform IME or not. For FME, just two options are used as in [8]: the most complex FME option for Mode 7 and the simplest FME option for Mode 1. The early skip mode decision is somewhat similar to the pre-skip in [8]. The resulting graph is shown in Fig. 7 with its graph denoted by "TEST". The R-D performance is very similar to the result in [8] for both Akiyo and Foreman videos. Figure 8 compares the number of blocks encoded in the IP modes. The result shows that the modified version similar to [8] results in a significant increase of intra blocks which degrades the R-D performance. This simulation draws a conclusion that the skip of IME may cause a significant degradation of R-D performance so that it is important to have a power-scaling algorithm that gradually decreases the complexity of IME without a complete removal of IME operation.

Acknowledgments This work is supported by the Center for Integrated Smart Sensors funded by the Ministry of Science, ICT and Future Planning as the Global Frontier Project.

References

1. Chang X, Zhang M, Zhang G, Zhang Z, Wang J (2007) Adaptive clock gating technique for low power IP core in SoC design. In: Proceedings of international symposium on circuits and systems, pp 2120–2123, May 2007
2. Tsai C, Chen T, Chen L (2006) Low power entropy coding hardware design for H.264/AVC baseline profile encoder. In: Proceedings of international conference on multimedia and expo, pp 1941–1944, July 2006
3. Kuo C, Lei S (2006) Design of a low power architecture for CABAC encoder in H.264. In: Proceedings of Asia Pacific conference on circuits and systems, pp 243–246, Dec 2006

4. Chen T, Chen Y, Tsai S, Chien S, Chen L (2007) Fast algorithm and architecture design of low-power integer motion estimation for H.264/AVC. IEEE Trans Circ Syst Video Technol 17 (5):568–577
5. Chen T, Chen Y, Chen L (2006) Low power and power-aware fractional motion estimation of H.264/AVC for mobile application. In: Proceedings of international symposium on circuits and systems, pp 5331–5334, May 2006
6. Chen Y, Chen T, Chen L (2006) Power-scalable algorithm and reconfigurable macro-block pipelining architecture of H.264 encoder for mobile application. In: International conference on multimedia and expo, pp 281–284, July 2006
7. Lian C, Tseng P, Chen L (2006) Low-power and power-aware video codec design: an overview. China Commun 2006:45–51
8. Chang W, Li G, Chang T (2009) Power-aware coding for H.264/AVC video encoder. In: Proceedings of international symposium on VLSI design/CAD
9. Kannur A, Li B (2009) Power-aware content-adaptive H.264 video encoding. In: Proceedings of international conference on acoustics, speech and signal processing, pp 925–928, April 2009
10. He Z, Cheng W, Chen X (2008) Energy minimization of portable video communication devices based on power-rate-distortion optimization. IEEE Trans Circ Syst Video Technol 18 (5):596–608
11. Kim J, Kim J, Kim G, Na S, Kyung C (2010) Event statistics and criticality-aware bitrate allocation to minimize energy consumption of memory-constrained wireless surveillance system. In: Proceedings of international conference on multimedia and expo
12. Kim J, Kim J, Kim G, Kyung C (2011) Power-rate-distortion modeling for energy minimization of portable video encoding devices. In: Proceedings of international midwest symposium on circuits and systems, pp 1–4, Aug. 2011
13. Chen T, Huang Y, Chen L (2004) Fully utilized and reusable architecture for fractional motion estimation of H.264/AVC. In: Proceedings of international conference on acoustics, speech and signal processing, pp. 9–12, May 2004
14. Kim H, Rhee C, Kim J, Kim S, Lee H (2011) Power-aware design with various low-power algorithms for an H.264/AVC encoder. In: Proceedings of international symposium on circuits and systems
15. Berkel C (2009) Multi-core for mobile phones. In: Proceedings of DATE
16. Xu N, Zhang F, Luo Y, Jia W, Xuan D, Teng J (2009) Stealthy video capturer: a new video-based spy-ware in 3G smartphones. In: Proceedings of second ACM conference on wireless network security
17. Chen T, Chien S, Huang Y, Tsai C, Chen C, Chen T, Chen L (2006) Analysis and architecture design of an HDTV720p 30 frames/s H.264/AVC encoder. IEEE Trans Circ Syst Video Technol 16(6):673–688
18. Su L, Lu Y, Wu F, Li S, Gao W (2007) Real-time video coding under power constraint based on H.264 co-dec. In: Proceedings of VCIP
19. Jung J, Moon D, Lee H (2008) Computation reduction of H.264/AVC motion estimation by search range adjustment and partial cost evaluation. In: Proceedings of international conference on electronics, information and communication, pp 229–233, June 2008
20. Rhee C, Jung J, Lee H (2010) A real-time H.264/AVC encoder with complexity-aware time allocation. IEEE Trans Circ Syst Video Technol 20(12):1848–1862
21. Bjontegaard G (2001) Calculation of average PSNR differences between R-D curves. In: VCEG-M33 of ITU-T Q6/16

Part II
Event/Object Detectors for Smart Sensing

Low-Power Operation for Video Event Data Recorder

Jinyoung Yang, Jongpil Jung and Chong-Min Kyung

Abstract Due to limited battery capacity, reducing power consumption of mobile surveillance camera like a video event data recorder is important to extend surveillance time. In this chapter, we propose a design of low-power video event data recorder which records events such as movement of objects, or impact to the camera itself. Duty-cycling of two different encoders, which are a low-power encoder and a high-compression encoder, are employed to implement the low-power video event data recorder. Operating time of the proposed system is considerably extended by duty-cycling of the two encoders in the event-driven operation; the system mainly stays in event detection mode and wakes up only when an event is detected. Because the most valuable information in the event is right before or at the moment of event detection, the proposed system records video from 10 s before the event detection. According to experiment, the energy consumption of the proposed system is decreased up to 25.1 % (by 33.2 % on average) of conventional video event data recorder. As energy consumption of the proposed system is reduced by 66.8 % on average, the surveillance time of the proposed system can be increased by three times consequentially.

Keywords Video event data recorder · Duty-cycling of heterogeneous codecs · Low-power operation · Event-driven mode · Low-energy surveillance camera · DCT coefficient · Event occurrence rate

1 Introduction

With the growth of the surveillance camera market, demand on portable surveillance camera grows due to increasing needs in places where power line is not available. The portable surveillance camera usually operates in event-driven mode for reducing energy consumption. However, focus of conventional researches based

J. Yang (✉) · J. Jung · C.-M. Kyung
Center for Integrated Smart Sensors, ITC Building(N1), Daehak-ro 291, Yuseong-gu, Daejeon 305-701, Republic of Korea
e-mail: jyyang0308@kaist.ac.kr

C. Kyung (ed.), *Theory and Applications of Smart Cameras*,
KAIST Research Series, DOI 10.1007/978-94-017-9987-4_6

on event-driven operation has been to raise operators' attention rather than reducing energy consumption. Video surveillance system [1] is proposed to detect abandoned or removed objects. Traffic surveillance framework [2] is proposed for detection, classification, and tracking of vehicles. Pedestrian detection [3] is still another active research topic. The objective of research is to automatically detect abnormal events and inform the operators of the events [4].

In [5], energy consumption of the surveillance camera is considerably decreased due to the event-driven operation. The system is mostly in sleep mode and wakes up only when an event is detected. However, the system in [5] is not practical from the viewpoint of video event data recorder (VEDR), because it takes some time for the system to wake up. As the system starts recording after an event is detected, it cannot record the most valuable scenes which is the moment of event detection or scene prior to the detection. According to [6], it is required for VEDR to save video data starting from 10 s before the time of event and ending 10 s after the time of event. The 10 s video prior to the event is crucial to find out how the event occurred.

Optimizing video encoders [7, 8] and adopting low-resolution image sensors [1, 9] were proposed to reduce power consumption of surveillance camera. However, because the video encoder is just one of power consuming components, system-level optimization beyond codec is more effective to reduce power consumption of surveillance camera. The resolution of cameras in commercial smartphone has already reached 2 K QHD, i.e., 2560 × 1440. High definition video is increasingly required in surveillance applications as well.

In this chapter, we propose a design of low-power and event-driven surveillance camera which records events. The objective of the proposed design is to maximize the operating time of battery-based surveillance camera system by minimizing the energy consumption. Duty-cycling of two different video encoders is employed to record the events starting from 10 s before the event.

In order to verify energy reduction of the proposed design, we applied the design to VEDR and compared energy consumption of the proposed system with that of conventional VEDR. VEDR started from car black box system which records data from various sensors of vehicle such as speed sensor, water sensor, brake sensor, and light sensor [10, 11]. Recently, high-end VEDRs, which record high definition video as well as the values of sensors on vehicle, have been released. Since the VEDR is operated by battery in the vehicle, it is critical to save energy. When the weather is cold, performance of car battery is significantly degraded [12]. If the power consumption of the event data recorder of the parked vehicle is excessive, it can lead to a failure to start engine in such a situation.

2 Operation of the Video Event Data Recorder

The VEDR shown Fig. 1 is an electronic device installed in a vehicle to sort out the cause of accident. It applies the concept of a blackbox in an airplane which works as the most important tool to find causes of aircraft crash. In the event of car accident,

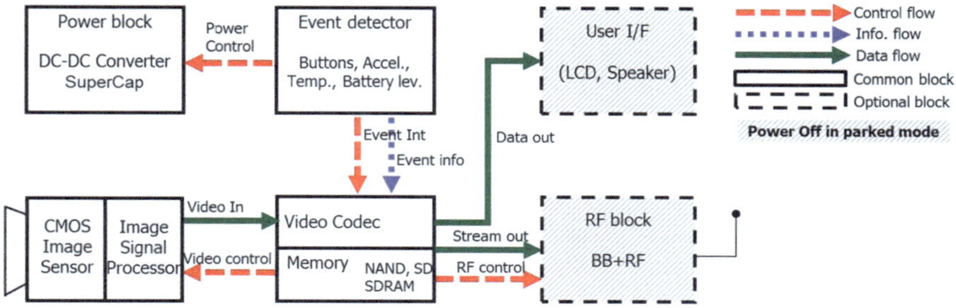

Fig. 1 Block diagram of an event-driven VEDR

it facilitates judgment on arguing liability for the accident. The main function of the VEDR is to record video data of the pre-accident and post-accident. The market of the VEDR in Korea has grown up more than 30 times; from 47,000 devices in 2008 to 1.5 million devices in 2012.

The VEDR operates as follows: When a car engine is on, the VEDR is operated in driving mode. When a car engine is off, the VEDR is operated in parked mode. In the driving mode, the VEDR records video continuously and saves them in AlwaysMovie folder. If AlwaysMovie folder becomes full, the oldest file of AlwaysMovie folder is overwritten by the newest one. When an impact is applied to the vehicle or user presses 'record' button in the driving mode, recorded file in AlwaysMovie folder is copied in the EventMovie folder. In the parked mode, the VEDR continuously records video data and stores them for 10 s in temporary buffer, which are discarded unless an event occurs. If the VEDR detects an event, which either motion event in the camera or impact event on the car, the VEDR starts to save the video in MotionMovie or EventMovie folder according to the type of events.

In early VEDR, video from camera is always saved even during parked mode. The VEDR in parked mode should monitor car's surroundings to detect car's damage and potential threat as long as possible. According to [13, 14], the average driving time in Korea and Europe is under 2 h and the average daily car trips in Europe is 2.5 trips. It means the average daily parking time is more than 22 h. According to [14], parking time can be split in two parts. One part can be named 'active parking' which is the time when the car is parked after a trip. The other part can be named 'inactive parking' which is the time when the car is parked before the first trip of the day or after the last trip of the day. The average daily total active parking time is $4 \sim 7$ h and the average daily total inactive parking time is $16 \sim 19$ h. From surveillance camera's point of view, the VEDR is at least required to satisfy the daily maximum monitoring time which is 19 h. However, if the VEDR consists of Full HD camera, 32 GB storage, and video encoder with 60 MB/min compression rate, the VEDR stores recent 8 h 52 min video depicted as *A* point in Fig. 2. If the VEDR adopts 16 GB storage, it only stores recent 4 h 26 min depicted as *B* point in Fig. 2. Based on the actual storage capacity requirement of the VEDR

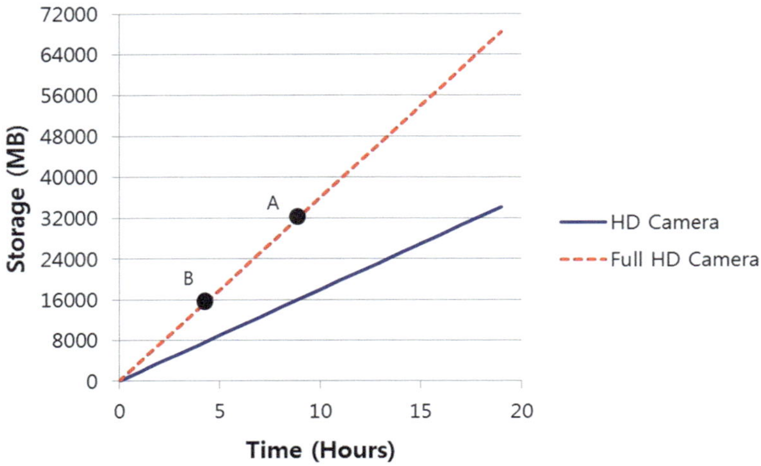

Fig. 2 Storage capacity of VEDR

in Fig. 2, the VEDR operation which saves data continuously in parked mode is not acceptable. To satisfy the basic requirement of surveillance camera, i.e., monitoring time, the VEDR needs to operate in event-driven scheme in parked mode. In event-driven scheme in parked mode, the VEDR stores video in the storage only at the occurrence of the event.

2.1 Power Consumption of Conventional Event-Driven System

To analyze power consumption of the VEDR, we measured power consumption of a commercial VEDR in driving mode and parked mode shown in Tables 1 and 2, respectively. The VEDR consumes 4.05 W in driving mode with Full HD input (1920 × 1080) and 30 frames per second. If we change input video quality to HD or VGA in Table 1, then the average power consumption of the VEDR is slightly reduced. Not surprisingly, the power consumption of a commercial VEDR in parked mode shown in Table 2 is only about 10 % lower than the power consumption in driving mode shown in Table 1. This results from the fact that the operation in driving mode is the same as operation in parked mode except for the event-driven scheme. Event-driven scheme in parked mode consists of three phases: standby phase, motion detection phase, and impact detection phase. In Table 2, standby phase (phase 1) is normal operation state before event occurrence. Phase 2 and 3 are states after impact detection and motion detection, respectively. Phase 2 and 3 are considered as event phase.

Table 1 Power consumption of VEDR in driving mode

Video mode		Average power consumption (W)
Resolution	Frame rate (frames per second)	
1920 × 1080	30	4.05
1280 × 720	30	3.88
640 × 480	30	3.77
1920 × 1080	15	3.90
1280 × 720	15	3.80
640 × 480	15	3.75
1920 × 1080	10	3.86
1280 × 720	10	3.77
640 × 480	10	3.72

Table 2 Power consumption of VEDR for each phase of the event-driven scheme in parked mode

	Phase in the event-driven scheme		
	Phase 1	Phase 2	Phase 3
Average power consumption (W)	3.68	3.75	3.75

From this measurement, we can get information about total power consumption in each mode. In order to measure detailed-power consumption of major parts of the VEDR, we developed an evaluation board using a cortex-A8 application processor (AP) [15]. Key features of the AP we used for our VEDR are cortex-A8 processor embedded, 2D/3D graphic engine, dual camera interface, Full HD encoding/decoding, and advanced low-power technology. Typical VEDR consists of CMOS image sensor (CIS), image signal processor (ISP), video encoder, microprocessor, memory system, video data storage, and communication module. We measured actual power consumption of major parts of the VEDR as shown Fig. 3. Major parts of the VEDR in Fig. 3 can be categorized into three parts. Core-related part including H.264 codec, DRAM with core DDR, core I/O, ARM processor constitutes one part, while camera part and LCD part are other parts. The core-related part and the camera part are always turned on in each mode, but LCD part can be turned off in parked mode. The core part consumes around 60 % of total power consumption. So, in order to reduce total power consumption of the VEDR, we need to reduce power consumption of the core-related part first. The main functions of the core-related part are event detection and video compression. If we have low-power event detector and video codec (LPEDVC) module as a substitute for the core-related part in standby phase of the parked mode, we can reduce average power consumption in parked mode.

In wireless sensor networks, MAC protocols [16–19] were developed to conserve energy consumption of the sensor nodes through duty-cycling scheme. By using duty-cycling of the sensor node, the sensor node extends its lifetime. Similar concept can be used in the proposed VEDR design. If the LPEDVC module

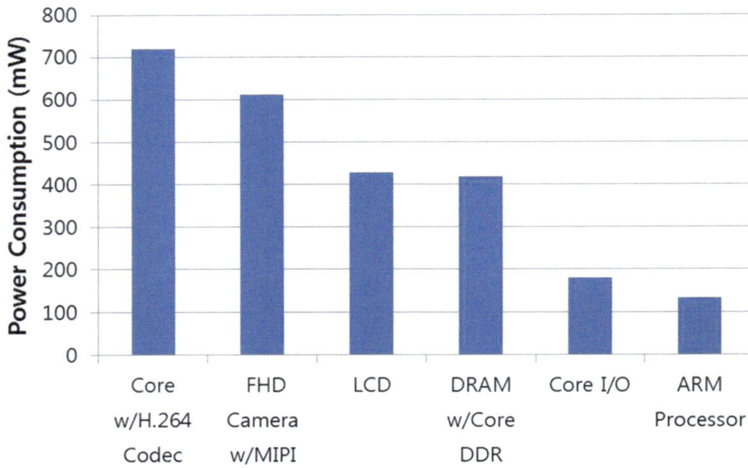

Fig. 3 Power consumption breakdown of conventional VEDR

implemented in FPGA is added to the conventional VEDR as shown in Fig. 4 and duty-cycling of the LPEDVC module and the core-related parts are employed based on the event-driven scheme, average power consumption of the system in the parked mode can be significantly reduced to P_{park} as shown in Fig. 5b. In standby phase with no event detected, i.e., in T_{stbyi}, only the event detector and CIS/ISP are turned on. Since power consumption of event detector is much less than that of video encoder or microprocessor, system power consumption in standby phase, i.e., P_{stby}, is much less than that of conventional VEDR. When an event is detected, power consumption of event-driven VEDR, i.e., P_{evt}, is slightly more than that of conventional VEDR because of the event detector. Figure 5 shows how power consumption of both systems varies with time. Average power consumption of event-driven VEDR is much less than power consumption of conventional VEDR unless events are detected too frequently.

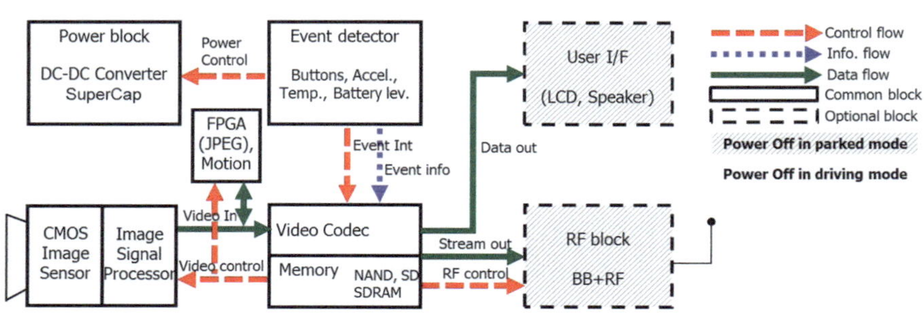

Fig. 4 System diagram of a proposed VEDR

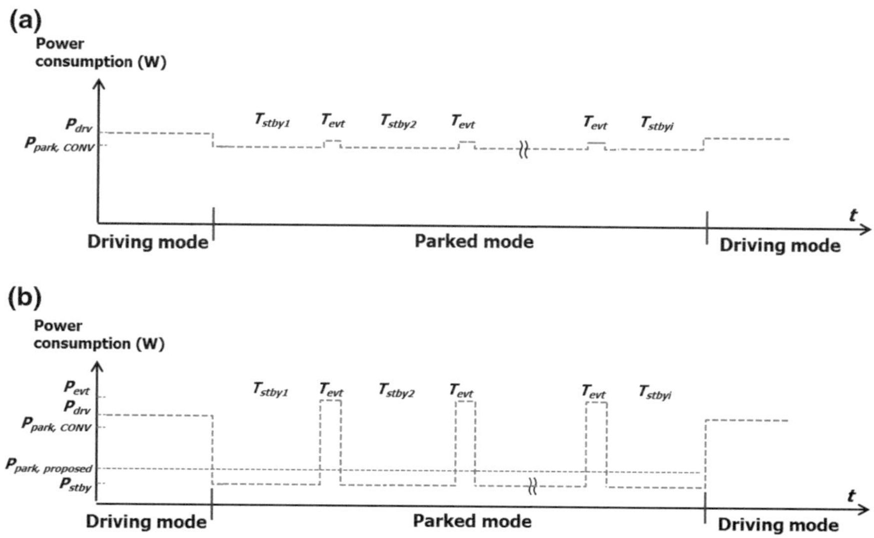

Fig. 5 Comparison of power consumption (**a** conventional VEDR, **b** proposed VEDR)

For event-driven VEDR, type of events must be defined. Motion in captured scene is the most widely used type of event for surveillance camera. If there is no moving object in video, then the video does not provide any events. Another important event for vehicle is acceleration. If a vehicle is physically damaged by another vehicle or pedestrian, momentary acceleration is detected. In this chapter, we assume that event has occurred when a change of motion or acceleration is detected.

2.2 Proposed Method

As mentioned in the introduction, the VEDR is required to start recording 10 s before the detection of an event. In conventional video surveillance systems [1, 2, 4], it is possible to have video before an event. Because they do not have storage and energy constraints, they record every video regardless of event. While the VEDR has storage and energy constraints, the VEDR must operate with event-driven scheme and low power consumption. However, as shown in Table 2, event-driven scheme itself does not provide substantial lifetime gains. Our approach to satisfy storage constraint and relax energy constraint simultaneously in event-driven scheme is as follows: We add an LPEDVC module in front of the core-related parts, and turn on the LPEDVC module and turn off the core-related part. After detecting an event, the LPEDVC module turns on the core-related part. Then AP boots itself as quickly as possible and the core-related part compresses the video data delivered from the LPEDVC module and AP saves the compressed video data into storage. After completing recording, the core-related part is turned off again.

Similar approaches [20–22] like ours have been proposed to save energy consumption in wireless sensor networks and surveillance camera system. If we replace the LPEDVC module with motion detector using Passive Infrared (PIR) sensor like in [20], this system may save energy consumption dramatically but cause other problem. Because this system wakes up after an event is detected, it can only have video after the event. Considering that it takes some time for this system to wake up, it starts recording few seconds after an event is detected. In such a case, the most important portion of video will not be recorded. In [21, 22], they can reduce power consumption in event-driven scheme of the surveillance system and record video for 10 s before an event as well. However, they use buffer memory instead of the LPEDVC module to record temporary video. After an event, they compress the temporary video in the buffer memory and save compressed video data into storage. In order to record video for only 10 s, they need 829.4 MB and 1.87 GB buffer memory to record HD and Full HD video, respectively. These buffer memory sizes are too big to be used practically.

In order to implement system based on our approach, we need to resolve practical issues as follows: (1) How to send pre-event data to AP, (2) How to boot AP as soon as possible after detecting an event, (3) How to implement event detection function, i.e., motion detection, using low-power image or video codec. The issue 2 can be resolved using Linux fast booting technology [23]. For simplicity, we assume AP's booting time is zero. To resolve the issue 3, we have to decide which codec will be used first. We have considered two candidates as low-power image or video codec, which are light-weight compression [24] and JPEG [25]. The two candidates are comparable from the viewpoint of low-power image codec, i.e., gate count and power consumption. However, according to [26], JPEG encoder can be used as motion detector as well. On the other hand, we need to devise motion detection algorithm if we adopt light-weight compression [24]. In the proposed design, we choose JPEG as low-power image codec and motion detector in the LPEDVC module. Here we describe the issue 1 in detail.

We devise three methods to transfer pre-event data to AP as shown in Figs. 6, 7, 8. First, we consider scheme 1 shown in Fig. 6. During T_{stbyi}, only CIS/ISP, JPEG Encoder, and DRAM for JPEG are in operation. After detecting an event, H.264 encodes post-event video for a while, saves the post-event video file and pre-event JPEG file into storage simultaneously shown in Fig. 6b. Additional data path from DRAM for JPEG to storage is required and two different file formats coexist in storage. To overcome the problems, we devise scheme 2 as shown Fig. 7. During T_{stbyi}, only CIS/ISP, JPEG Encoder, and DRAM for JPEG are working. After detecting an event, H.264 encodes post-event video for a while and saves the post-event video file first as shown in Fig. 7b. After saving the post-event video file, JPEG decodes pre-event JPEG file and transfers the decoded format into H.264 input as shown in Fig. 7c. Then H.264 part encodes pre-event video and AP saves the pre-event video file. However, if several events occur with $10 \sim 20$ s inter-event interval, control logic processing event-driven operation got to be

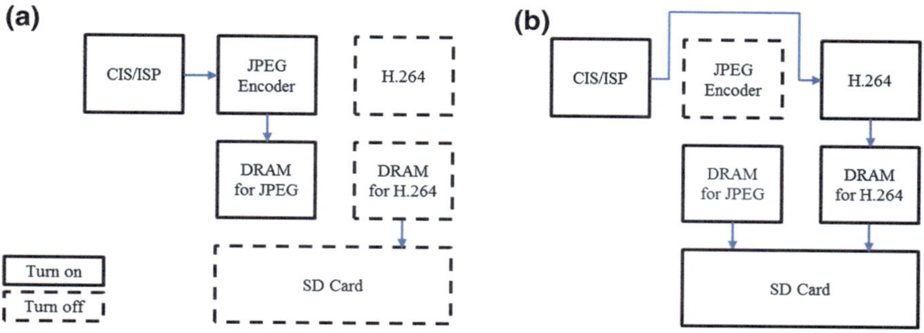

Fig. 6 Pre-event data transfer idea 1 **a** Operation in T_{stbyi}, **b** Operation in T_{evt}

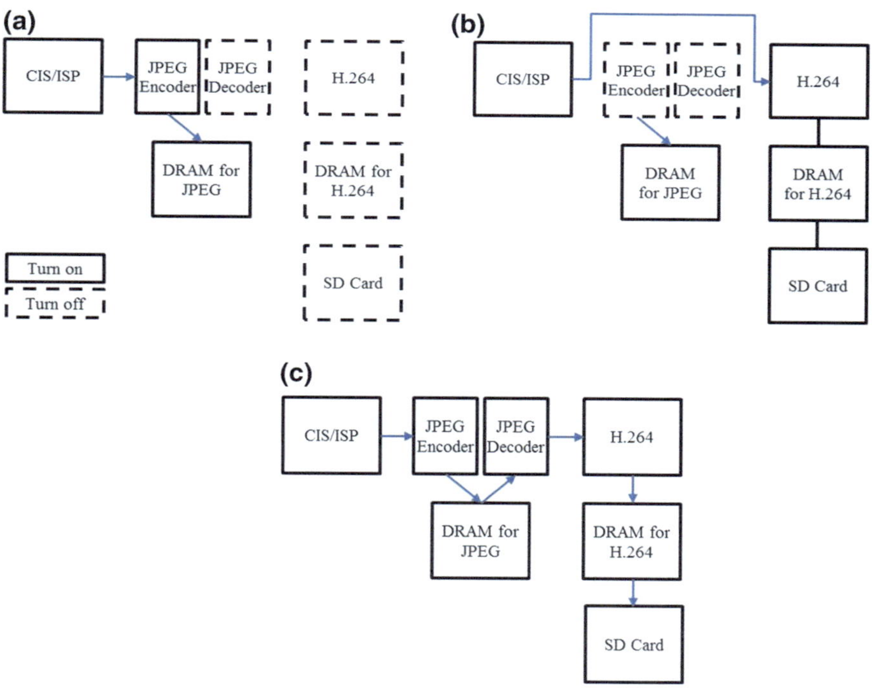

Fig. 7 Pre-event data transfer idea 2 **a** Operation in T_{stbyi}, **b** Operation in the first half of T_{evt}, **c** Operation in the second half of T_{evt}

complicated. For simple and robust operation, we devised scheme 3 shown in Fig. 8. During T_{stbyi}, the operation is the same as scheme 1 and 2. After detecting an event, AP is turned on. JPEG encodes post-event image into DRAM for JPEG, JPEG decodes pre-event image from DRAM for JPEG, and transfers the decoded image into H.264 input as shown in Fig. 8b. Then H.264 part encodes

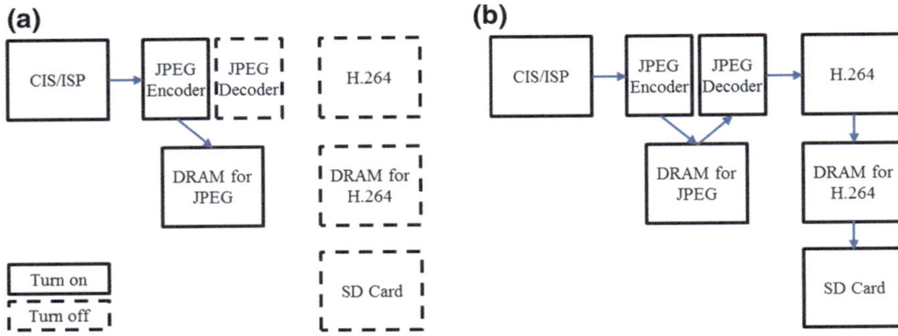

Fig. 8 Pre-event data transfer idea 3 **a** Operation in T_{stbyi}, **b** Operation in T_{evt}

pre/post-event video and AP saves the pre/post-event video file. A problem with scheme 3 is time difference between event detection time and record starting time. This can be solved if we can delay video data for a few seconds. If the few seconds of video are temporarily saved in volatile memory, the system is capable of having all required data even after the event is detected. In conventional VEDR, this is done by using buffer memory between H.264 encoder and data storage. Video captured by CIS/ISP is encoded in H.264 format and saved in the buffer at all time. If an event happens, the data in the buffer is written to data storage. In such a system, however, energy consumption is wasted because all system components including power-hungry core-related part including H.264 encoder must stay on.

To save power consumption of VEDR, we need to turn off as many components as possible. The most power consuming parts such as H.264 encoder and processor must be turned off for considerable energy reduction. However, CIS/ISP must remain on for following reasons even though there is no event detected: (1) pre-event video data is important, (2) nobody knows when the event happens, and (3) the event detector needs video data for event detection. In order to turn off H.264 encoder, uncompressed data from CIS/ISP must be temporarily saved. However, data size of uncompressed video data from CIS/ISP is too large to be saved in memory practically. The size of uncompressed 10 s Full HD video with 30 frames per second is 1.7 GB.

In the proposed design, power consumption of the proposed system can be reduced by adding light-weight image codec to conventional system with H.264 video encoder. The light-weight codec, i.e., JPEG in this case, is added between CIS/ISP and H.264 encoder. The video data from CIS/ISP is transferred to JPEG encoder, and JPEG encoder compresses the video data and temporarily saves the data in DRAM which is directly connected to the JPEG encoder. Other power-consuming blocks such as H.264 encoder, processor, memory, storage, and communication module can all be turned off in this state. After an event is detected, H.264 encoder, processor, memory, storage, and JPEG decoder are turned on. JPEG decoder decodes JPEG-compressed data in DRAM, H.264 encoder encodes the video data, and AP saves them in storage. Details of system design are explained in Sect. 3.

Because most components in conventional VEDR are turned off at standby phase, power consumption of the proposed system is significantly lower than otherwise. However, since event detector, JPEG codec, and memory are added to the proposed system, power consumption during the event phase in parked mode is slightly more than conventional VEDR. If events happen infrequently, average power consumption of the proposed system in parked mode is much less than conventional system. Results on energy reduction in various situations are explained in Sects. 4 and 5.

3 Low-Power System Design

Block diagram of the proposed system is shown in Fig. 9. Like a conventional VEDR system, the proposed system uses CIS/ISP, H.264 encoder, main processor, event detector, accelerometer, memory, storage, and communication module. Micro SD card is used as storage of video data for user convenience. The addition of the LPEDVC module which includes JPEG encoder, decoder, and DRAM for JPEG is the main difference of the proposed design compared to conventional system. The video data flow starts from CIS and ends at the storage in typical video surveillance system. In the proposed system, the video data flow has two different paths: (1) one passing through JPEG encoder, and (2) the other bypassing the JPEG encoder. These paths are decided by operation modes of VEDR, which are driving mode and parked mode. To decide operation modes of VEDR, VEDR takes ACC signal (on/off signal for accessories) of the vehicle as an input. If a driver turns on

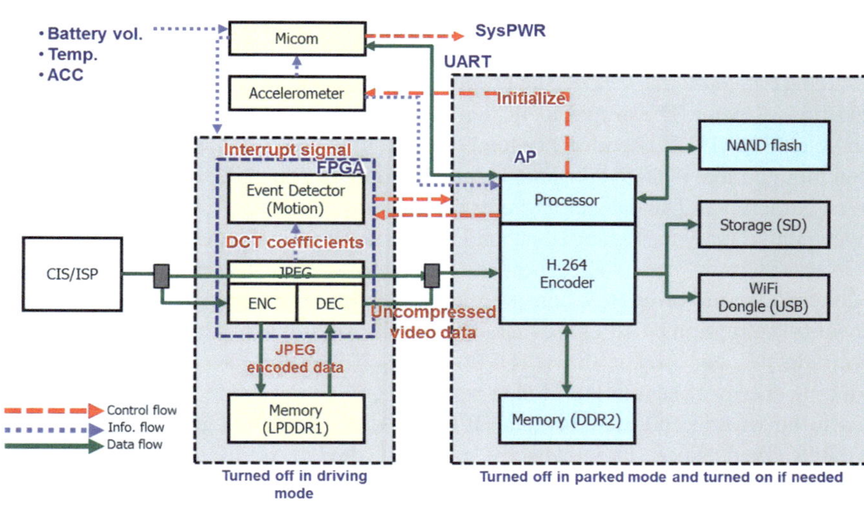

Fig. 9 Proposed system overview

the engine, ACC signal rises high and the VEDR operates as driving mode. When a driver stops engine, ACC signal falls and VEDR changes its operation mode to parked mode.

In addition to the change of video data flow, information flow is also changed. In conventional VEDR, always-on AP deals with the information from accelerometer and video data for event detection. However, in the proposed design, the event detector in parked mode deals with the information from accelerometer and video data. For motion detection, the event detector uses DCT coefficients [26]. To detect physical impact on the vehicle, the event detector utilizes values from the accelerometer. When an event is detected, the event detector passes the information to the processor.

3.1 Parked Mode

The proposed design featuring the addition of the LPEDVC module in Sect. 2.2 is employed during parked mode. Depending on the detection of events, the system switches its mode between standby phase and event phase. Since the video data path which goes through JPEG encoder is selected during parked mode, mux and demux between CIS/ISP and H.264 encoder are switched accordingly.

If there is no event detected, the system operates in standby phase which is shown in Fig. 10a. In the standby phase, the system is in the state of waiting for an event. When the system is in standby phase, we can turn off many components in the system. As shown in Fig. 5b, the power consumption of system in standby phase significantly affects total energy consumption of the system. Only the minimally required components are turned on for event detection and temporary video saving: CIS/ISP, event detector, micom, accelerometer, JPEG encoder, and low-power memory for DRAM for JPEG-encoded data.

Like a conventional VEDR, the proposed system should record video prior to the detection of event. The proposed system saves the 10 s of data in the memory which is connected to JPEG encoder. For temporary saving, CIS/ISP always takes a picture even though there is no event. Then a series of images from CIS/ISP are temporarily saved in memory. The data structure in memory for the images is like circular buffer conceptually. Only the recent 10-second images are kept in the memory, and old data is overwritten by new data. The size of uncompressed data from CIS/ISP is a problem in terms of energy consumption and cost. According to [27], read/write power consumption of LPDDR memory is quite significant, while the power consumption of deep power down (DPD) mode of LPDDR is very small. So, for reducing energy consumption of data read/write, it is important to reduce the size of data to be written and read, so that DPD mode of LPDDR is utilized as much as possible. And since large-capacity memory device is more expensive than low-capacity memory device, it is required to reduce the data size. This is one of the reasons why JPEG encoder is employed in the proposed design. Uncompressed data from CIS/ISP is transferred to JPEG encoder. After the JPEG encoding, data size is

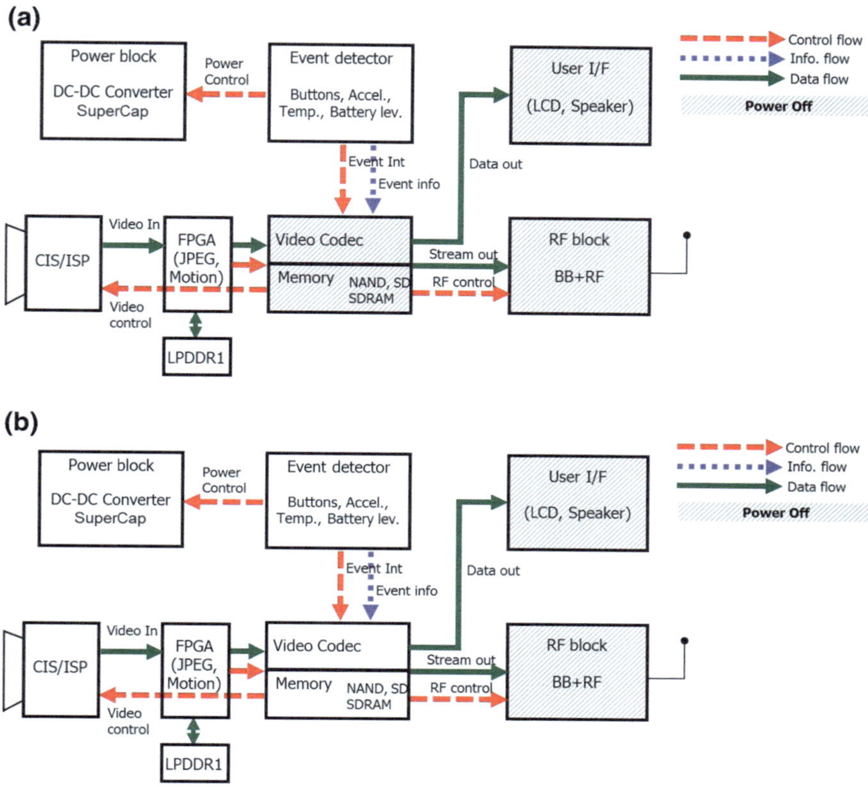

Fig. 10 Operation of a proposed system in parked mode **a** Operation in T_{stbyi}, **b** Operation in T_{evt}

reduced by up to 95 % with quality of PSNR 36 dB. Thus, the size of 10-second Full HD video can be reduced from 13.6 to 0.7 Gb. If we consider margin for variable compression rate, 1 Gb LPDDR1 memory is enough for implementation.

Since the event detector always consumes power during the standby phase, it is also important to design the event detector as simple as possible. We apply basic motion detection method based on inter-frame difference [28, 29]. Motion is detected when difference between two consecutive frames exceeds defined threshold. Previous frame data is saved to get inter-frame difference. If all uncompressed data from CIS/ISP is passed on to the even detector, power consumption of event detector is enormous due to huge data size. In the proposed design, movement within captured scene is detected by using DCT coefficients in compressed domain, not data in pixel domain. DC coefficient from 8 × 8 block DCT represents average intensity of the block. This has additional effect of downsizing image from 1920 × 1080 to 240 × 135. Since only luminance information is sufficient for motion detection (i.e., color information can be discarded), data size required for a frame is reduced down to 32.4 KB. Conventional motion detection is usually implemented using luminance information in the processor. For example, the processor scales down input image into QVGA (320 × 240) grade using scaler block and compares inter-frame difference, and data size required for a frame is 76.8 KB.

The event detector also detects physical impact on a vehicle using accelerometer. The accelerometer is initialized by the processor at the very beginning so that it operates stand alone. It operates even when processor is in sleep or power-off mode. The accelerometer triggers an interrupt signal when acceleration at any direction exceeds threshold. When any kind of events is detected by the event detector, the system switches to event phase. In this phase, the system starts encoding in H.264 format and saves encoded file in the storage. All components that were turned off during the standby phase are turned on except for communication module. When the event detector issues enable signal for wake-up, processor wakes up. Then other blocks including H.264 encoder and storage become ready for H.264 encoding. After the system is ready for H.264 encoding, JPEG decoder starts decoding data at low-power memory which are a series of images prior to the event detection. The 10-second delayed images are delivered to the H.264 encoder and finally it is saved in the storage. In the meanwhile, a series of images from CIS/ISP are still saved in memory which will be saved in storage through H.264 encoder at the end. At 20 s after event, the system switches back to the standby phase. Timing diagram between two phases is described in the Fig. 11.

3.2 *Driving Mode*

In driving mode, the system operates as shown in Fig. 9. The system operates similar to conventional VEDR which does not have the LPEDVC module. When a driver turns on the engine, the proposed system reads ACC signal from vehicle (which

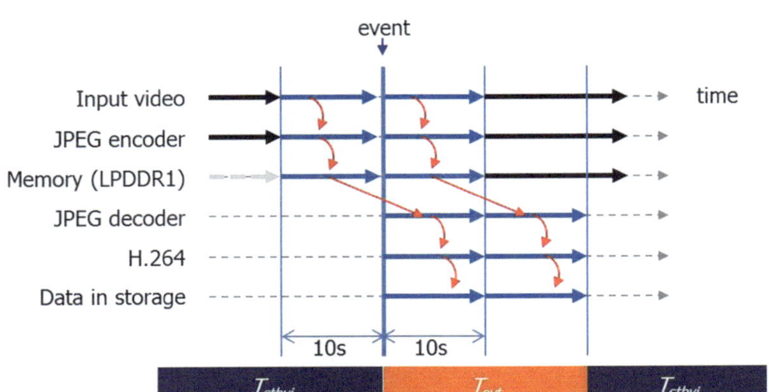

Fig. 11 Timing diagram between standby phase and event phase

switches from low to high at the moment) and the mux after CIS/ISP is set to skip the LPEDVC module. The components for the LPEDVC module such as JPEG encoder, JPEG decoder, and low-power memory, are turned off during the driving mode. Other components such as CIS/ISP, H.264 encoder, processor, event detector, and storage are always turned on. The function of this mode is very simple because it is just required to capture and save. The video data from CIS/ISP module is directly delivered to the H.264 encoder, and they are saved at the storage in H.264 format.

Since the vehicle moves in the driving mode, motion detection is meaningless for an event. The accelerometer is used as the only input of event detector in the driving mode. As mentioned earlier, the main function of VEDR is to record video data of the pre-accident and post-accident. To detect accident, the accelerometer seems to be enough. However, any threat to the vehicle deserves to be recorded. Nobody knows when accident or threat happens. So, saving important data in more secure place is more important than saving energy consumption in the driving mode. The role of event detection in this mode is quite different from the parked mode. Every video in the driving mode is saved in AlwaysMovie folder. If an event is detected, the video of the event is duplicated and saved in separate space of storage. It helps to be prepared for memory failure.

The power consumption of the driving mode is significant because the most power-consuming parts, AP with H.264 encoder and processor, are always turned on. The only event-driven feature in this mode is the storage. Considering energy consumption of micro SD card is very small compared with total energy consumption, it can be said that the system constantly consumes a lot of energy.

4 Power Analysis in Parked Mode

In the proposed system, we apply power-gating to minimize the power consumption of system. Total energy consumption of the proposed system depends on which components are turned on for how long. And total power consumption is the sum of power consumption of components which are turned on at the time. We assume that power consumption of each component is constant.

In the proposed design of VEDR, there are two different operating modes: driving mode and parked mode. As shown in Fig. 9, most components including CIS/ISP, H.264 encoder, processor, main memory, micro SD card, and other peripherals are turned on during driving mode. Since power consumption in the driving mode (P_{drv}) is constant according to our assumption, energy consumption in the driving mode (E_{drv}) is simply equal to power consumption multiplied by duration of the driving mode (T_{drv}) as shown in (1).

$$E_{\text{drv}} = T_{\text{drv}} \cdot P_{\text{drv}} \tag{1}$$

Power consumption of whole system in the driving mode is the sum of power consumption of components that are turned on in the mode as expressed in (2)

$$P_{\mathrm{drv}} = P_{\mathrm{CIS}} + P_{\mathrm{H264}} + P_{\mathrm{PROC}} + P_{\mathrm{DRAM}} + P_{\mathrm{SD}} + P_{\mathrm{PERI}}, \tag{2}$$

where P_{CIS}, P_{H264}, P_{PROC}, P_{DRAM}, P_{SD}, and P_{PERI} are power consumption of CIS/ISP, H.264 encoder, processor, main memory, micro SD card, and other peripherals, respectively.

Energy consumption in the parked mode is different from the driving mode because some components or blocks are turned on or off according to the event detection. Total energy consumption in the parked mode (E_{park}) can be divided into energy consumption in the standby phase (E_{stby}) and energy consumption in the event phase (E_{evt}). Total energy consumption also depends on the number of detected events (N_{evt}). Total energy consumption in the parked mode can be shown in (3).

$$E_{\mathrm{park}} = E_{\mathrm{stby}} + N_{\mathrm{evt}} \cdot E_{\mathrm{evt}} \tag{3}$$

Energy consumption of the standby phase (E_{stby}) is equal to power consumption of the standby phase (P_{stby}) multiplied by duration of the standby phase (T_{stby}) as expressed in (4).

$$E_{\mathrm{stby}} = T_{\mathrm{stby}} \cdot P_{\mathrm{stby}} \tag{4}$$

Equation (5) shows power consumption in the standby phase, which is the sum of power consumption of CIS/ISP (P_{CIS}), JPEG encoder (P_{JPEGENC}), low-power DRAM connected to JPEG encoder (P_{LPDRAM}), and event detector (P_{ED}). Other components are turned off in the standby phase.

$$P_{\mathrm{stby}} = P_{\mathrm{CIS}} + P_{\mathrm{JPEGENC}} + P_{\mathrm{LPDRAM}} + P_{\mathrm{ED}} \tag{5}$$

Energy consumption for recording one event video (E_{evt}) is equal to power consumption of the event phase (P_{evt}) multiplied by average duration of an event (T_{evt}) as shown in (6).

$$E_{\mathrm{evt}} = T_{\mathrm{evt}} \cdot P_{\mathrm{evt}} \tag{6}$$

Power consumption of the event phase is the sum of power consumption of all block including CIS/ISP, event detector, JPEG encoder, JPEG decoder (P_{JPEGDEC}), low-power DRAM connected to JPEG encoder (P_{LPDRAM}), H.264 encoder, processor, main memory, micro SD card, and other peripherals.

$$\begin{aligned} P_{\mathrm{evt}} = & P_{\mathrm{CIS}} + P_{\mathrm{ED}} + P_{\mathrm{JPEGENC}} + P_{\mathrm{JPEGDEC}} + P_{\mathrm{LPDRAM}} \\ & P_{\mathrm{H264}} + P_{\mathrm{PROC}} + P_{\mathrm{DRAM}} + P_{\mathrm{SD}} + P_{\mathrm{PERI}} \end{aligned} \tag{7}$$

If we divide Eq. (3) by T_{park}, we can get P_{park}. T_{stby} can be expressed using T_{park} and T_{evt}. If we define f as event occurrence rate (i.e., total event duration divided by total parked mode duration), power consumption in the parked mode can be expressed in (8).

$$\begin{aligned}
P_{\text{park}} &= \frac{T_{\text{stby}}}{T_{\text{park}}} \cdot P_{\text{stby}} + \frac{N_{\text{evt}} \cdot T_{\text{evt}}}{T_{\text{park}}} \cdot P_{\text{evt}} \\
&= \frac{T_{\text{park}} - N_{\text{evt}} \cdot T_{\text{evt}}}{T_{\text{park}}} \cdot P_{\text{stby}} + \frac{N_{\text{evt}} \cdot T_{\text{evt}}}{T_{\text{park}}} \cdot P_{\text{evt}} \\
&= P_{\text{stby}} + \frac{N_{\text{evt}} \cdot T_{\text{evt}}}{T_{\text{park}}} (P_{\text{evt}} - P_{\text{stby}}) \\
&= P_{\text{stby}} + \frac{N_{\text{evt}} \cdot T_{\text{evt}}}{T_{\text{park}}} (P_{\text{H264}} + P_{\text{PROC}} + P_{\text{DRAM}} + P_{\text{SD}} + P_{\text{PERI}} + P_{\text{JPEGDEC}}) \\
&\leq P_{\text{stby}} + f \cdot P_{\text{drv}}
\end{aligned} \tag{8}$$

To reduce power consumption in the parked mode, we should decrease P_{stby} as much as possible and decrease the event occurrence rate as well according to Eq. (8). Using Eq. (8), we can draw Fig. 12. Lower bound of power consumption is P_{stby} at $f = 0$, upper bound is the sum of P_{drv} and P_{stby}, and power consumption depending on f is somewhere under solid line. If P_{stby} is designed as a quarter of P_{drv}, P_{park} is equal to P_{drv} at $f = 3/4$. Assuming that we design P_{stby} as a quarter of P_{drv}, the system applying our approach seems to be effective until three quarters of f, which is depicted in gray area in Fig. 12. Conventional VEDR in the parked mode consumes around 90 % of power consumption in the driving mode shown in Eq. (9). If the LPEDVC module consumes 25 % of power consumption of the driving mode, we can get effective range of f as Eq. (10) using Eq. (8). This means our approach with P_{stby} as a quarter of P_{drv} remains competitive with conventional VEDR until 65 % of event occurrence rate in terms of power consumption.

$$P_{\text{park,CONV}} \cong 0.9 \cdot P_{\text{drv}} \tag{9}$$

$$\begin{aligned}
&0.9 \cdot P_{\text{drv}} \geq 0.25 \cdot P_{\text{drv}} + f \cdot P_{\text{drv}} \Big|_{P_{\text{stby}} = 0.25 \cdot P_{\text{drv}}} \\
&f \leq 0.65
\end{aligned} \tag{10}$$

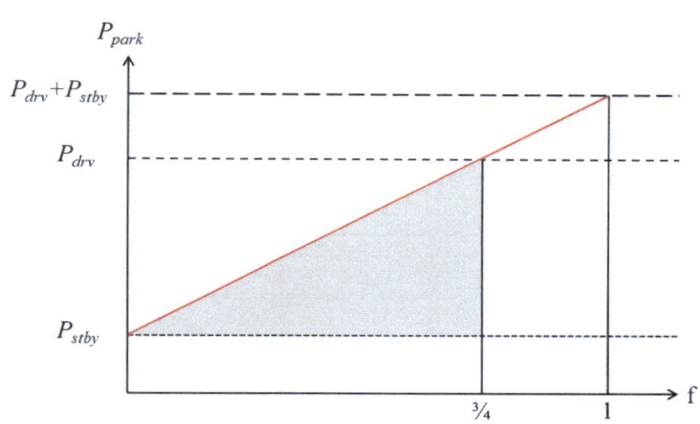

Fig. 12 Power consumption in parked mode according to event occurrence rate

5 Performance Evaluation

Experimental results on how much energy can be saved by the proposed method are given in this section. The energy consumption is calculated by the energy model in the previous section. With the model, we compared the energy consumption of the proposed system with the energy consumption of conventional VEDR on various event occurrence scenarios. The value of each component's power consumption is based on the measurement from the prototype shown in Fig. 13. In our prototype, commercially available AP [15] is used. The AP includes ARM Cortex-A8 processor which operates at 800 MHz and H.264 encoder which encodes 1080p video at 30 frames per second. In order to implement the LPEDVC module, we utilize low-power FPGA [30]. According to power simulation of the FPGA and other measurements shown in Fig. 3, P_{stby} is estimated around 820 mW which is about a quarter of P_{drv}. The energy consumption of the proposed system depends on the frequency of event occurrence, while energy consumption of conventional VEDR is almost constant regardless of event occurrence frequence rate.

To obtain the statistics of frequency of event occurrence, we need to know the distribution of parking place. According to [14], however, the distribution of parking places looks different in each country. For fair comparison, scenarios for parking places are carefully selected considering following parameters. The frequency at night is much lower than the frequency during daytime. Type of location

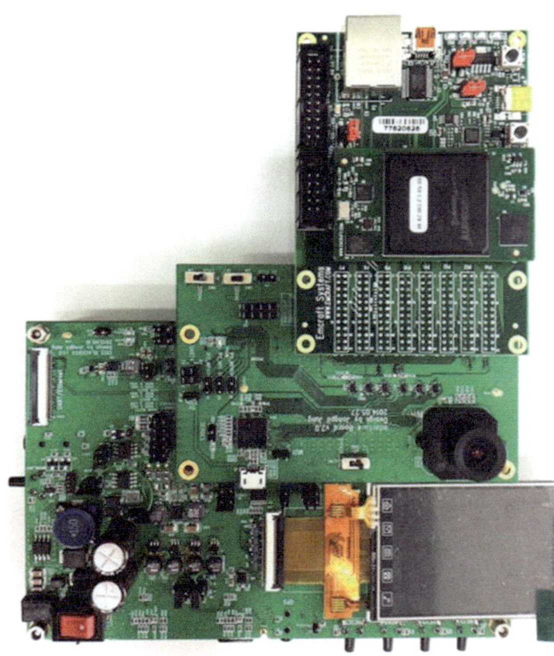

Fig. 13 Prototype design

Table 3 The frequency of event occurrence in scenario

Scenario ID	Day/night	Indoor/outdoor	Parking place type	Events per hour
1	Night	Outdoor	Public area[1]	0.6
2	Day	Outdoor	Own place at Home[1]	27.1
3	Night	Outdoor	Kerbside unregulated[1]	1.5
4	Day	Indoor	Reserved at work[1]	1.0
5	Day	Indoor	Shopping mall	89.5
6	Day	Outdoor	Kerbside unregulated[1]	22.4
7	Night	Indoor	Reserved at work[1]	7.9
8	Day and night	Outdoor	Own place at Home[2]	19.6
9	Day	Indoor	Reserved at work[2]	2.2
10	Night	Outdoor	Public area[2]	2.4
11	Night	Outdoor	Kerbside regulated[2]	17.0

is another important factor. In crowded areas such as parking lots at mall, stadium, and main street, the frequency of event occurrence is extremely high. The frequency in residential area is relatively low compared with crowded area. It also depends on whether the parking lot is indoor or outdoor. Considering all, 11 scenarios were selected for the experiment, which are described in Table 3. Superscript on parking place type in the table is used to discriminate different locations of the same type.

Experimental results are shown in Fig. 14. To compare energy consumption of the proposed system with that of conventional VEDR, each value for the proposed system is normalized to the value for conventional VEDR. As shown in Fig. 14, energy consumption is significantly reduced in every scenario. The energy consumption of proposed design is reduced by up to 74.9 % in scenario ID 1, and by 66.8 % on average. Reduction of the energy consumption is most significant in the case of public area during night because event occurrence rate is very low as 0.6 events per hour. The energy consumption is reduced even in the crowded area. However, the degree of reduction is 30.6 % which is not as significant as residential area at night. Using the results, we can get normalized power consumption by

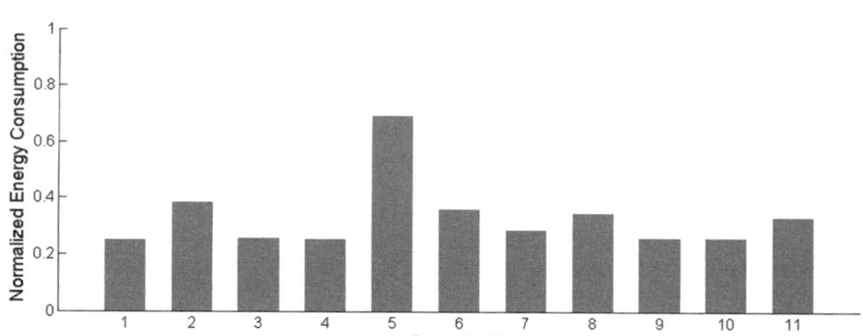

Fig. 14 Normalized energy consumption in selected scenarios

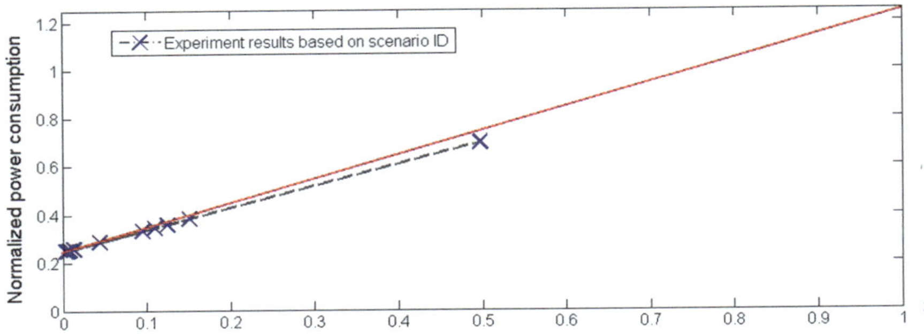

Fig. 15 Normalized power consumption versus f

dividing total parked mode duration. We can calculate f for each scenario ID from Table 3, e.g., $89.5 \times 20/3{,}600 = 0.4972$ for scenario ID 5. The normalized power consumption based on f is shown in Fig. 15. The power consumption of the proposed system is below the solid line which is the analytical upper bound given in Eq. (8).

Among detected events in the crowded area, not every event was meaningful. Important video in this application is about moving objects which can damage the vehicle of interest. Most of events detected in the crowded area were pedestrians at a distance, which are not a threat to the vehicle. If we design better event detector which eliminates false positive effectively based on region of interest, total energy consumption will be reduced further.

6 Conclusion

In this chapter, an energy-aware low-energy VEDR is proposed. It is crucial for video surveillance system like the VEDR to extend its operating time under limited storage and energy constraint. To overcome the limited storage constraint, the proposed system records video only when defined events are detected. Like conventional VEDR, the proposed system always catches important moment, because it records the video of defined event starting from 10 s prior to the event. The energy consumption of the proposed system under selected scenarios is up to 25.1 and 33.2 % on average of that of conventional VEDR due to duty-cycling of the LPEDVC and the core-related part based on event occurrence. Since energy consumption of the proposed system is reduced by 67 % on average, the monitoring time of the proposed system can be extended by three times. The energy consumption of proposed system is expected to be reduced even more when less power consuming LPEDVC module and an event detector with less false positive are incorporated.

The proposed design can also be used in wearable streaming cameras [31] and the blackbox camera for bike [32]. In these applications, event-driven scheme can be very helpful in extending products' operating time.

Acknowledgments This work is supported by the Center for Integrated Smart Sensors funded by the Ministry of Science, ICT and Future Planning as the Global Frontier Project.

References

1. Mango M et al (2009) Multimodal abandoned/removed object detection for low-power video surveillance systems. In: IEEE international conference on advanced video and signal based surveillance (AVSS), pp 188–193, Sep 2009
2. Robert K (2009) Night-time traffic surveillance: A robust framework for multi-vehicle detection, classification and tracking. In: IEEE international conference on advanced video and signal based surveillance (AVSS), pp 1–6, Sep 2009
3. Belongie S, Dollar P, Perona P (2010) The fastest pedestrian detector in the west. In: British machine vision conference
4. Remagnino P, Shihab A, Jones G (2004) Distributed intelligence for multi-camera visual surveillance. Pattern Recogn 37(4):675–689 (Agent Based Computer Vision)
5. Kim G et al (2012) Energy-aware operation of black box surveillance cameras under event uncertainty and memory constraint. In: IEEE international conference on multimedia and expo (ICME), pp 782–787, July 2012
6. Video data recording systems for road vehicle accidents, Technical report KS C 5078:2013R, Korean Agency for Technology and Standards, May 2013
7. Chien S-Y et al (2012) Power optimization of wireless video sensor nodes in m2 m networks. In: Design automation conference (ASP-DAC), Asia and South Pacific, pp 401–405, Jan 2012
8. Jin X, Goto S (2011) Encoder adaptable difference detection for low power video compression in surveillance system. Sig Process Image Commun 26(3):130–142
9. Fernandez-Bemi J et al (2011) Wi-flip: a wireless smart camera based on a focal-plane low-power image processor. In: ACM/IEEE international conference on distributed smart cameras (ICDSC), pp 1–6, Aug 2011
10. Jung SM, Lim M-S (2007) System on chip design of embedded controller for car black box. In: International symposium on information technology convergence (ISITC), pp 217–221, Nov 2007
11. Kassem A et al (2008) Vehicle black box system. In: Annual IEEE systems conference, pp 1–6, Apr 2008
12. Linden D, Reddy TB (2001) A handbook of batteries, 3rd edn. McGraw-Hill, New York
13. Korea Transportation Safety Authority (2012) A study on survey of the 2012 car mileage report. http://www.ts2020.kr/
14. European Commission, Driving and parking patterns of European car drivers—a motility survey. http://publications.jrc.ec.europa.eu/repository/handle/111111111/26994
15. CoreLogic, CLM9722 datasheet. http://www.corelogic.co.kr/down_pb/110525_CLM9720CLM9721_PB_v5.0_noDRM.pdf
16. Ye W, Heidemann J, Estrin D (2002) An energy-efficient mac protocol for wireless sensor networks. In: 21st international annual joint conference of the IEEE computer and communications societies (INFOCOM'02), New York, NY, USA
17. van Dam T, Langendoen K (2002) An adaptive energy efficient mac protocol for wireless sensor networks. In: 1st ACM conference on embedded networked sensor systems (SenSys), pp 53–64

18. Polastre J, Hill J, Culler D (2004) Versatile low power media access for wireless sensor networks. In: 2nd ACM conference on embedded networked sensor systems (SenSys), pp 95–107, Nov 2004
19. El-Hoiydi A, Decotignie J (2005) Low power downlink mac protocol for infrastructure wireless sensor networks. ACM Mob Netw Appl 10(5):675–690
20. Jung D, Teixeira T, Barton-Sweeney A, Savvides A (2007) Model-based design exploration of wireless sensor node lifetimes. In: Proceedings of 4th European conference EWSN 2007, pp 277–292, Jan 2007
21. Masashi M et al Image recording apparatus. Japan Patent JP2006-127206A
22. Takajawa N et al Recording information generating device and recording information generating program and information recording medium. Japan Patent JP2014-036428A
23. Park H The fast booting of embedded linux. The 3th CE Linux Forum Korea Technical Jamboree. http://tree.celinuxforum.org/CelfPubWiki/KoreaTechJamboree3
24. Kim H et al (2013) A low-power video recording system with H.264/AVC and light-weight compression. In: Proceedings of 2013 IEEE workshop on signal processing systems (SiPS), pp 183–188, Oct 2013
25. Pennebaker WB Mitchel JL JPEG: Still Image Data Compression Standard. Springer, New York
26. Kim W (2014) Low complexity, high accuracy event detection using DCT coefficients for parked mode vehicle surveillance camera. Master's thesis, Department of Electrical Engineering, KAIST
27. Mobile dram power-saving features and power calculations. Technical report TN-46-12, Micron technology, May 2009
28. Benezeth Y et al (2010) Comparative study of background subtraction algorithms. J Electron Imaging 19(3):033003-033003-12
29. Ghidary S et al (2000) Human detection and localization at indoor environment by home robot. IEEE Int Conf Syst Man Cybern 2:1360–1365
30. Microsemi, IGLOO2 FPGA datasheet. http://www.microsemi.com/products/fpga-soc/fpga/igloo2docs#documents
31. Looxcie, Looxcie HD user manual. http://www.looxcie.com
32. Rideye, Rideye 32 GB. http://www.rideye.com

Low-Power Face Detection for Smart Camera

Hyung-Il Kim, Seung Ho Lee and Yong Man Ro

Abstract Recently the development of intelligent surveillance system increasingly requires low power consumption. For power saving, this chapter presents an event detection function based on automatically detected human faces, which adaptively changes from low-power camera mode to high performance camera mode. We propose efficient face detection (FD) method being operated under the low-power camera mode. By employing two-stage structure (i.e., region-of-interest (ROI) selection and false positive (FP) reduction), the proposed FD method requires very low computational complexity and memory requirements without sacrificing the face detection robustness. Experimental results demonstrated that the proposed FD could be implemented in low-power video cameras with promising performance.

Keywords Face detection · Intelligent surveillance system · Low power hardware architecture · Event detection · Smart camera · Two-stage structure

1 Overview of Face Detection for Smart Camera

The advances in computing, communication, and sensor technology are recently pushing the development of many new applications in pervasive computing, sensor networks, and embedded systems [1]. As one example of the innovation, smart

H.-I. Kim · S.H. Lee · Y.M. Ro (✉)
School of Electrical Engineering, KAIST, Daehak-Ro, Yuseong-Gu, Kaist 305-701 Republic of Korea
e-mail: ymro@ee.kaist.ac.kr

H.-I. Kim
e-mail: hyungil.kim@kaist.ac.kr

S.H. Lee
e-mail: leesh09@kaist.ac.kr

C. Kyung (ed.), *Theory and Applications of Smart Cameras*,
KAIST Research Series, DOI 10.1007/978-94-017-9987-4_7

cameras are equipped with high-performance onboard computing and communication functionalities by combining video sensing, processing, and communications in a single embedded device. Thanks to the functionalities, the smart cameras can support more complex and challenging applications including smart rooms, intelligent surveillance, tracking, and motion analysis [1].

Among the applications, intelligent surveillance system has become increasingly important for the purpose of public safety and security [2]. For example, a main function of the system is to detect and analyze specific events such as suspicious actions or an unidentified object [3]. The surveillance system, where sensor nodes are equipped with low-power cameras, can be utilized in an intrusion detection scenario [4]. In the application, it is crucial to get visual information about an event area [4] under consideration. For that purpose, face detection (FD) can be one of the key techniques in intelligent surveillance systems [2]. FD is a necessary step for all the face related applications such as access control, person-specific identification, etc. In addition, FD is useful for reliable surveillance because FD is widely known as a relatively mature technique [5].

There have been several works [6–10] on FD. In [6, 7], a support vector machine (SVM) classifier [11] based FD was proposed. Due to the exhaustive computation for comparison between test feature vector and support vectors in every scaled image of the input frame, this method is inefficient in terms of computational complexity. In order to speed-up the computation time, the methods in [12, 13] incorporated a skin-color filtering. However, the memory cost increased due to the requirement of additional use of color information.

Since early 2000's, considerable research efforts have been dedicated to development of FD methods [14]. In [8], Viola and Jones proposed the first real-time FD based on Haar-like features and the adaptive boosting (AdaBoost)-based learning algorithm [8]. Due to its effectiveness, boosting learning-based FD methods have been regarded as de facto standard of FD in real-world applications [14]. A number of researchers have addressed challenges in boosting learning [9, 10, 12, 13]. To speed-up testing of FD, the number of features (or weak classifiers) was reduced [9] or skin-color filter was incorporated into FD, which could significantly reduce the search area in an input image [12, 13]. To deal with large variation in face pose, multi-view FD was proposed with a "divide and conquer" strategy. To enable multi-view FD, individual classifier was learned for each view. Given a scan window was passed to all classifiers (i.e., parallel cascade [10]) or sequentially passed to classifiers of pyramid structure [15]. However, this approach needs a quite large memory space for storing training models (e.g., feature information and thresholds [16]), which is not preferred for use in surveillance system. Furthermore, since the training process is too complicated, it is not easy to update the training model according to the change of surveillance environment. In addition, a common limitation of the previous works [8, 12, 13, 15] is that they only consider faces larger than 20×20 pixels or 24×24 pixels although much smaller faces are often encountered in surveillance systems [17].

Modern surveillance systems require a compact design and low power operation [3, 18] (e.g., in terms of the memory cost which is known to be dominant portion of power consumption in a digital system [19]). In reality, various surveillance environments introduce large variations in illumination and camera view angles, which require adaptive change of training models. The previous FD methods in [6–10] could not be appropriate for the surveillance system requirements mentioned above.

In this chapter, we describe a new FD method aiming to cope with the above-mentioned limitations in surveillance system. For efficient and smart surveillance system, we introduce a new event detection function based on detected human faces in surveillance system. The function remains in a low-power standby mode (i.e., FD operating on an extremely low-power camera) until an intrusion alarm is made by a detected face. The intrusion alarm immediately wakes up the surveillance system so that the system can perform a more sophisticated analysis such as person-specific identification on the intruder in a high performance camera. This structure allows surveillance system to operate on a minimal power budget (refer to the Fig. 1).

In particular, we propose a two-stage FD framework for the low-power camera mode in the introduced event detection. Since a strong classifier in the second stage is conducted for only selected region-of-interests (ROIs) from the first stage with low computational cost, the FD algorithm can be efficiently operated in real-time. In addition, the described FD framework is well suited for operating on a low-power camera (e.g., detecting faces in small sized low-power monochrome camera). Further, the detector explained in this chapter requires pre-training for only a single classifier in the second stage, which is much more efficient than most existing face detectors. Furthermore, the hardware architecture of the proposed algorithm is implemented and validated in the gate level simulation. In particular, the unified static random access memory (SRAM) structure reduces gate counts for memory usage. In addition, the pipelined structure contributes to speed-up the process considerably.

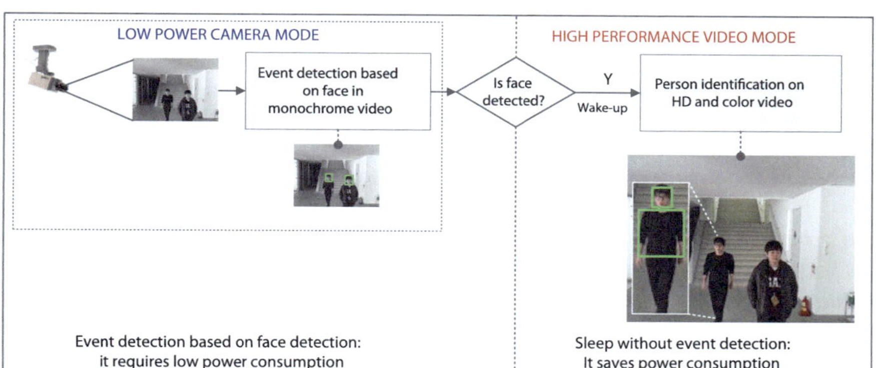

Fig. 1 Operation scenario using the face-based event detection two camera modes [22]. © 2014 IEEE

Experimental results show that the proposed face detector is useful for event detection function of surveillance, which requires both low-computation and memory complexities. It is also shown that the proposed method is robust to variations in video acquisition condition (e.g., pose or illumination variation). Further, the feasibility of adopting the proposed method in hardware has been successfully validated by using hardware design verification. Note that this chapter is restructured based on the paper published in [22]. In addition, some of the figures and tables listed in this chapter are derived from [22].

The rest of the chapter is organized as follows: Sect. 2 presents human face-based event detection for intelligent surveillance system. Section 3 describes the proposed ROI-based two-stage FD algorithm. In Sect. 4, the comparative experiments and hardware architecture are present. Finally, Sect. 5 provides some concluding remarks.

2 Human Face-Based Event Detection for Low Power Operation

There are two essential steps in intelligent surveillance systems [2]: (1) object detection and (2) object recognition. Object detection is basically a very time-consuming task due to exhaustive search in every scaled image of input image. Furthermore, object detection requires high memory consumption because it is performed on whole and scaled image regions.

In this section, we introduce an event detection function which aims to reduce the power consumption required for object detection in intelligent surveillance systems. Figure 1 shows an operating scenario of the proposed event detection. It contains low-power camera mode and high performance camera mode. The low-power camera mode (with QVGA [20] monochrome frames) is always turned on for continuously sensing human intrusions while keeping power consumption low. Only when a face is detected by the FD module, an intrusion alarm is made to wake up the high performance camera mode. The object recognition (i.e., person identification) is performed on this camera mode for further analysis on the intruder. The facial feature and other useful personal attributes, such as clothing information, are extracted. The extracted features are then used to match the unknown intruder with persons in a human database, determining the identity label. For reliable analysis, video frames of sufficiently high resolution (e.g., HD [20]) are used in the high performance camera mode. Color information is also available in this mode because it can provide useful discriminative information for person identification (facial color, clothing color, etc. [12, 13]). The use of the two different camera modes (i.e., low power and high performance) is able to achieve significant power savings over traditional video surveillance systems which use a single camera mode.

In the next section, we present a very efficient and robust FD method suited for the low-power camera mode, which has not been deeply investigated in literature. Note that many effective methods developed for person identification can be found in [21].

3 Low-Power Face Detection Algorithms

To detect various sizes of faces from an image, the proposed FD method finds face-like patterns by sliding 16 × 16 pixels scanning window through every downscaled image from the original input image, where the image is quantized to 4-bit to reduce the power dissipated in image sensor. Herein, all of the generated scan windows are the input of the FD framework. As shown in Fig. 2, the proposed method is comprised of two stages, ROIs selection and false positive (FP) reduction. In the first stage, each scan window is rapidly checked whether to be a face candidate (ROI) or not. In the second stage, only the ROIs are further examined by using a strong classifier [22].

The proposed FD method is suitable for operating under the low-power camera mode in the following aspects: (1) the two-stage design enables to speed-up FD without using additional skin color information; (2) the proposed method deals with very small sized faces (e.g., 16 × 16 pixels) due to a robust feature extraction (refer to Sect. 3.1); (3) As demonstrated in our simulation results in Sect. 4, the proposed FD itself requires very low power consumption (especially in terms of memory). The detailed descriptions of the two stages (i.e., ROI selection and FP reduction) are given in the next subsections.

3.1 *The First Stage: Region-of-Interest (ROI) Selection*

This section describes the ROI selection method which is based on pre-filtering, feature extraction, and feature templates matching. For a given scan window, the pre-filtering step is to speed-up FD by rejecting many negative scan windows in early stage. Then, the feature extractor generates the facial features to be used for the classification using feature templates matching.

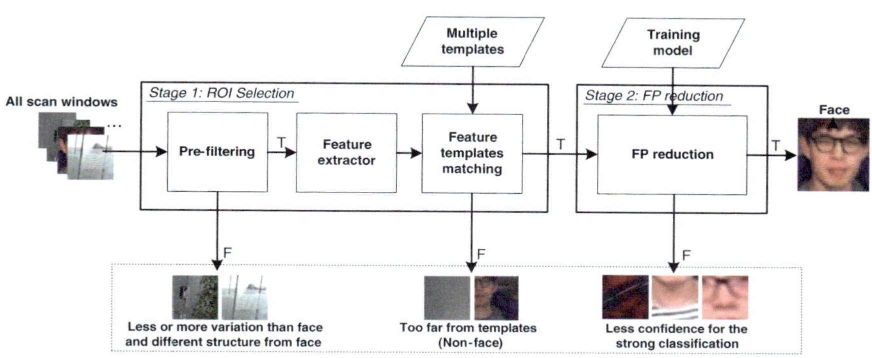

Fig. 2 Overview of the proposed FD method for low-power camera mode

3.1.1 Pre-filtering Step for Speed-up

A human face image contains a certain amount of variance due to the presence of the facial components (e.g., eyes, nose, mouth etc.). An image variance filter measures the variance of the pixel luminance values within a scan window to reject the ones that are unlikely to contain a face. For measuring image variance, we define a 6 × 6 pixels region $\mathbf{R}_{\text{var}}$ as shown in Fig. 3a. The region $\mathbf{R}_{\text{var}}$ contains the facial components but not the background region. As the image variance measure, we adopt a standard deviation σ as follows:

$$\sigma = \sqrt{\frac{1}{N}\sum_{k=1}^{N}(x_i - \bar{x})^2}, \quad (1)$$

where $\bar{x}$ is the mean of the all pixel intensity values $x_i \quad (i = 1, \ldots, N)$. In addition, N ($8 \times 8 = 64$ in this chapter) is the number of pixels within the region $\mathbf{R}_{\text{var}}$. In Eq. 1, σ represents how spread out the pixel intensity values are within the region $\mathbf{R}_{\text{var}}$. Using the σ value, we determine if a given scan window is face or non-face. In other words, if the σ value of the scan window is larger than $t_{\min}$ and smaller than $t_{\max}$, the scan window is passed, otherwise, the scan window is rejected. Herein, the $t_{\min}$ and $t_{\max}$ are a lower and an upper decision threshold values, respectively.

As the second filtering step, we use another facial characteristic that eye region in the human face is likely to be dark. This structural information can be used to reject negative scan windows that have erroneously passed the image variance filter. To capture facial structural information, we make use of the difference of pixel intensity values between the two rectangle regions, which is motivated by Haar-like

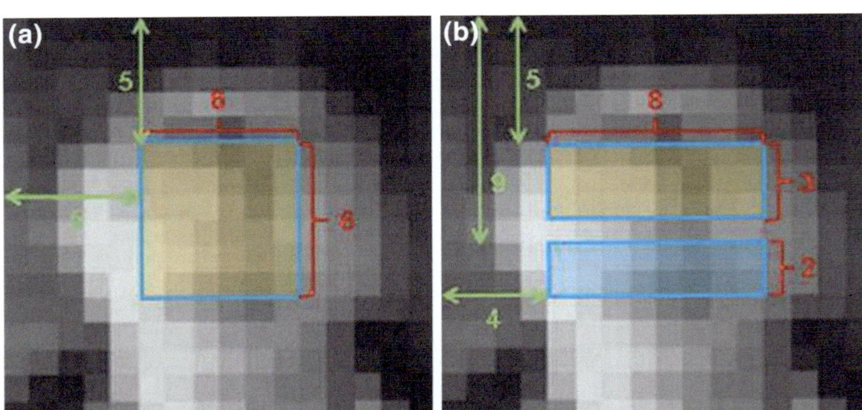

Fig. 3 **a** 6 × 6 facial region for measuring image variance. **b** Two rectangle regions for capturing facial structure

features in [8]. The two rectangles are defined as 8 × 3 pixels upper region $\mathbf{R}_u$ (that corresponds to two eyes) and the 8 × 2 pixels lower region $\mathbf{R}_l$ (that corresponds to cheeks and nose) for a given 16 × 16 scan windows (see Fig. 3b). To determine if a given scan window is face or non-face, we first compute the difference γ of mean pixel intensity value between the upper region and the lower region:

$$\gamma = \bar{x}_l - \bar{x}_u, \tag{2}$$

where $\bar{x}_l$ and $\bar{x}_u$ are the mean pixel intensity values of the lower region $\mathbf{R}_l$ and the upper region $\mathbf{R}_u$, respectively. Then, if the difference γ is larger than the predefined decision threshold value γ_{th}, the scan window is passed, otherwise, the scan window is rejected.

3.1.2 Feature Extraction

For a given scan window (passed window from the previous pre-filtering stage), the feature extractor generates the facial features to be used for the classification using feature templates matching. For effective face representation, we propose block texture features that emphasize facial components (i.e., two eyes and mouth). We divide each window region to three blocks ($\mathbf{R}_{\text{L\_eye}}, \mathbf{R}_{\text{R\_eye}}$, and $\mathbf{R}_{\text{mouth}}$) as shown in Fig. 4. The individual block regions are separately represented by histogram feature vectors. It is important to note that the histogram-based features are less susceptible to subtle rotation and translation in face representation [23]. Hence, histogram based features could increase the tolerance of the FD.

For the histogram feature vectors, we adopt a local micro texture like local binary pattern (LBP) [24] which can efficiently encode facial texture pattern. The main advantages of LBP include [24]: (1) low computational complexity and (2) invariance against monotonic illumination variation. Note that any other histogram-based texture features could be used for the proposed FD scheme.

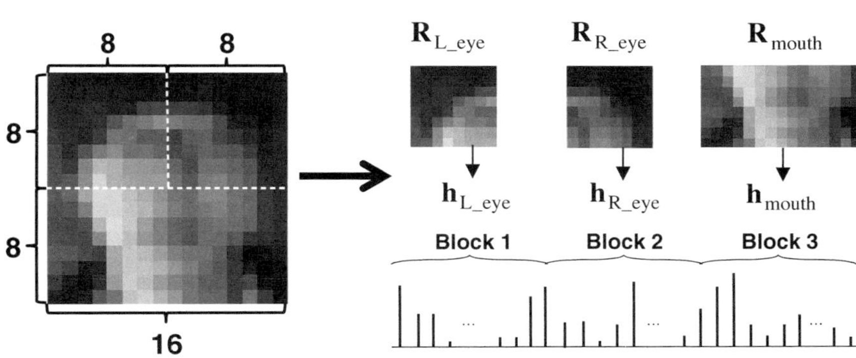

Fig. 4 Illustration of the block texture feature extraction [22]. © 2014 IEEE

For extracting LBP histogram, the uniform LBP operator [24] is used with parameters (P, R) = (4,1), where P is the number of neighboring pixels equally spaced on a circle of radius R [24] (In this chapter, horizontal and vertical neighboring pixels are considered.). It produces a 15-dimensional $(= P(P-1)+3)$ [25] feature vector of LBP histogram for each block region, where each value has an integer value. In order to eliminate the effect of different sizes of block, we normalize the histogram obtained from each block region. Herein, we multiply the number of pixels within each region for efficiently implementing the hardware by using only integer programming instead of dividing each value for the normalization.

Finally, in order to reflect facial structural information in face feature, the three histogram feature vectors $\mathbf{h}_{\mathrm{L\_eye}}$, $\mathbf{h}_{\mathrm{R\_eye}}$, and $\mathbf{h}_{\mathrm{mouth}}$, are concatenated resulting in 45-dimensional global histogram feature vector $\mathbf{h}_{\mathrm{SW}}$ for a scan window:

$$\mathbf{h}_{\mathrm{sw}} = [(\mathbf{h}_{\mathrm{L\_eye}})^T(\mathbf{h}_{\mathrm{R\_eye}})^T(\mathbf{h}_{\mathrm{mouth}})^T]^T, \quad (3)$$

where T denotes the transpose operator.

The feature vector is fed into the binary classification based on feature template matching, which will be explained in the next subsection.

3.1.3 Multiple Template Matching

Using the feature vector $\mathbf{h}_{\mathrm{SW}}$ with type of integer, the decision is made through the feature templates matching. In order to detect faces with various poses, we propose multiple feature templates, each of which represents a single face pose (e.g., frontal and 45 degrees in yaw). As shown in Fig. 5, the feature vector ($\mathbf{h}_{\mathrm{sw}}$) is compared with each feature template one by one. The comparison is repeated until the best-matched feature template is found (e.g., 3rd feature template in Fig. 5). Note that the number and types of feature templates can vary according to the image acquisition conditions or target applications. The method for feature templates generation is explained in Sect. 4. In order to classify the feature vector $\mathbf{h}_{\mathrm{sw}}$, we compute the distance (l^1-norm distance in this paper for computational efficiency) $\mathrm{dist}^{(k)}$ between $\mathbf{h}_{\mathrm{sw}}$ and the k-th feature templates $\mathbf{h}_{\mathrm{ft}}^{(k)}$ $(k = 1, \ldots, K)$ in order, where we can decrease the computational load by using Manhattan distance (integer operation) instead of using Euclidean distance (floating point operation). For each template, the computed distance $\mathrm{dist}^{(k)}$ is compared with the pre-determined decision threshold values. If the distance $\mathrm{dist}^{(k)}$ is smaller than s, the scan window is classified as a face candidate. Otherwise, the distance computation for the next feature templates $\mathbf{h}_{\mathrm{ft}}^{(k+1)}$ (for $k+1 \leq K$) is performed and the comparison is continued.

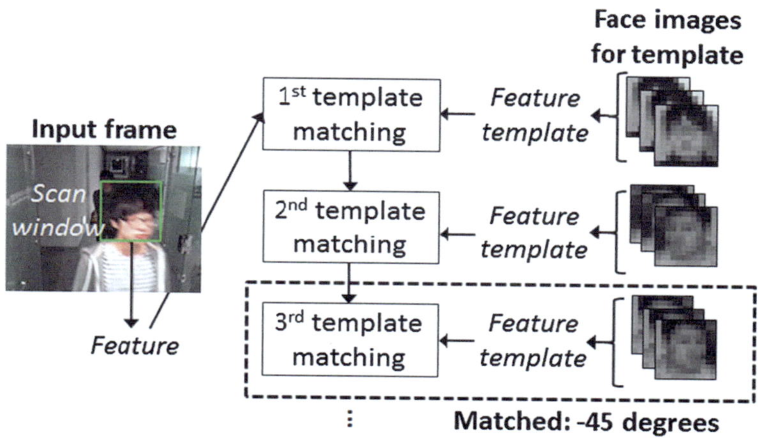

Fig. 5 Illustration for the proposed multi-view face detection using feature templates matching [22]. © 2014 IEEE

3.2 *False Positive (FP) Reduction*

To reduce unnecessary wake-up of the high performance camera mode, this section describes the FP reduction stage using a strong classifier. As the strong classifier, we adopt the linear support vector machine (SVM) [11] due to its robustness and high generalization capability [26]. The SVM is learnt by minimizing the classification error as well as maximizing a margin which represents a distance between support vectors and optimal hyperplane [11]. In order to obtain the SVM training model, the SVM learns the following quadratic optimization functions [27]:

$$\begin{aligned}&\min_{\alpha_i} \frac{1}{2}\sum\nolimits_{i=1}^{n}\sum\nolimits_{j=1}^{n}\alpha_i\alpha_j y_i y_j \mathbf{\Phi}(\mathbf{h}_i)^T\mathbf{\Phi}(\mathbf{h}_j)-\sum\nolimits_{i=1}^{n}\alpha_i,\\ &\text{subject to } 0\le\alpha_i\le C,\ \sum\nolimits_{i=1}^{n}\alpha_i y_i=0,\end{aligned} \tag{4}$$

where $\alpha_i, y_i, \mathbf{h}_i$, and C are the i-th Lagrangian multiplier, the i-th class label ($y_i \in \{-1, +1\}$, '+1' for positive and '−1' for negative), the feature vector of the i-th training sample, and the regularization parameter, respectively. In Eq. 4, the kernel function $\mathbf{\Phi}(\mathbf{h}_j)$ represents the mapping of feature space into a higher dimensional space. Note that in the linear SVM, for efficient and simple computations, $\mathbf{\Phi}(\mathbf{h}_j)^T\mathbf{\Phi}(\mathbf{h}_j)$ becomes the inner product operation, i.e., $\mathbf{h}_i^T\mathbf{h}_j$.

It is important to note that the training images used to train the SVM classifier correspond to the output scan windows from the ROI selection stage. Hence, huge variation contained in scan windows which the SVM classifier should handle can be significantly reduced. The reduced variation allows the SVM classifier to be trained more efficiently and effectively.

Using the obtained SVM training model (i.e., support vectors ($\mathbf{h}_i$, $i = 1, \cdots, n_{SV}$) and Lagrangian multipliers (α_i)), the SVM confidence value η for $\mathbf{h}_{SW}$ is computed by using the following equation:

$$\eta = \sum_{k=1}^{n_{SV}} \alpha_i \mathbf{h}_i^T \mathbf{h}_{sw} = \mathbf{w}^T \mathbf{h}_{sw}, \tag{5}$$

where the weight vector ($\mathbf{w}$) can be defined by the linear combination of support vectors with the Lagrangian multipliers. We can see that the computation of the SVM confidence value η in Eq. 5 requires only one inner product operation in the linear SVM. This means that the second stage of the proposed FD method needs to store only a 45-dimensional feature vector for $\mathbf{w}$, which leads to very low memory complexity. Furthermore, we calculate down to six places of decimals by rounding off for the computational efficiency. This contributes to minimize the cost of multiplier in hardware implementation. For the final decision on a given scan window, η is compared with the predefined confidence threshold value η_{thrs}. The scan window is determined as a face if the η is larger than η_{thrs}, otherwise, it is determined as a non-face.

4 Experiments

In this section, we evaluated the effectiveness of the proposed FD method for low-power camera. For the evaluation, videos were acquired by using a web camera (Microsoft LifeCam) and a closed-circuit television (CCTV) camera (NCD-2000P) in experiments. The information of the videos used in the experiment is summarized in Tables 1 and 2.

For the proposed FD method, the downscaled images for scanning face windows were obtained from the original input images with scale factor of 1.4 by a nearest-neighbor interpolation method. To detect faces in the image, the window was scanned left to right with two-pixel shift. To reduce FPs, the detection results overlapped at a location were merged to form a final detection result, i.e., FD was valid only when the number of overlapped detection results was more than four.

As discussed in Sect. 3.1, two simple pre-filtering methods were adopted to speed-up FD. First, we reject scan windows that have too large spatial variances or

Table 1 Test videos used in the experiments [22]. © 2014 IEEE

Video id	V1	V2	V3	V4
Place	Hall	Entrance	Lab	Stair
Frame size (pixels)	320 × 240	320 × 240	293 × 240	426 × 240
No. of frames	135	457	90	209
No. of subjects	5	3	1	1

Table 2 Test videos used in the experiments with different types of challenges [22]. © 2014 IEEE

Video id	V5	V6
Frame size (pixels)	320 × 240	426 × 240
No. of frames	135	209
Type of challenge	Illumination variation	Pose variation

too small spatial variances compared to face. Standard deviation is used as an image spatial variance measure. Second, we reject scan windows that have different structure from face in which the eye region is brighter than the cheek region [8].

To this end, we subtract mean pixel value of upper rectangle region (8 × 3 pixels corresponding to two eyes) from that of lower rectangle region (8 × 2 pixels corresponding to cheek and nose). If the subtraction value of a scan window is smaller than a predefined value (21 is used in this paper), the scan window is rejected. The feature extraction step takes only the scan windows that have passed the pre-filtering step.

In order to construct the feature templates, we collected 64 face images of various styles, which were not present in the test video. To deal with variation in facial pose, we considered the three different poses, i.e., frontal, +45° in yaw, and −45° in yaw. In each pose, the block texture feature was extracted from every face image, using the method in Sect. 3.1. After that, the feature templates were obtained by averaging the block texture features.

We conducted the SVM training using the publicly available LIBSVM library [27], based on the quadratic optimization function in Eq. 4. For training data, we collected the face and non-face images from random web images, video frames captured by the web camera under CCTV environments, and images from the public face databases (e.g., FERET DB [28]). The number of face samples for training is 1,084 images and the number of non-face samples is 7,120 images.

For the comparative evaluation, we presented the two existing well-known FD methods: (1) Viola-Jones face detector (termed as Viola-Jones) [8] and (2) AdaBoost based on LBP feature (termed as LBP + AdaBoost) [29]. Both methods were executed with OpenCV (open computer vision library) version 2.4.2. The FD parameters, i.e., scale factor, minimum neighbor, and minimum face size were set to 1.1, 3, and 24 × 24 pixels, respectively. As the measures of FD performance, we used a recall [30], precision [30], and F1-score [31]. The ground truth of a video frame was annotated manually (only faces larger than 16 × 16 pixels have ground truth values assigned).

In addition, we designed the FD hardware for a low-power camera mode. For the evaluation, we performed the gate level simulation by using DongBu 0.11um cell library. The details on the design are described in Sect. 4.2.

4.1 Face Detection Performance Evaluation

In this section, we performed comparative experiments using four videos (V1 to V4 in Table 1) shown in Fig. 6. In each video, the subjects move towards camera. The videos contained very challenging frames with occlusion, blurring, highlighting, head tilting, etc. Because the minimum detectable size of the two comparison methods (i.e., LBP + Adaboost and Viola-Jones method) was 24 × 24 pixels, the comparison results were obtained only for faces larger than 24 × 24 pixels for fair comparisons. As shown in Table, 3, the proposed FD outperforms the other two methods. In addition, the proposed FD method could achieve an acceptable performance (87.62 % of F1-score) for faces larger than 16 × 16 pixels.

To examine the robustness of the proposed FD method to illumination variations and pose variations, we performed FD on the V5 (for the illumination variation) and the V6 (for the pose variation). The information for the V5 and V6 can be found in Table 2. V5 included frames acquired from the different illumination conditions ranging from 5 lux to 250 lux with camera exposure value set to '−8'. In V6, the subject performed a slow head rotation ranging from +60° (looking at right) to −60° (looking at left). Table 4 shows the comparative results for faces with varying illumination. We observe that the proposed FD method clearly outperforms the Viola-Jones method. Some FD results are shown in Fig. 7a (the face was detectable

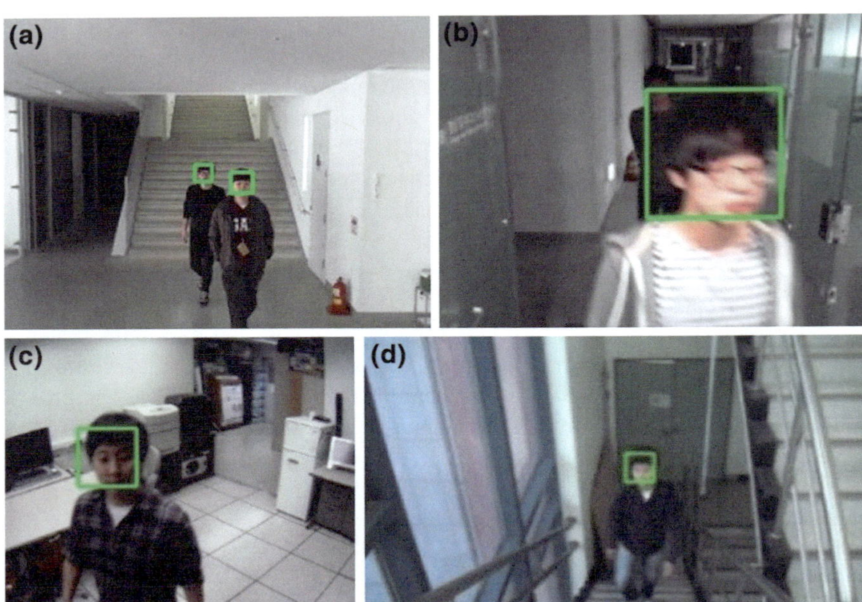

Fig. 6 Example of video frames with the results of the proposed method [22]. **a** For V1 with occlusion between multiple subjects. **b** For V2 with blurred and flashed face. **c** For V3 with shaded face. **d** For V4 with tilted face by high camera angle. © 2014 IEEE

Table 3. Experimental result for V1–V4. **a** For faces larger than 24 × 24 pixels. **b** For faces larger than 16 × 16 pixels [22]. © 2014 IEEE

(a)

Method	Recall	Precision	F1-score
Viola-Jones	72.48 %	79.02 %	75.61 %
LBP + AdaBoost	43.18 %	96.50 %	59.66 %
Proposed	**90.07 %**	**94.01 %**	**92.00 %**

(b)

	Recall	*Precision*	*F1-score*
Proposed	**81.75 %**	**94.40 %**	**87.62 %**

Table 4. Experimental result for V5 with illumination variation [22]. © 2014 IEEE

Method	Recall (%)	Precision (%)	F1-score (%)
Viola-Jones	17.68	93.59	29.74
Proposed	**57.14**	**100.00**	**72.73**

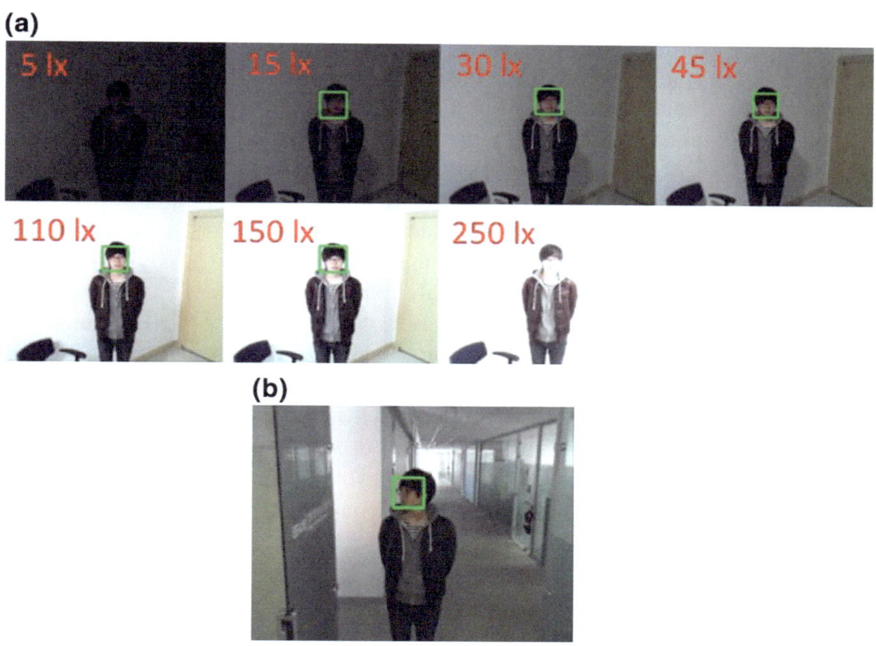

Fig. 7 Example of video frames with the results of the proposed method [22]. **a** For V5. **b** For V6. © 2014 IEEE

Table 5. Experimental result for V6 with pose variation [22]. © 2014 IEEE

Method	Recall (%)	Precision (%)	F1-score (%)
Viola-Jones	29.47	25.45	27.32
Proposed	**75.26**	**95.97**	**84.37**

in the illuminations between 15 lux and 150 lux). Table 5 shows that the proposed FD method is also robust to pose variations achieving much higher performance than Viola-Jones method. This stems from the fact of the relative robustness of texture feature to illumination variation and the use of multiple templates with the different poses. These results demonstrate that the proposed FD method is tolerable in terms of moderate pose and illumination variations.

4.2 Face Detection Simulation for Hardware Implementation

In this section, we present a hardware implementation for the proposed FD method and its simulation results. In general, most of power consumption in a digital system is directly affected by memory usage, i.e., data transfer and storing a large amount of data in a memory [19]. In particular, digital systems with image processing and computer vision techniques are likely to conduct memory-based operation. They unavoidably require a lot of memory usage. In order to reduce memory usage, our designed FD system optimizes the system architecture and reduces the number of required gate counts. Especially, the image scaler for obtaining multi-scale images for the FD task is implemented by a line memory (e.g., SRAM)-based design unlike the previous implementations based on a frame memory (e.g., dynamic RAM (DRAM)). In this way, we can considerably reduce the number of gate counts for low-power consumption. In addition, our FD system can process all of the scaled data simultaneously through merging line memories for each scaled image. Moreover, to increase the speed of the operation and parallelism, our FD system uses a pipelined structure for the block texture feature extraction, and the computations of the mean and the standard deviation used in the pre-filtering.

Figure 8 shows the block diagram for the hardware implementation of the proposed FD method. The system is largely comprised of four modules: image pyramid scaler, block texture feature extractor, feature templates matching, and FP reduction. Herein, the input signal is "PI[7:0]" with 8-bit monochrome QVGA image and the output signals are "PO_X[6:0]" and "PO_Y[6:0]" with 7-bit, where "PO_X" and "PO_Y" are upper left horizontal and upper left vertical positions of the detected window. For the input image signal, the first module generates image pyramids that consist of multi-scale images with the scale factor of 1.4. In this module, the unified SRAM based scaler generates four scaled images with the integer scale factor for the original image and 1.4 times scaled image. Then, the line mergers combine the line data stored in line memories into one data stream via

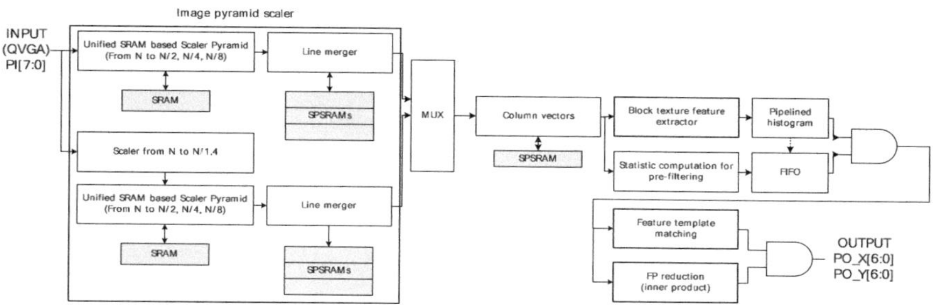

Fig. 8 Overall block diagram of the implemented hardware architecture [22]. © 2014 IEEE

SPSRAMs (single ported SRAMs). It enables to process the line data of multi-scale at the same time. Using a multiplexer (MUX), the data stream is controlled. The data stream per line enters the block texture feature extractor and statistic computation blocks. By pipelining the two blocks and data flows, we can obtain the pipelined histogram vector and statistics (i.e., mean and standard deviation for pre-filtering). Then, the extracted feature vector is matched with the feature templates in the feature templates matching module. Finally, FP reduction (i.e., inner product) using the linear SVM is conducted sequentially.

To verify the effectiveness of the hardware implementation of the proposed FD method, the designed FD system was evaluated by gate level simulation using DongBu 0.11um cell library. From the simulation result, we can see that the FD system can be implemented with 83 K gates. In particular, only 7 K gates are required for generating a 320 × 240 pixels image scaled pyramid, which is much more efficient than the FD system of [32] where 75 K gates are required for processing a 160 × 120 pixels input image. In [33], the authors implemented QVGA input image-based FD system with edge feature using Sobel filtering and a Naïve Bayes classifier. In the FD system [33] 268 KB were used for SRAM. On the other hand, our FD system requires only 10 KB for SRAM. These simulation results demonstrated that our FD system is very efficiently designed.

5 Conclusion

In this chapter, we introduced an event detection function for low-power intelligent surveillance system. The function remains in a very low-power camera mode except when human intruder is detected by automatic face detection. When a face is detected, a higher performance camera mode becomes available for further reliable analysis on the human intruder. We also proposed an efficient ROI-based two-stage face detection (FD) method suitable for operating on low-power camera mode. The comparative experiments showed that the proposed FD framework outperformed the widely used FD methods for the challenging videos in surveillance environments. Moreover, as the implemented FD hardware uses 83 K gate counts, the proposed FD method could be adopted for low-power event detector. As future work, we will further investigate the FD hardware implementation by using field programmable gate array (FPGA).

Acknowledgments This work is supported by the Center for Integrated Smart Sensors funded by the Ministry of Science, ICT & Future Planning as Global Frontier Project.

References

1. Bramberger M, Doblander A, Maier A, Rinner B, Schwaback H (2006) Distributed embedded smart cameras for surveillance applications. IEEE Comput Mag 39(2):68–75
2. Hu W, Tan T, Wang L, Maybank S (2004) A survey on visual surveillance of object motion and behaviors. IEEE Trans Syst Man Cybern Part C Appl Rev 34(3):334–352
3. Kim G, Kim J, Jung J, Kyung C. –M. (2012) Energy-aware operation of black box surveillance cameras under event uncertainty and memory constraint. In: IEEE international conference multimedia and Expo (ICME), pp 782–787
4. Irgan K, Ünsalan C, Baydere S (2014) Low-cost prioritization of image blocks in wireless sensor networks for border surveillance. J Netw Comput Appl 38:54–64
5. Huang C, Ai H, Li Y, Lao S (2007) High-performance rotation invariant multiview face detection. IEEE Trans Pattern Anal Mach Intell 29(4):671–686
6. Osuna E, Freund R, Girosi F (1997) Training support vector machines: an application to face detection. In: IEEE computer society conference computer vision and pattern recognition (CVPR), pp 130–136
7. Li Y, Gong S, Liddell H (2000) Support vector regression and classification based multi-view face detection and recognition. In: IEEE international conference automatic face and gesture recognition (FG), pp 300–305
8. Viola P, Jones MJ (2004) Robust real-time face detection. Int J Comput Vision 57(2):137–154
9. Jang J–S, Kim J–H (2008) Fast and robust face detection using evolutionary pruning. IEEE Trans Evol Comput 12(5):562–571
10. Wu B, Ai H, Huang C, Lao S (2004) Fast rotation invariant multi-view face detection based on real AdaBoost. In: IEEE international conference automatic face and gesture recognition (FG), pp 79–84
11. Vapnik VN (1998) Statistical Learning Theory. Wiley
12. Tabatabaie ZS, Rahmat RW, Udzir NIB, Kheirkhah E (2009) A hybrid face detection system using combination of appearance-based and feature-based methods. Int J Comput Sci Network Secur 9(5):181–185
13. Kim B, Ban S–W, Lee M (2008) Improving AdaBoost based face detection using face-color preferable selective attention. Intell Data Eng Autom Learn LNCS5326:88–95
14. Zhang C, Zhang Z (2010) A survey on recent advances in face detection. Technical report, MSR-TR-2010–66, Microsoft Research
15. Li S, Zhu L, Zhang Z, Blake A, Zhang H, Shum H (2002) Statistical learning of multi-view face detection. In: European conference computer vision (ECCV), pp 67–81
16. Wei Y, Bing X, Chareonsak C (2004) FPGA implementation of AdaBoost algorithm for detection of face biometrics. In: IEEE international workshop on biomedical circuits and systems, pp S1–6
17. Zou WWW, Yuen PC (2012) Very low resolution face recognition problem. IEEE Trans Image Proc 21(1):327–340

18. He T, Krishnamurthy S, Stankovic JA, Abdelzaher T, Luo L, Stoleru R, Yan T, Gu L (2004) Energy-efficient surveillance system using wireless sensor networks. In: ACM international conference mobile systems, applications, and services, pp 270–283
19. Coumeri SL, Thomas DE (2000) Memory modeling for system synthesis. IEEE Trans Very Large Scale Integr (VLSI) Syst 8(3):327–334
20. http://en.wikipedia.org/wiki/Graphics_display_resolution
21. Tan X, Chen S, Zhou Z-H, Zhang F (2006) Face recognition from a single image per person: a survey. Pattern Recogn 39(9):1725–1745
22. Kim H–I, Lee SH, Moon JI, Park H–S, Ro YM (2014) Face detection for low power event detection in intelligent surveillance system. In: IEEE international conference digital signal processing (DSP), pp 562–567
23. Chan CH, Kittler J (2010) Sparse representation of (multiscale) histograms for face recognition robust to registration and illumination problems. In: IEEE international conference image processing (ICIP), pp 2441–2444
24. Ahonen T, Hadid A, Pietikäinen M (2006) Face description with local binary patterns: application to face recognition. IEEE Trans Pattern Anal Mach Intell 28(12):2037–2041
25. Zhu C, Bichot C, Chen L (2011) Color orthogonal local binary patterns combination for image region description. Rapport technique RR-LIRIS-2011-012, LIRIS UMR, 5205, 15
26. Fauvel M, Chanussot J, Benediktsson JA (2006) Evaluation of kernels for multiclass classification of hyperspectral remote sensing data. In: IEEE international conference acoustics, speech, and signal processing (ICASSP)
27. Chang C–C, Lin C–J (2011) LIBSVM: a library for support vector machines. ACM Trans Intell Syst Technol 2(3):article 27
28. Phillips PJ, Moon H, Rizvi SA, Rauss PJ (2000) The FERET evaluation methodology for face-recognition algorithms. IEEE Trans Pattern Anal Mach Intell 22(10):1090–1104
29. http://docs.opencv.org/doc/tutorials/objdetect/cascade_classifier/cascade_classifier.html
30. http://en.wikipedia.org/wiki/Precision_and_recall
31. http://en.wikipedia.org/wiki/F1-score
32. Hori Y, Kuroda T (2007) A 0.79-mm^2 29-mW real-time face detection core. IEEE J Solid-State Circuits 42(4):790–797
33. Nguyen D, Halupka D, Aarabi P, Sheikholeslami A (2006) Real-time face detection and lip feature extraction using field-programmable gate arrays. IEEE Trans Syst Man Cybern Part B Cybern 36(4):902–912

Accurate Face and Human Detection Using Hybrid Local Transform Features

Daijin Kim and Bongjin Jun

Abstract We propose two novel local transform features: local gradient patterns (LGP) and binary histograms of oriented gradients (BHOG). LGP assigns one if the neighboring gradient of a given pixel is greater than the average of eight neighboring gradients and zero otherwise, which makes the local intensity variations along the edge components robust. BHOG assigns one if the histogram bin has a higher value than the average value of the total histogram bins, and zero otherwise, which makes the feature computation time fast due to no further post-processing and SVM classification. We also propose a hybrid feature that combines several local transform features by AdaBoost feature selection method where the best local transform feature among several local transform features (LBP, LGP, and BHOG), which has the lowest classification error, is sequentially selected until we obtain the required classification performance. This hybridization makes the face and human detection robust to the global illumination change by LBP, the local intensity change by LGP, and the local pose change by BHOG, which improves the detection performance considerably. We apply the proposed local transform features and the hybrid feature to the face detection problem using MIT+CMU and FDDB face database and the human detection problem using INRIA human database. The experimental results show that the proposed LGP and BHOG features attain accurate detection performance and fast computation time, respectively, and the hybrid feature provides a considerable improvement of face detection and human detection.

Keywords Local binary pattern · Local gradient pattern · Binary histograms of oriented gradients · Feature hybridization · Face detection · Human detection

D. Kim (✉)
Department of Computer Science and Engineering, Pohang University of Science and Technology, San31, Hyoja-dong, Nam-gu, Gyeongbuk, Pohang 790-784, Korea
e-mail: dkim@postech.ac.kr

B. Jun
StradVision, Rm 106, Ji-Gok Research Building, San 31, Hyoja-dong, Nam-gu, Gyeongbuk, Pohang 790-784, Korea
e-mail: simple21@postech.ac.kr

C. Kyung (ed.), *Theory and Applications of Smart Cameras*,
KAIST Research Series, DOI 10.1007/978-94-017-9987-4_8

1 Introduction

Face and Human detection is one of the important topics in computer vision. It has been widely used for the practical and real-time applications in many areas such as digital media (cell phone, smart phone, smart TV, digital camera), intelligent user interfaces (Wii, MS Kinnect), intelligent visual surveillance, and interactive games. Conventional face and human detection methods usually take the pixel color (or intensity) [37] directly as the information cue. However, these cues are sensitive to the illumination changes and noises [29]. To tackle this problem, many researchers have introduced the transform features that convert the pixel color (or intensity) by a certain nonlinear transformation function. They can be categorized into two transform features: the intensity-based transform features and the gradient-based transform features.

First, the intensity-based transform features convert the pixel color (or intensity) into the encoded value by comparing the pixel value with the neighboring pixel value. Papageorgiou and Poggio [26] introduced the Haar-like features that encoded the differences in average intensities between two rectangular regions and they applied to extract the textures irrespective of pixel color (or intensity). Viola and Jones [39, 40] used the Haar-like features to detect the faces. They used an integral image [40] to compute the Haar-like features efficiently and an efficient scheme for constructing a strong classifier by cascading several weak classifiers using AdaBoost training. Yan et al. [41] proposed the binary Haar feature that kept only the directional relationship in the Haar feature computation. However, the discriminating power of a single binary Haar feature was too weak to construct a robust classifier. They also proposed the assembled binary Haar (ABH) feature that integrated three binary Haar features to improve the discriminative power of the binary Haar feature. However, the dimensionality of ABH feature is very huge. Furthermore, they proposed the locally assembled binary (LAB) Haar feature that combined 8 locally adjacent 2-rectangle to reduce the size of feature dimensionality. The LAB Haar feature represented the local intensity differences at various locations, scales, and orientations. Ojala et al. [24] proposed the local binary patterns (LBP) feature that was derived from a general definition of texture in a local neighborhood of the image. They encoded an image pixel into a 8-bit binary pattern that compared the intensity of center pixel within the 3×3 block with the intensity values of 8 boundary pixels with the 3×3 block and representing the comparison result as 1 or 0. One important advantage of the LBP feature was that it was invariant to the monotonic change of illumination. Zabin and Woodfill [42] proposed the census transform (CT) that is similar to the LBP feature. The LBP feature and its variants have been widely used in many applications: face detection [19, 43], face recognition [1, 44], facial expression recognition [12, 33], gender recognition [36], face authentication [16], gait recognition [21], image retrieval [38], texture classification [14, 25], shape localization [17], and object detection [15].

Second, the gradient-based transform features convert the pixel color (or intensity) into the gradient magnitude and orientation. Lowe [22] proposed the SIFT descriptor that extracted distinctive invariant features from images and was invariant to image scale and rotation. The SIFT descriptor computed a histogram of local oriented gradients around the key point and represented the histogram in a 128 dimensional vector. It was obtained by computing the gradient magnitude and orientation on the key points, where the key points were obtained by finding the maxima and minima of the difference of Gaussian (DOG) images among three adjacent layers. It also required an image pyramid to make the SIFT descriptor scale invariant. Ke and Sukthankar [20] proposed the PCA-SIFT that used the principal component analysis (PCA) instead of histogram to normalize gradient patch. The feature vector was significantly smaller than the SIFT feature vector. They showed that PCA-based local descriptors were distinctive and robust to image deformations but it took a long computation time to extract the local descriptors. Bay et al. [2] proposed the speeded up robust features (SURF) that was an efficient implementation of SIFT by using the integral image. The SURF descriptor was obtained by computing the gradient magnitude and orientation on the key points, where the key points was obtained by finding the maxima of the Haar-like box filtered images. It did not require the image pyramid because it used many different sized box filters using integral image. Dalal and Triggs [4] proposed the histogram of oriented gradients (HOG) that divided the object into many fixed sized blocks, computed the HOG of each block, and represented the object by a concatenation of the block's HOG vectors. The HOG feature has been widely used in many applications: human detection [4, 6, 46], face recognition [5], object detection [10, 11] and emotion recognition [3]. Many researchers [9, 34, 45, 46] have also extended the original HOG to use variable-sized blocks, which improved the detection performance greatly.

In this chapter, we take two representative local transform features: local binary patterns (LBP) and histogram of oriented gradients (HOG) because LBP is robust to the global illumination change and HOG is robust to the local pose change. However, the local transform features have some limitations such that LBP is sensitive to local intensity changes due to makeup, wearing of glasses, and a variety of background and HOG requires a long processing time to compute the feature transformation.

To overcome these limitations, we propose two new local feature transforms: LGP and BHOG. LGP assigns one if the neighboring gradient of a given pixel is greater than the average of eight neighboring gradients, and zero otherwise, which makes the local intensity variations along the edge components robust. We show that LGP has a higher discriminant power than LBP in both the difference between face histogram and non-face histogram and the detection error based on face/face distance and face/non-face distance. BHOG assigns one if the histogram bin has a higher value than the average value of the total histogram bins, and zero otherwise, which makes the feature computation time fast due to no further post-processing and SVM classification.

We also propose a hybrid feature that combines several local transform features by AdaBoost feature selection method where the best local transform feature among several local transform features (LBP, LGP, and BHOG), which has the lowest classification error, is sequentially selected until we obtain the required classification performance. This hybridization makes the face and human detection robust to the global illumination change by LBP, the local intensity change by LGP, and the local pose change by BHOG, which improves the detection performance considerably.

This chapter is organized as follows. Section 2 describes the LGP feature to overcome the limitation of the LBP feature. Section 3 describes the BHOG feature to speed up the computation of the HOG feature. Section 4 describes a hybridization of several local transform features that combines them by AdaBoost feature selection method. Section 5 describes the experimental results of face and human detection that demonstrates the usefulness of the proposed local transform features and the hybrid feature. Finally, Sect. 6 presents conclusions.

2 Local Gradient Patterns

Many variants of LBP have been applied to tasks such as face detection, face recognition, facial expression recognition, gender recognition, face authentication, gate recognition, image retrieval, texture classification, shape localization, and object detection. However, they are sensitive to local intensity variations that occur commonly along edge components such as eyes, eyebrows, noses, mouths, whiskers, beards, or chins due to internal factors (eye glasses, contact lenses, or makeup) and external factors (different backgrounds). This sensitivity generates many different patterns of local intensity variations and makes training of the face and human detection by AdaBoost difficult. To overcome this problem, we propose a novel face and human representation method called Local Gradient Patterns (LGP), which generates constant patterns irrespective of local intensity variations along edges.

The LGP operator uses the gradient values of the eight neighbors of a given pixel, which are computed as the absolute value of intensity difference between the given pixel and its neighboring pixel. Then, the average of the gradient values of the eight neighboring pixels is assigned to the given pixel and is used as the threshold value for LGP encoding as follows. A pixel is assigned a value of 1 if the gradient value of a neighboring pixel is greater than the threshold value, and a value of 0 otherwise. The LGP code for the given pixel is then produced by concatenating the binary 1s and 0s into a binary code (See Fig. 1).

The LGP operator is extended to use different sizes of neighborhoods. We consider a circle of radius r centered on a specified pixel and take p sampling points along on the circle (See Fig. 2). To obtain the values of pixel positions in the neighborhood for r and p, bilinear interpolation is necessary. It uses a $2 \times r + 1$ by $2 \times r + 1$ kernel that summarizes the local structure of an image. At a given center pixel position (x_c, y_c), it takes the $2 \times r + 1$ by $2 \times r + 1$ neighboring pixels surrounding of the center pixel. Here, we define the gradient value between the center pixel i_c and its neighboring pixel i_n as $g_n = |i_n - i_c|$, and set the average of p gradient values as $\bar{g} = \frac{1}{p}\sum_{n=0}^{p-1} g_n$. Then, $\mathrm{LGP}_{p,r}(x_c, y_c)$ can be expressed as

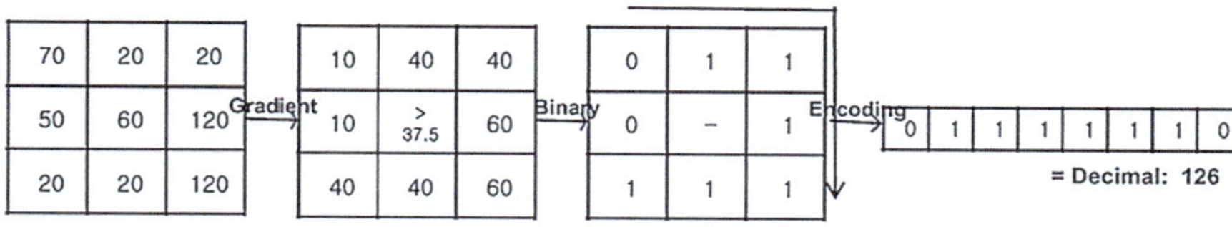

Fig. 1 The original LGP operator. © 2013 IEEE

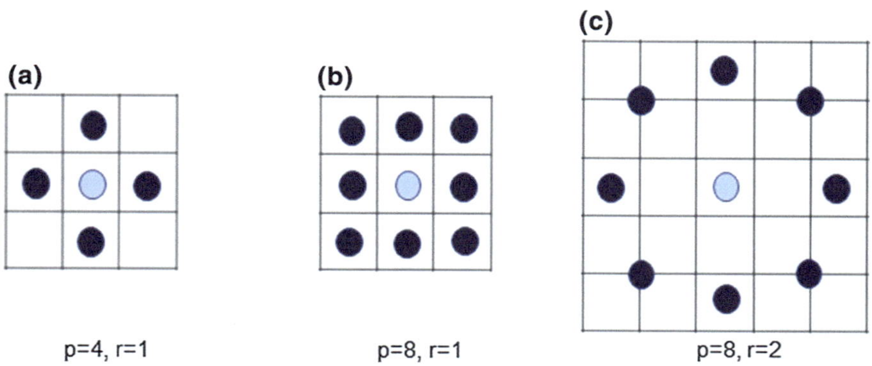

Fig. 2 Three examples of neighboring pixels: $LGP_{4,1}$, $LGP_{8,1}$ and $LGP_{8,2}$. © 2013 IEEE

$$\mathrm{LGP}_{p,r}(x_c, y_c) = \sum_{n=0}^{p-1} s(g_n - \bar{g})2^n, \tag{1}$$

where

$$s(x) = \begin{cases} 0, & \text{if } x < 0, \\ 1, & \text{otherwise.} \end{cases} \tag{2}$$

Figure 3 illustrates that LBP and LGP generate the same codes and the different codes depending on the global and local intensity changes. When the intensity levels of both the background and the foreground are changed together (globally), LGP and LBP both generate invariant patterns (See Fig. 3a). However, when the intensity level of the background or the foreground is changed locally, LGP generates invariant patterns but LBP generates variant patterns (See Fig. 3b, c). This difference occurs because LGP generates patterns using the gradient difference ($s(g_n - \bar{g})$), whereas LBP generates patterns using the intensity difference ($s(i_n - i_c)$). For the nearly uniform color region, there exist the small variations of absolute intensity differences between two neighboring pixels. We can suppress these small variations of absolute intensity differences by setting the threshold as a predefined value that is a little greater than the average absolute difference.

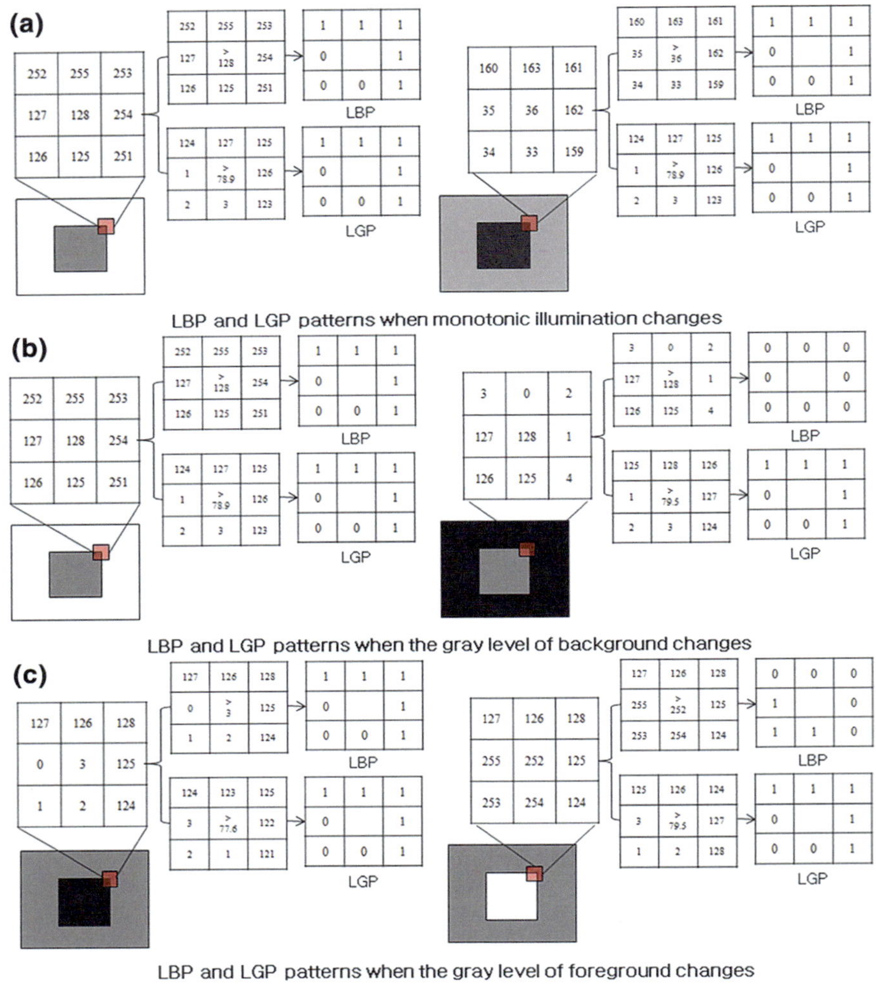

Fig. 3 LBP and LGP patterns when the intensity levels are changed globally or locally. © 2013 IEEE

3 Binary Histograms of Oriented Gradients

Dalal and Triggs [4] showed that the HOG feature combined with a linear SVM was a good detection performance of human beings. They took the overlapped block division method, the 1-D centered mask [−1, 0, 1], and the L2-Hys normalization method. However, it showed a slow processing speed of 1 fps for the 320 × 240 image although it took very small number of search windows (800 windows per image).

Q. Zhu et al. [46] used a cascade of rejectors and AdaBoost training to select the features which needed to be evaluated in each stage. This method could process 320 × 240 images over the speed of 5 fps, while maintaining an accuracy level similar to the existing HOG methods. However, it was still not enough to run in real-time, because each HOG feature consisted of 36 dimensional histogram vectors for each block and the weak classifiers of AdaBoost were the linear SVMs with HOG features.

To overcome this problem, we propose a novel face and human representation method called the binary histograms of oriented gradients (BHOG) that assigns one if the histogram bin has a higher value than an average value of the total histogram bins and zero otherwise, where threshold is just . Therefore, the BHOG feature for a given block is represented by concatenating the binary 1s and 0s into a binary code (See Fig. 4). While the HOG feature represents each block by the 256 bit vector (8 bins × 32 bits), the BHOG feature represents each block by the 8 bits, which makes the processing time efficient.

The BHOG feature is generated as follows. First, we compute the square of gradient magnitude and orientation of all pixels within the block. Second, we build the orientation histogram $\mathrm{HOG}(b)$, $b = 0, 1, \ldots, 7$ in the same way of generating the HOG feature. Third, we encode the orientation histogram into 8 bit vector, where each bit is determined by thresholding: If the histogram bin has a higher value than a given threshold, the 1 bit is assigned. Otherwise, the 0 bit is assigned. The decimal form of the 8 bit BHOG feature for a given block is expressed as

$$\mathrm{BHOG} = \sum_{n=0}^{7} s(\mathrm{HOG}(n) - \mathrm{Th})2^n, \tag{3}$$

where Th denotes the average of HOG as $\mathrm{Th} = \frac{1}{8}\sum_{n=0}^{7} \mathrm{HOG}(n)$ and a sign function $s(\cdot)$ is defined as

$$s(x) = \begin{cases} 1, & \text{if } x > 0, \\ 0, & \text{otherwise.} \end{cases} \tag{4}$$

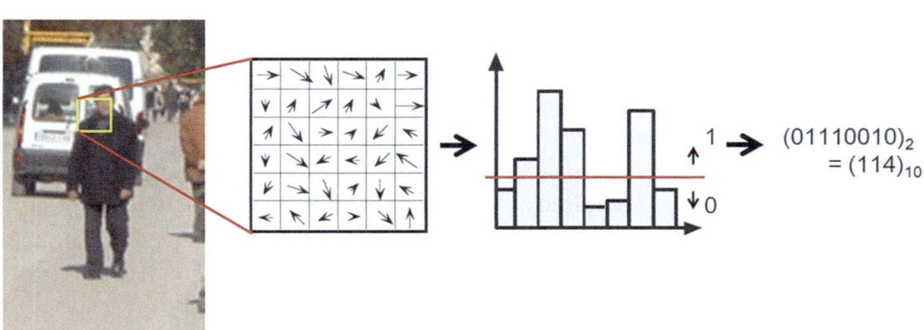

Fig. 4 Binary histograms of oriented gradients. © 2013 IEEE

The BHOG feature has several advantages over the HOG feature as follows. First, the BHOG featuer does not require the square root operation in computing the gradient magnitude because it just compares the value of histogram bin with a given threshold. Second, the BHOG feature does not perform normalization of the orientation histograms which is the essential part in the original HOG, since it just requires the relative comparison between the value of histogram bin and a given threshold value. Third, the BHOG feature can be obtained by the AdaBoost training because it can be represented as one dimensional scalar value.

However, the HOG feature cannot use the Adaboost training because it is represented by a $N \times M$ dimensional vector that is obtained by concatenating N blocks, where each block is M dimensional vector. Therefore, the HOG feature is obtained by applying the linear SVM to the vector and then applying the Adaboost training to the scalar value of the SVM result. Finally, the BHOG feature uses the variable-sized blocks from 3×3 to $W \times H$, where W and H denote the width and height of the image, which can capture a lot of useful information that is spread over different scales and it can capture a large sized part of the human body (e.g. head, arm, leg).

4 Hybridization of Local Transform Features

We propose a hybridization of local transform features that combines them by AdaBoost feature selection method, where the best local transform feature among several local transform features (LBP, LGP, and BHOG), which has the lowest classification error, is sequentially selected until we obtain the required classification performance. The pool of feature candidates consists of a large set of point features in the case of LBP and LGP features and a huge number of block features with a variety of sizes from 3×3 to $W \times H$ in the case of BHOG feature. The selected features should not be redundant and characterize both intra-class variability and inter-class variability well. This hybridization makes the face and human detection robust to the global illumination change by LBP, the local intensity change by LGP, and the local pose change by BHOG, which improves the detection performance considerably. To select discriminative features from LBP, LGP, and BHOG, we use AdaBoost based on LBP, LGP, and BHOG.

The overall procedure of selecting the hybrid feature using the AdaBoost training is given below. First, we prepare the positive and the negative training images. Second, we initialize the weight values of the positive and the negative training images. Third, we obtain the positive and the negative training feature images of three different local transform features: LBP, LGP and BHOG. Fourth, we compute the classification errors for all feature images. Fifth, we select the best local transform feature that has the minimum classification error. Finally, we update the weight values of all the training images such that the training images incorrectly classified by the selected feature have large weight values and the training images correctly classified by the selected feature have small weight values in the subsequent iterations. We prevent to re-select the previously selected feature by the other feature type by sharing the weight values among LBP, LGP and BHOG features.

After AdaBoost training, we obtain a strong classifier $H(\mathbf{C})$, where $\mathbf{C}$ includes LBP, LGP, and BHOG feature images. Then, it is represented by the sum of weak classifiers as

$$H(\mathbf{C}) = \sum_{\mathbf{x} \in S_T^{\mathrm{LBP}}} h_{\mathbf{x}}(\mathbf{L}(\mathbf{x})) + \sum_{\mathbf{x} \in S_T^{\mathrm{LGP}}} h_{\mathbf{x}}(\mathbf{G}(\mathbf{x})) + \sum_{\mathbf{x} \in S_T^{\mathrm{BHOG}}} h_{\mathbf{x}}(\mathbf{B}(\mathbf{IH}(\mathbf{x})), \quad (5)$$

where $\mathbf{L}$ is an LBP feature, $\mathbf{G}$ is an LGP feature, $\mathbf{IH}$ is an integral histogram [27] of the HOG feature whose size is $w \times h$ of one detection window, $\mathbf{B}(\cdot)$ is a binary HOG feature value computed from HOG feature vector, S_T^{LBP}, S_T^{LGP}, and S_T^{BHOG} are the sets of selected LBP, LGP and BHOG features at the final iteration, respectively, $\mathbf{x}$ denotes the selected feature as $\mathbf{x} = (\mathrm{type}, x, y, w, h)$ (If type is LBP or LGP, x and y represents feature location, while w and h has no meaning, if type is BHOG, x and y represent the center position of the selected block, while w and h represent the width and height of the selected block.), and $h_{\mathbf{x}}(\cdot)$ is the weak classifier that consists of a lookup table with a dimensionality of 2^N $\{0, 2^N - 1\}$, N is bit length of LBP, LGP, and BHOG) whose index is just the LBP, LGP, or BHOG value.

The value at each index of the lookup table indicates that the smaller it is, the more positive training images have the index and the larger it is, the more negative training images have the index. The weak classifiers are constructed using AdaBoost training [13], which updates the weight of each training sample such that misclassified instances are given a higher weight in the subsequent iteration. Table 1 shows an overall procedure of selecting the hybrid feature using the AdaBoost training procedure and Table 2 shows a detailed sub-procedure of selecting the best feature.

5 Experimental Results and Discussion

5.1 Face Detection

5.1.1 Data Preparation

We prepared 30,000 images from the FDD06[1] database, which contained the faces with the race, illumination, color and texture variations. We detect the faces in the image manually and normalized the detected faces to the face images with a fixed size of 22 × 24 pixels using the manually marked both eye's center positions. We generated 300,000 training face images by shifting slightly the face images, scaling

---

[1]See database(http://imlab.postech.ac.kr/faceDB/FDD06/FDD06.html.

Table 1 Hybrid feature selection using AdaBoost training. © 2013 IEEE

1. Prepare the training images $\{(T_i, c_i)\vert i = 1, 2, \ldots, N_p + N_n\}$,
where N_p and N_n denote the number of positive and negative training images, respectively, $c_i = 0$ for $T_i \in P$ and $c_i = 1$ for $T_i \in N$, where P and N denote positive and negative training images, respectively.
2. Initialize the weights of the positive and negative training images as $w_i = \begin{cases} \frac{1}{N_p} & \text{for } c_i = 0, \\ \frac{1}{N_n} & \text{for } c_i = 1, \end{cases}$
define the set of selected features $S_1 = \{\}$, set the number of selected features to N_s, and set the values of the weak classifier $h_{\mathbf{x}_t}(\gamma) = 0$, where $\mathbf{x}_t$ denotes one of LBP, LGP, and BHOG features, $t = 1, 2, \ldots, N_s$ and the feature index $\gamma = 0, \ldots, 2^N - 1$.
3. Apply LBP, LGP, and HOG to all positive and negative training images.
Let L_i, G_i, and $\mathbf{IH}_i$ be the positive and negative training LBP, LGP and integral histogram of HOG feature images, respectively.
4. For $t = 1, 2, \ldots, T$
(a) Select the best feature $\mathbf{x}_t$ with the classification error *epsilon*$_t$, by performing the tasks in Table 2.
(b) Update the weak classifier at the selected feature $\mathbf{x}_t$ as $h_{\mathbf{x}_t}(\gamma) = h_{\mathbf{x}_t}(\gamma) + \alpha_t z_t(\gamma)$,
where $\gamma = 0, \ldots, 2^N - 1$ and $\alpha_t = \frac{1}{2}\ln\left(\frac{1-\varepsilon_t}{\varepsilon_t}\right)$.
(c) Update the weights of positive and negative training images as
if the type of $\mathbf{x}_t$ is LBP, $w_{t+1}(i) = w_t(i) \cdot \begin{cases} e^{-\alpha_t}, & \text{if } z_t(L_i(\mathbf{x}_t)) = c_i, \\ e^{\alpha_t}, & \text{if } z_t(L_i(\mathbf{x}_t)) \neq c_i, \end{cases}$
if the type of $\mathbf{x}_t$ is LGP, $w_{t+1}(i) = w_t(i) \cdot \begin{cases} e^{-\alpha_t}, & \text{if } z_t(G_i(\mathbf{x}_t)) = c_i, \\ e^{\alpha_t}, & \text{if } z_t(G_i(\mathbf{x}_t)) \neq c_i, \end{cases}$
if the type of $\mathbf{x}_t$ is BHOG, $w_{t+1}(i) = w_t(i) \cdot \begin{cases} e^{-\alpha_t}, & \text{if } z_t(\mathbf{B}(\mathbf{IH}_i(\mathbf{x}_t))) = c_i, \\ e^{\alpha_t}, & \text{if } z_t(\mathbf{B}(\mathbf{IH}_i(\mathbf{x}_t))) \neq c_i, \end{cases}$
(d) Normalize the weights of positive and negative training images as $w_{t+1}(i) = \frac{w_{t+1}(i)}{\sum_{i=1}^{N_p+N_n} w_{t+1}(i)}$.
5. The final strong classifier is the sum of weak classifiers as
$H(\mathbf{C}) = \sum_{\mathbf{x}\in S_T^{\text{LBP}}} h_{\mathbf{x}}(\mathbf{L}(\mathbf{x})) + \sum_{\mathbf{x}\in S_T^{\text{LGP}}} h_{\mathbf{x}}(\mathbf{G}(\mathbf{x})) + \sum_{\mathbf{x}\in S_T^{\text{BHOG}}} h_{\mathbf{x}}(\mathbf{B}(\mathbf{IH}(\mathbf{x})))$,
where S_T^{LBP}, S_T^{LGP}, and S_T^{BHOG} are the set of selected feature positions at the final iteration T.

the face images with 0.95, 1.0, and 1.05 scale-factors, and rotating the face images by −15, 0, and 15 degrees in order to detect the faces irrespective of positions and scales. In addition, we mirrored the training face images to make them doubled. Figure 5 shows some typical training face images that were normalized by two eyes.

We prepared 17,000 non-face images from the FDD06 database, which did not contain the faces and generated 300,000 training non-face images by resizing the non-face images and taking the image patches with a fixed size of 22 × 24 pixels from the resized non-face images at random positions. These non-face images were used to train only the 1st stage of the cascade of face detectors, which will be explained later. From the 2nd stage of the cascade of face detectors, only the non-face images that were classified as false positives in the previous stage, were used to train the current stage face detector.

Table 2 A sub-procedure of selecting the best feature. 2013 IEEE

1. Generate the weight tables from the positive and negative training LBP, LGP, and BHOG feature images as $W_t^{k,\mathrm{LBP}}(\mathbf{x},\gamma)=\sum_{i,\mathbf{x},\gamma} w_t(i)I(L_i(\mathbf{x})=\gamma)I(c_i=k)$, $W_t^{k,\mathrm{LGP}}(\mathbf{x},\gamma)=\sum_{i,\mathbf{x},\gamma} w_t(i)I(G_i(\mathbf{x})=\gamma)I(c_i=k)$, $W_t^{k,\mathrm{BHOG}}(\mathbf{x},\gamma)=\sum_{i,\mathbf{x},\gamma} w_t(i)I(\mathbf{B}(\mathbf{IH}_i(\mathbf{x}))=\gamma)I(c_i=k)$,
where $k = 0$ or 1 for positive or negative training samples, respectively, and $I(\cdot)$ is an indicator function that takes a value of 1 if the argument is true, and 0 otherwise.
2. Compute the error $\varepsilon_t(\mathbf{x})$ for each lookup table as $\varepsilon_{\mathrm{LBP}} = \sum_\gamma \min\left\{W_t^{0,\mathrm{LBP}}(\mathbf{x},\gamma), W_t^{1,\mathrm{LBP}}(\mathbf{x},\gamma)\right\}$, $\varepsilon_{\mathrm{LGP}} = \sum_\gamma \min\left\{W_t^{0,\mathrm{LGP}}(\mathbf{x},\gamma), W_t^{1,\mathrm{LGP}}(\mathbf{x},\gamma)\right\}$, $\varepsilon_{\mathrm{BHOG}} = \sum_\gamma \min\left\{W_t^{0,\mathrm{BHOG}}(\mathbf{x},\gamma), W_t^{1,\mathrm{BHOG}}(\mathbf{x},\gamma)\right\}$, $\varepsilon_t(\mathbf{x}) = \min\{\varepsilon_{\mathrm{LBP}}, \varepsilon_{\mathrm{LGP}}, \varepsilon_{\mathrm{BHOG}}\}$.
3. Select the best feature position $\mathbf{x}_t$ as $\mathbf{x}_t = \begin{cases} \mathbf{x} = \min_{\mathbf{x}} \varepsilon_t(\mathbf{x}), & \text{if } \lvert S_t \rvert < N_s, \\ \mathbf{x} = \min_{\mathbf{x}\in S_t} \varepsilon_t(\mathbf{x}), & \text{otherwise,} \end{cases}$
where N_s is the allowed number of selected feature positions.
4. Update the set of selected features as
if the type of $\mathbf{x}_t$ is LBP, $S_{t+1}^{\mathrm{LBP}} = \{S_t^{\mathrm{LBP}} \cup \mathbf{x}_t\}$,
if the type of $\mathbf{x}_t$ is LGP, $S_{t+1}^{\mathrm{LGP}} = \{S_t^{\mathrm{LGP}} \cup \mathbf{x}_t\}$,
if the type of $\mathbf{x}_t$ is BHOG, $S_{t+1}^{\mathrm{BHOG}} = \{S_t^{\mathrm{BHOG}} \cup \mathbf{x}_t\}$, $S_{t+1} = \{S_{t+1}^{\mathrm{LBP}} \cup S_{t+1}^{\mathrm{LGP}} \cup S_{t+1}^{\mathrm{BHOG}}\}$.
5. Determine the dominant class indicator $z_t(\gamma)$ of the feature value γ at the selected feature $\mathbf{x}_t$ as $z_t(\gamma) = \begin{cases} 0, & \text{if } W_t^0(\mathbf{x}_t,\gamma) > W_t^1(\mathbf{x}_t,\gamma), \\ 1, & \text{otherwise.} \end{cases}$

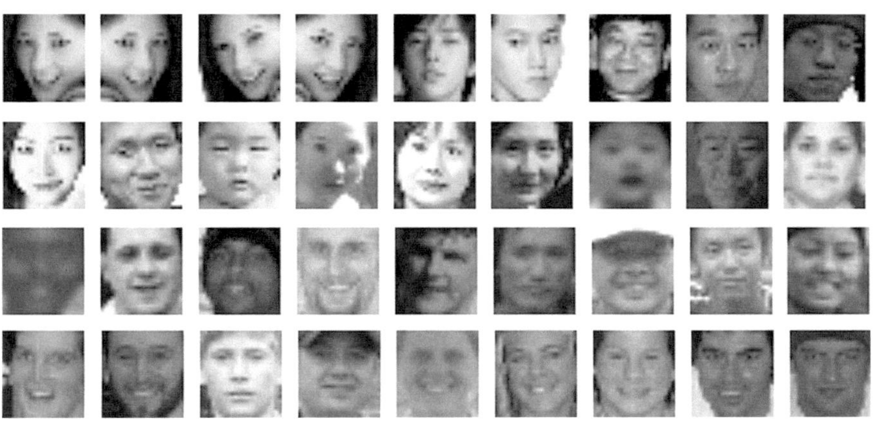

Fig. 5 Normalized training face images. © 2013 IEEE

We also prepared 15,000 images from the internet, which were not used for training and generated 150,000 validation face images in the same way of generating the training face images. We also prepared 15,000 non-face images from the internet, which were not used for training and generated 250,000 validation non-face images in the same way of generating the training non-face images.

5.2 Training Procedure

We have three different face detectors that use different features such as LBP, LGP and LBP+LGP+BHOG hybrid features, respectively. The AdaBoost training procedure of three face detectors is explained below.

First, we transform the training face and non-face images into the training face and non-face LBP, LGP, and BHOG feature images. Second, we compute the classification errors of all features. Third, we select one best feature with the minimum classification error at the current iteration. Fourth, we update the weight values of the training face and non-face feature images. Fifth, we check the stop condition that we achieve 99 % detection rate and 4 % false positive error rate using the validation face and non-face feature images. If the stop condition is satisfied, then we stop and obtain the selected features: the positions features in the case of LBP and LGP and the position and block features in the case of hybrid feature. Otherwise, we normalized the weight values of the training face and non-face feature images and go to the second step.

5.2.1 Cascade of Face Detectors

Since the proposed face detection method is based on classifying every possible window in the image as positive images or negative images, it takes long computation to detect the face in the high resolution image. To make the detection fast, we can employ the cascade of face detectors using the AdaBoost training method used by Viola and Jones [39].

In the real experiments, we trained three different cascades of face detectors using the LBP, LGP, and the hybrid feature images. However, we failed to train the cascade of face detectors using the BHOG feature images because the BHOG feature has only 8 different patterns in the case of 3×3 size of block. We set the maximum number of selected features of stage 1, 2, 3, and 4–26, 60, 120, and 400, respectively.

Figure 6 shows the selected features of three different cascade of face detectors using the LBP, LGP, and hybrid features, where white dots denote the positions of the selected point features in case of the LBP and LGP features and the center positions of the selected block features in the case of BHOG feature, and the rectangular boxes denote the sizes of the selected block features. We represent the center points of all the selected block features but did not represent the sizes of all the selected block features because it is very difficult to draw the boxes of all the selected block features within the face image. From Fig. 8, we know that (1) the LBP features are mostly selected from eye and mouth endpoints because they capture the common characteristics to all training face images, (2) the LGP features are widely selected from all face regions because they capture the locally changing gradient information and (3) the BHOG features are mostly selected from the eye, nose and mouth regions because they capture the common block information to all training face images.

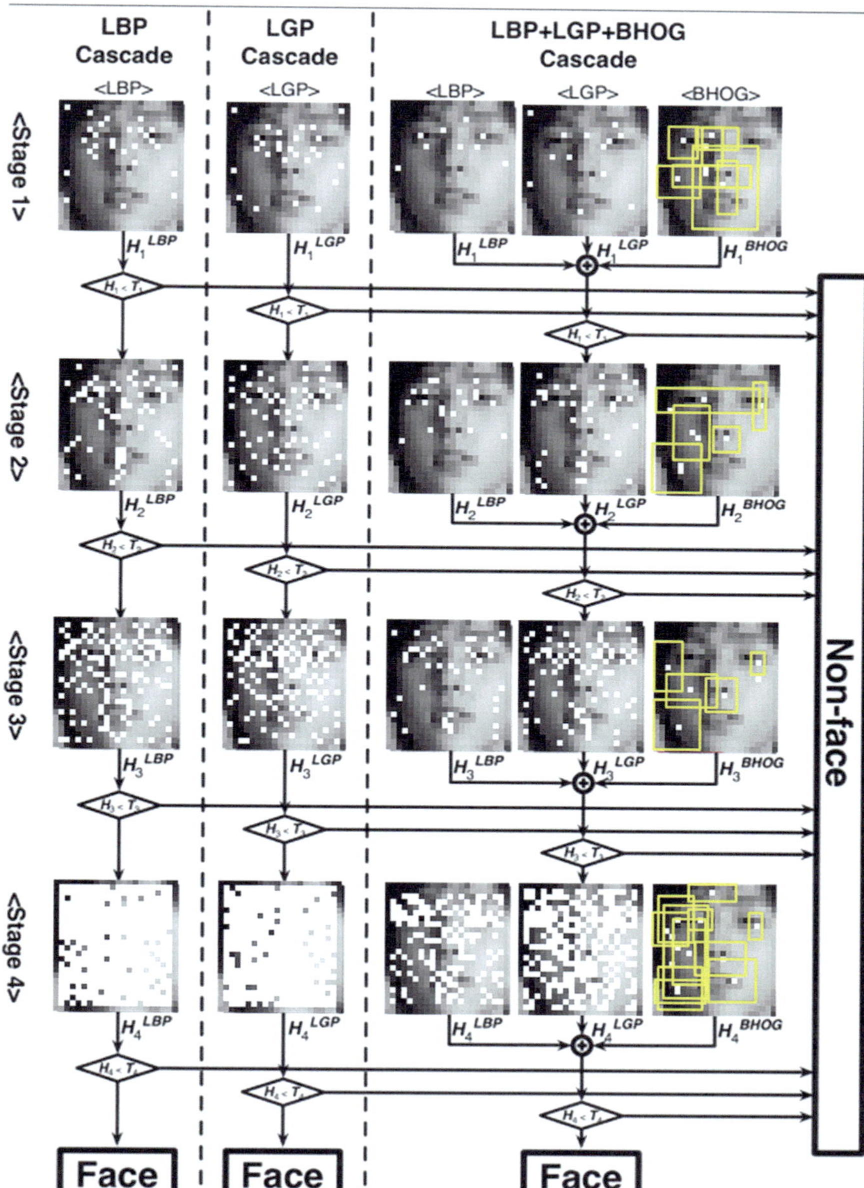

Fig. 6 Selected features of three cascades of face detectors. © 2013 IEEE

Table 3 shows the number of selected features in each stage that is determined from the training of the cascade of face detectors using the hybrid feature images. From Table 3, we know that (1) the LGP features are selected more than the LBP

Table 3 The number of selected features in each stage. 2013 IEEE

Feature	Stage 1	Stage 2	Stage 3	Stage 4	Total
# of LBP	8	19	39	146	212
# of LGP	12	34	76	242	364
# of BHOG	6	7	5	12	30
Total	26	60	120	400	606

Table 4 The computation time in each stage for training three different cascades of face detectors. 2013 IEEE

Cascade	Training time (LBP or LGP)	Training time (Hybrid)
Stage 1	≈1 min	≈6 min
Stage 2	≈5 min	≈40 min
Stage 3	≈30 min	≈4 h
Stage 4	≈23 h	≈3 days

and BHOG features because they are widely distributed over the all face region and (2) the BHOG features are rarely selected because they cover the large face components such as eyes, nose, and mouth.

Table 4 shows the computation time in each stage that is executed for the training of three different cascades of face detectors using the LBP, LGP and hybrid feature images, which run on the 2.83 GHz Intel Pentium IV PC system with 8 GB RAM. From Table 4, we know that the training time for the cascade of face detectors using the LBP and LGP feature images takes about one day while the training time for the cascade of face detectors using the hybrid feature images takes about four days.

5.2.2 Detection Performance

After training the proposed four-stage cascaded face detector, we evaluated the face detection accuracy using two kinds of face databases: the MIT+CMU database [30] (130 images with 483 faces), the Face Detection Data Set and Benchmark (FDDB[2]) database [18] (2,845 images with 5,171 faces). The face images in the MIT+CMU database are easy to detect because they are frontal and upright, and have mild illumination variations. The face images in the FDDB database are very difficult to detect because they include many occluded images and have large pose/illumination variations.

We considered six face detection methods for performance evaluation: the LBP feature-based face detector (LBP), the LGP feature-based face detector (LGP),

[2]See http://vis-www.cs.umass.edu/fddb/results.html.

the LBP+LGP feature-based face detector (LBP+LGP), the hybrid feature-based face detector (HYBRID). We compared four face detection methods (LBP, LGP, LBP+LGP, and HYBRID) with other existing face detection methods: Rowley-Baluja-Kanade [31], Viola-Jones [39], Mikolajaczyk et al. [23], Subburaman et al. [35].

Figure 7a, b show two receiver operating characteristic (ROC) curves that are obtained from several different face detection methods using the MIT+CMU database and the FDDB database, respectively. From Fig. 7a using the MIT+CMU database, we know that (1) the detection rate of the proposed HYBRID face detection method was the highest among all face detection methods by 0.959 when the false positive per image (FPPI) is one and (2) the number of false positives of the HYBRID, LBP+LGP, LGP, LBP, Viola-Jones, and Rowley-Baluja-Kanade methods at the 0.9 detection rate is 4, 7, 26, 67, 78, and 166, respectively,

From Fig. 7b using the FDDB database, we know that (1) the detection rates using the FDDB database are lower than those using the MIT+CMU database because the face images in the FDDB database has higher variations in the pose,

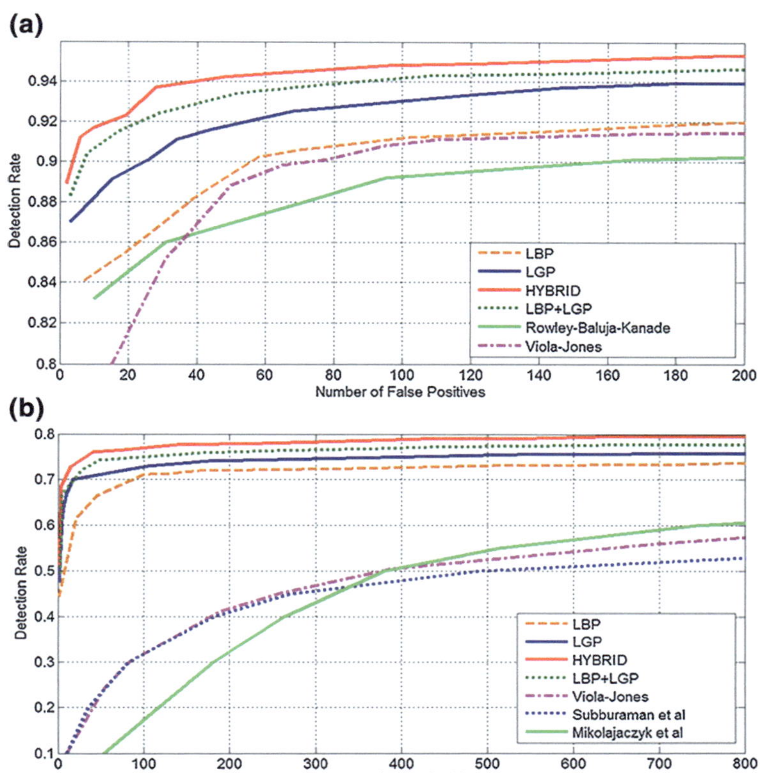

Fig. 7 ROC curves using (**a**) the MIT+CMU database and (**b**) the FDDB database. © 2013 IEEE

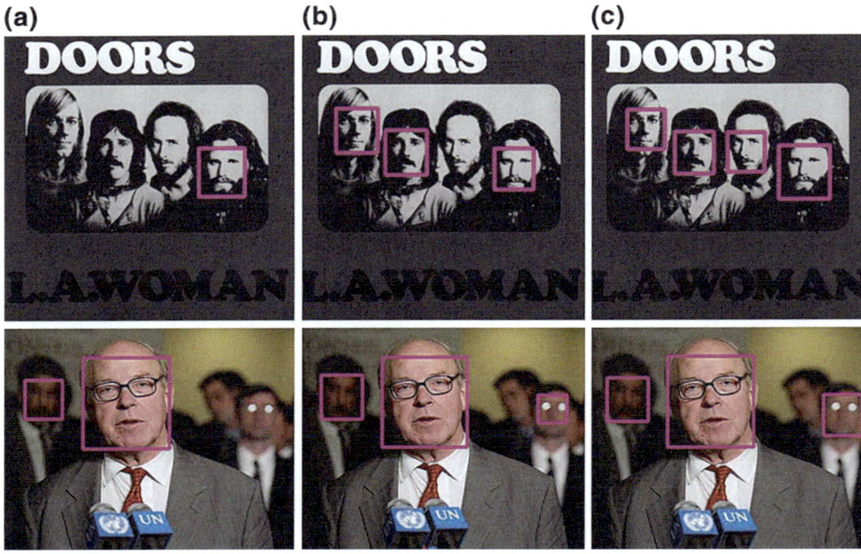

Fig. 8 Comparison of face detection results from (**a**) the LBP feature-based face detector, (**b**) the LGP feature-based face detector, and (**c**) the hybrid feature-based face detector. © 2013 IEEE

illumination, expression, and occlusion than those in the MIT+CMU database, (2) the detection rate of the proposed HYBRID method was the highest among all face detection methods by 78.9 % when the false positive per image (FPPI) is 0.1, (3) the detection rate of the HYBRID, LBP+LGP, LGP, LBP, Viola-Jones, Mikolajaczyk et al., and Subburaman et al. methods at the 0.1 FPPI are 78.2, 76.3, 74.2, 72.1, 46.2, 45.6, and 42.3 %, respectively.

Figure 8 shows the face detection results using the MIT+CMU database (top row) and the FDDB database (bottom row), where (a), (b) and (c) are obtained from the LBP feature-based face detector, the LGP feature-based face detector and the hybrid feature-based face detector, respectively. From Fig. 8, we know that the HYBRID feature-based face detector succeeds to find most of faces, even tiny faces with a size of 22 × 24, but the LBP and LGP feature-based face detectors fail to find them occasionally.

Figure 9 shows several face detection results using the hybrid feature-based face detector on the MIT+CMU and FDDB database, respectively.

5.2.3 Memory Size

Each weak classifier must store the confidence value at each LBP, LGP, and BHOG value in the lookup table, where the confidence value is represented by a real number, which consists of 8 bytes. Therefore, each weak classifier requires a memory space of 2,048 bytes (= 256 LGP patterns × 8 bytes). Because stages 1–4

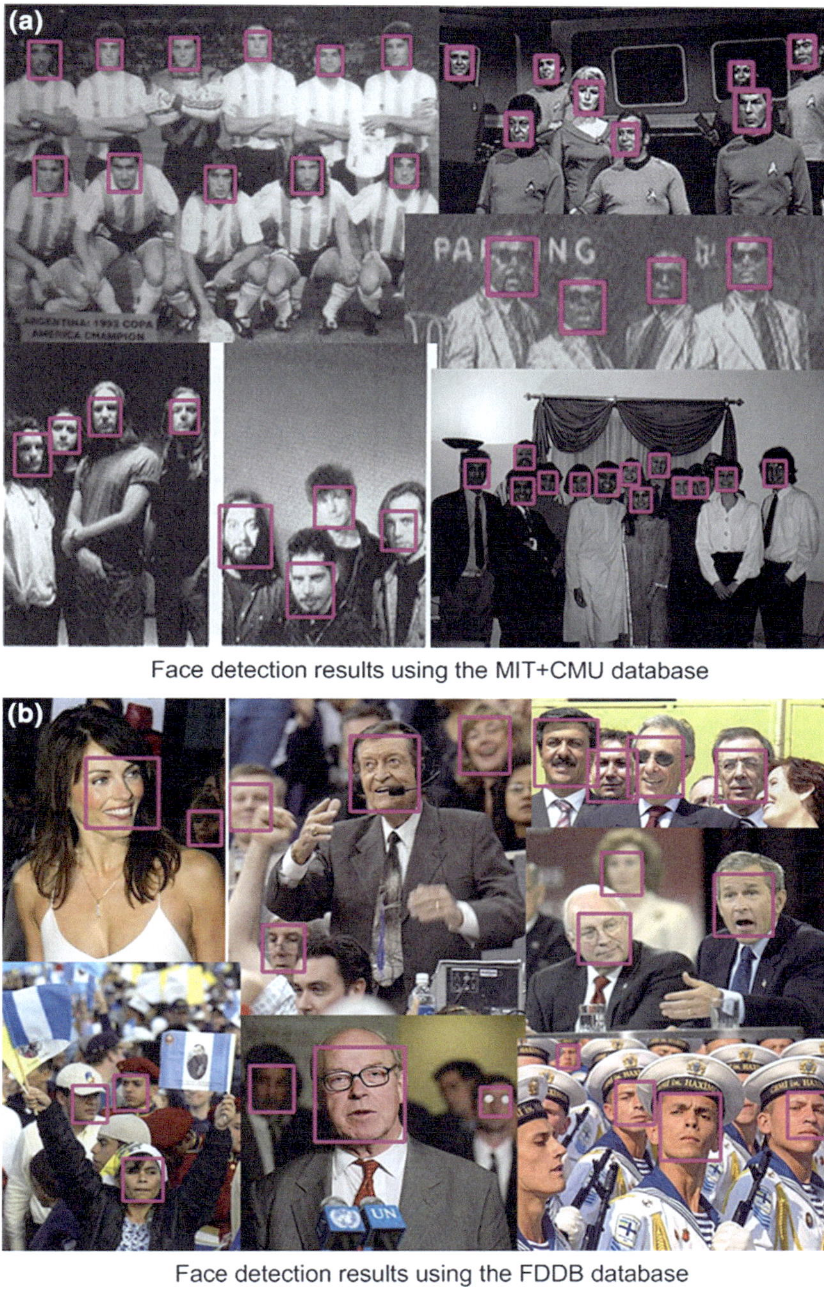

Fig. 9 Face detection results. 2013 IEEE

consist of 26, 60, 120 and 400 weak classifiers respectively, the total required memory space is 1.2 Mbytes (= 606 × 2,048 bytes), which is a burden for low-performance embedded systems. Furthermore, most low-performance embedded systems do not support the floating point operation. To overcome this limitation, we propose an encoding scheme of reducing the required memory space that quantizes the confidence value into 256 intervals and represents it as one byte value from 0 to 255. This encoding reduces the required memory size to 152 Kbytes (= 606 × 256 LGP patterns × 1 byte).

5.2.4 Computation Time

We represent the computation time of our face detector as a linear function $T(t) = N \times t + C$, where N is the number of possible detection windows in the image, t is the average computation time to process one detection window, and C is a constant time that includes the image loading time, the preprocessing time (the time for transforming the input image into the LBP, LGP, BHOG feature image, the time for making integral histogram of HOG in the case of the hybrid-based face detector, the time for making the integral image in the case of Viola-Jones face detector, the time for constructing the pyramid image).

We measured the computation time on a 2.83 GHz Intel Pentium IV PC system with 8 GB RAM. Table 5 shows the preprocessing time and the average computation time of several face detectors, where it is the average of the computation time of 10,000 320 × 240 input images.

The average computation times of the Rowley-Baluja-Kanade face detector [31] and the Schneiderman-Kanade face detector [39] were referred from [39], which stated that their face detector was roughly 15 times faster than the Rowley-Baluja-Kanade face detector and roughly 600 times faster than the Schneiderman-Kanade face detector. The proposed LGP feature-based face detector is slightly slower than the LBP-based face detector due to the gradient computation for LGP feature transformation. However, the LGP feature-based face detector is seven times faster than the Viola-Jones face detector because the LGP feature-based

Table 5 Comparison of average computation time among several face detectors (unit: 10^{-3} s). 2013 IEEE

Detector	Pyramid	Feature	Face	Total
	Image	Transform	Detection	Time
LBP feature-based [24]	1.7	1.76	6.20	9.66
LGP feature-based	1.7	2.05	6.07	10.12
HYBRID feature-based	1.7	9.78	25.78	37.26
Viola-Jones [39]	0.0	0.16	70.06	70.22
Rowley-Baluja-Kanade [31]	–	–	–	1053.3
Schneiderman-Kanade [32]	–	–	–	42132.0

face detector computes the weak classifier by one array reference to the lookup table, whereas the Viola-Jones face detector computes the weak classifier by more than six array references even with integral image.

The proposed hybrid feature-based face detector is roughly 2 times faster than the Viola-Jones face detector. Since most of the features of hybrid feature-based face detector consist of LBP and LGP features, there are a few number of BHOG features. Accordingly, hybrid feature-based face detector requires a few number of integral histogram computations which take much computation time. In contrast, all the weak classifiers of Viola-Jones face detector consist of Haar-like features which require high number of integral image computation.

5.3 Human Detection

5.3.1 Data Preparation

We prepared 618 images from the INRIA database [4], which contained 1,208 humans with the pose, illumination, appearance, and occlusion variations. We detect the human in the image manually and normalized the detected humans to the human images with a fixed size of 32×64 pixels using the manually marked head and toe positions. We generated 59,180 training human images by shifting slightly the human images and scaling the human images with 0.95, 1.0, and 1.05 scale-factors in order to detect the humans irrespective of positions and scales. In addition, we mirrored the training human images to make them doubled. Figure 10 shows some typical training human images that were normalized by the head and toe.

We prepared 1,218 nonhuman images from the INRIA database [4], which did not contain humans and generated 100,000 training nonhuman images by bootstrapping and resizing the nonhuman images and taking the image patches with a fixed size of 32×64 pixels from the resized nonhuman images at random positions. These nonhuman images were used to train only the first stage of the cascade of human detectors. From the 2nd stage of the cascade of human detectors, only the nonhuman images that were classified as false positives in the previous stage, were used to train the current stage human detector.

5.3.2 Training Procedure

We have two different human detectors that use different features such as BHOG and LBP+LGP+BHOG hybrid features, respectively. The BHOG feature uses the variable size of blocks from 4×4 to $W \times H$, where W and H denote the width and height of the window image, which it can capture a lot of useful information that is spread over different scales and it can capture a large sized part of the human body

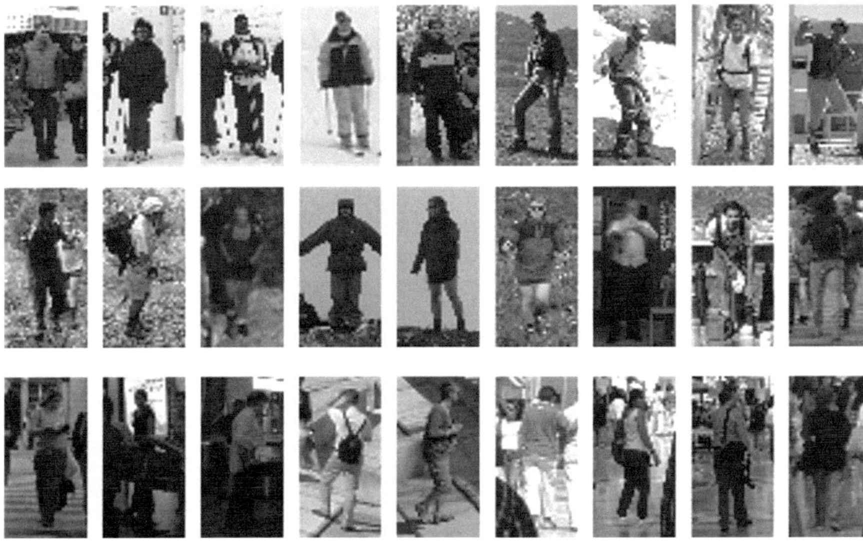

Fig. 10 Normalized training human images. © 2013 IEEE

(e.g. head, arm, leg). The AdaBoost training procedure of two human detectors is explained below.

First, we transform the training human and nonhuman images into the training human and nonhuman LBP, LGP, and BHOG feature images. Second, we compute the classification errors of all feature images. Third, we select one best feature with the minimum classification error at the current iteration. Fourth, we update the weight values of the training human and nonhuman feature images. Fifth, we check the stop condition that we achieve 96 % detection rate and 8 % false positive error rate using the validation human and nonhuman feature images. If the stop condition is satisfied, then we stop and obtain the selected features: the position features in the case of LBP and LGP and the position and block features in the case of hybrid feature. Otherwise, we normalize the weight values of the training human and nonhumane feature images and go to the second step.

5.3.3 Cascade of Human Detectors

We also take the cascade of human detectors to make the human detection fast. In real experiments, we trained two different cascades of human detectors using BHOG and LBP+LGP+BHOG hybrid feature images because the LBP and LGP features failed to train the human detectors. We set the maximum number of selected features of stage 1, 2, 3, 4, and 5–40, 80, 160, 320, and 1,600, respectively.

Figure 11 shows the selected features of two different cascade of human detectors using the BHOG and hybrid features, where white dots denote the positions of the selected point features in the case of the LBP and LGP features and the center positions of the selected block features in the case of BHOG feature, and the rectangular boxes denote the sizes of the selected block features. We represent the center points of all the selected block features but did not represent the sizes of all the selected block features because it is very difficult to draw the boxes of all the selected block features. From Fig. 11, we know that (1) the LBP features are mostly selected from the shoulder because they capture the common characteristics to all training human images, (2) the LGP features are mostly selected from the arms and legs with high variations because they capture the locally changing gradient information, (3) the BHOG features are widely selected from all-human regions such as head, arms, legs, and torso because they capture the common block information to all training human images.

Table 6 shows the number of selected features in each stage that is determined from the training of the cascade of human detectors using the hybrid feature images. From Table 6, we know that (1) the LGP features are selected more than the LBP features because they are widely distributed over the all- human region and (2) the BHOG features are most widely selected over the whole body region because they cover the large body part components such as arms, legs and torso.

Table 7 shows the training time of two different cascade of human detectors using the BHOG and hybrid features, which runs on the 2.83 GHz Intel Pentium IV PC system with 8 GB RAM. From Table 7, we know that the training of the cascade of human detectors using the BHOG feature images takes about seven days while the training of the cascade of human detectors using the hybrid feature images takes about nine days.

5.3.4 Detection Performance

After training the proposed five-stage cascaded human detector, we evaluated the human detection accuracy using the INRIA database [4] that contained 288 test images with 1,132 humans.

We considered four human detection methods for performance evaluation: the BHOG feature-based human detector (BHOG), the LGP+BHOG feature-based human detector (LGP+BHOG) the hybrid feature-based human detector (HYBRID). We compared three human detection methods (BHOG, LGP+BHOG, and HYBRID) with other existing human detection methods: HOG [4] and VJ (Viola-Jones) [7] using the evaluation protocol based on Pascal measure [8].

Figure 12 shows the receiver operating characteristic (ROC) curve that is obtained from several different human detection methods using the INRIA database. From Fig. 12, we know that (1) the detection rate of the HYBRID, LGP +BHOG, BHOG, $HOG_{64\times128}$, $HOG_{32\times64}$, and VJ at the one false positive rate per images (FPPI) was 85.5, 83.5, 79.5, 78.9, 41, and 58 %, respectively, which means that the proposed HYBRID human detection method was the highest among all other human detection methods, and (2) the number of false positives of the

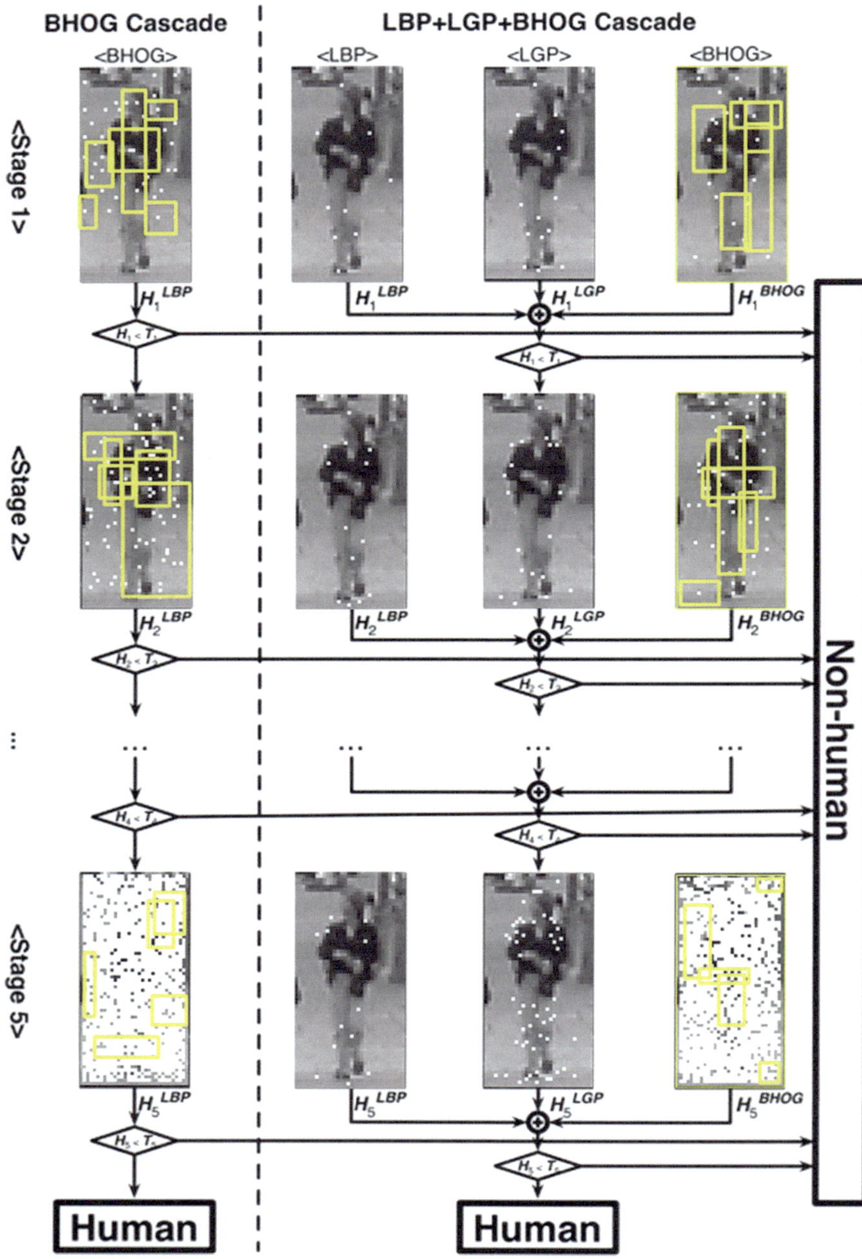

Fig. 11 Selected features of two cascades of human detectors. © 2013 IEEE

Table 6 The number of selected features in each stage. 2013 IEEE

Feature	Stage 1	Stage 2	Stage 3	Stage 4	Stage 5	Total
# of LBP	8	12	17	16	12	65
# of LGP	14	18	31	43	54	160
# of BHOG	18	50	112	261	1,534	1,975
Total	40	80	160	320	1,600	2,200

Table 7 The training time of two different cascades of human detectors. 2013 IEEE

Cascade	Training time (BHOG)	Training time (Hybrid)
1 stage	≈8 min	≈10 min
2 stage	≈30 min	≈50 min
3 stage	≈4 h	≈4 h
4 stage	≈1 day	≈2 days
5 stage	≈5 days	≈6 days

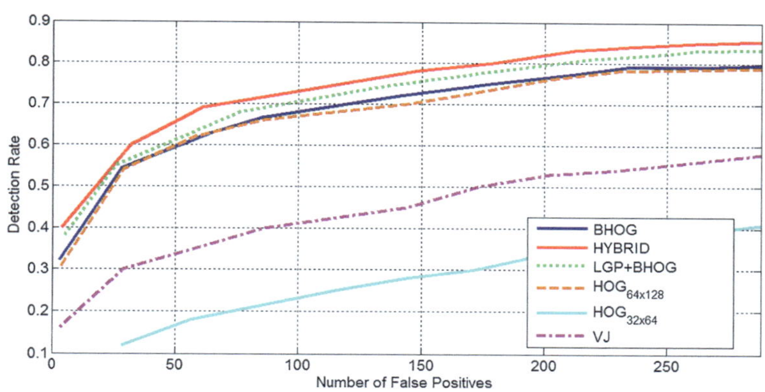

Fig. 12 ROC curves using the INRIA database. © 2013 IEEE

HYBRID, LGP+BHOG, BHOG, and HOG at the 70 % detection is 70, 92, 120, and 145, respectively.

Figure 13 shows the human detection results using the INRIA database, where (a), (b) and (c) are obtained from from the HOG-based human detector, the BHOG-based human detector, and the hybrid feature-based human detector, respectively. From Fig. 13, we know that the HYBRID feature-based human detector succeeds to find most of humans even small sized human with a size of 32 × 64, but the HYBRID and HOG-based human detectors fail to find them occasionally.

Figure 14 shows several human detection results using the hybrid feature-based human detector on the INRIA database.

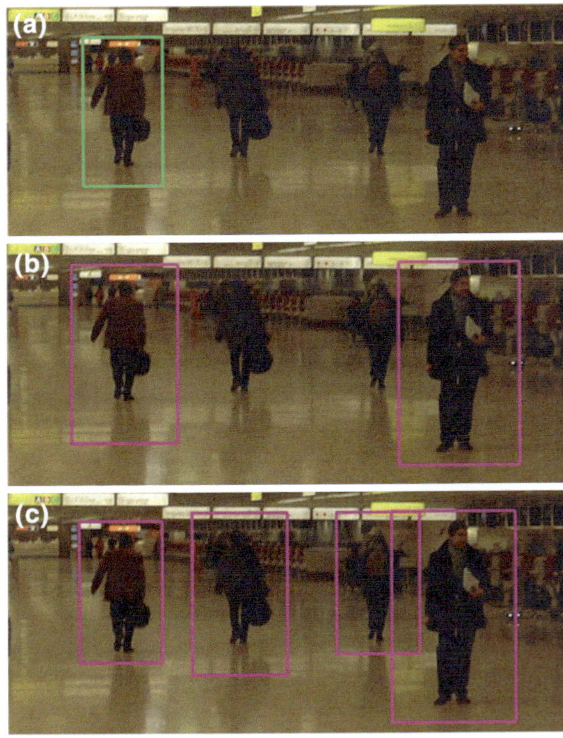

Fig. 13 Comparison of human detection results. © 2013 IEEE

We also evaluated the human detection accuracy using the MIT-CBCL[3] database that contained 924 front/back-view positive images (no negative images). Instead of training on the MIT-CBCL database, we use our trained detectors on the INRIA database and tested them on the MIT-CBCL database. We achieve that (1) the detection rate of the HYBRID, LGP+BHOG, BHOG, and HOG at the zero false positive rate per images (FPPI) was 93.1, 92.6, 90.2, and 84.5 %, respectively, which means that the proposed HYBRID human detection method was the highest among all other human detection methods, and (2) this indicates that our detectors have good generalization performance.

5.3.5 Computation Time

We measured the computation time of the HOG human detector [4], the proposed BHOG-based human detector, and the proposed hybrid-based human detector on a 2.83 GHz Intel Pentium IV PC system with 8 GB RAM. Table 8 shows the average

[3]See http://cbcl.mit.edu/software-datasets/PedestrianData.html.

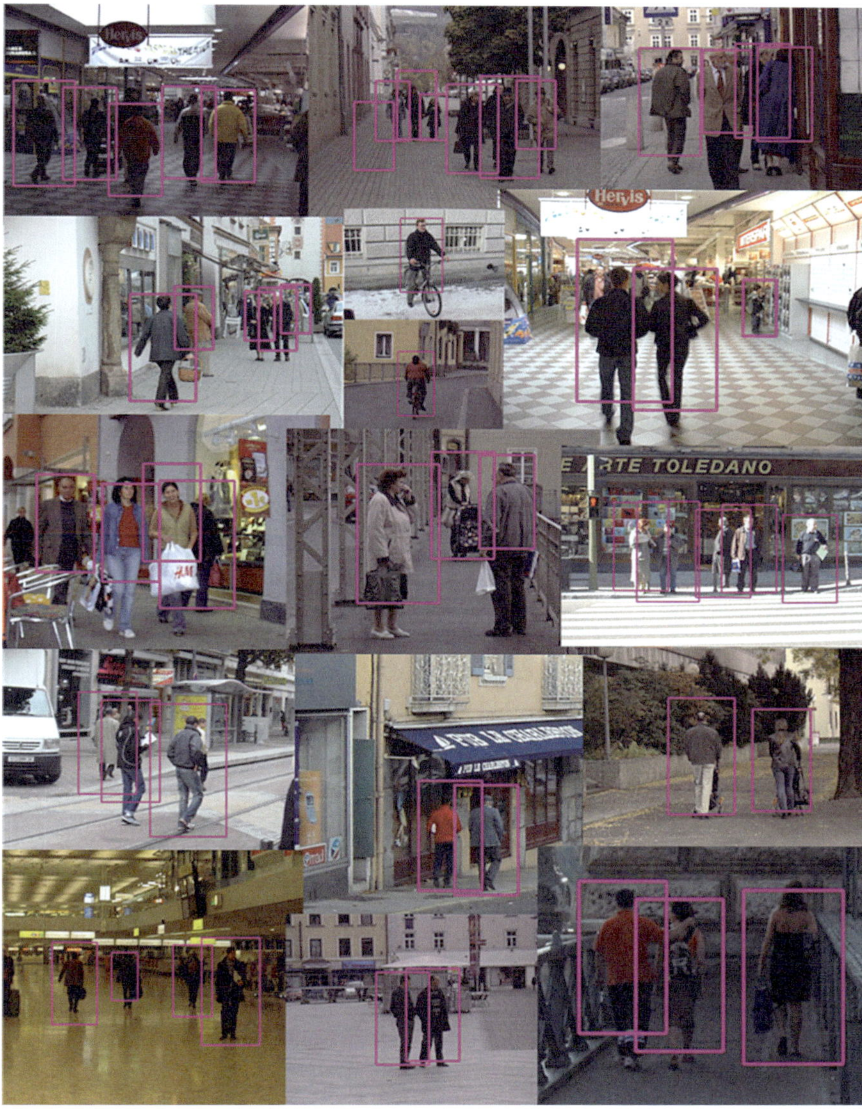

Fig. 14 Human detection results using the INRIA database. © 2013 IEEE

computation time of two human detectors, where it is the average of computation time of 1,000 320 × 240 input images. From Table 8, we know that (1) the existing HOG-based human detector works slowly in that it takes about 490 10^{-3} s (≈2 fps) and the BHOG-based human detector works fast in that it takes about 52 10^{-3} s (≈20 fps), which implies that the proposed BHOG-based human detector is about

Table 8 Comparison of average computation time among several human detectors (unit: 10^{-3} s). 2013 IEEE

Detector	Pyramid Image	Feature Transform	Human Detection	Total Time
HOG feature-based	1.7	87	401.3	490
BHOG feature-based	–	6.01	46	52.01
Hybrid feature-based	1.7	9.78	165.78	177.26
Cascade+HOG [46]	–	–	–	214
GPU implementation of HOG [28]	–	–	–	19

10 times faster than the existing HOG-based human detector, and (2) the hybrid feature-based human detector is roughly three times slower than the BHOG-based human detector because it uses the hybrid features. One interesting point is that the BHOG-based human detector shows 1 % higher detection rate than the HOG-based human detector in spite of its faster computation time.

6 Conclusion

The most commonly used face and human detection method was local transform feature-based method. Many researchers have introduced many different approaches using local transform features: specifically local binary patterns (LBP) and histograms of oriented gradients (HOG). Each approach had its own advantage in that LBP was robust to monotonic illumination variations and HOG was robust to local pose variations. However, these methods have some limitations such that LBP was sensitive to locally changing intensity changes and HOG required a huge computation time for the feature transformation.

To overcome the limitations of the previous approaches, we proposed two novel local feature transformation methods: local gradient patterns (LGP) and binary HOG (BHOG) and proposed a hybridization of local transform features that combined several local features (LBP, LGP, and BHOG or HOG) by AdaBoost feature selection method to improve the face and human detection performance given below.

LGP encoded an image pixel into a 8-bit binary pattern by comparing the gradient of the given pixel and the average of its 8 neighboring gradients. It was invariant to the local gradient variations that were caused by makeup, wearing of glasses, and a variety of background, and had higher discriminant power than LBP.

BHOG binarized the histogram values of HOG by thresholding them with the average value of the total histogram bins. It did not require the square root operation in computing the gradient magnitude and the normalization of the orientation histograms because it just compared the value of histogram bin with a given threshold and enabled to obtain the face and human detectors by the AdaBoost training because it was represented as one dimensional scalar value.

The hybridization of the multiple local transform features selected relevant features from the feature pool of LBP, LGP, and BHOG in order to improve the detection performance considerably. It took advantages of each local transform feature: LBP's robustness to local illumination change, LGP's robustness to locally changing intensity, and BHOG's robustness to local pose change.

We applied the proposed local transform features and its hybridization to face and human detection to validate the usefulness of the proposed methods First, the face detection rates of LBP, LGP and the hybridization of LBP, LGP, and BHOG features using MIT+CMU database were 90, 93, and 96 %, respectively, which showed that the LGP feature resulted in better face detection rate than the LBP feature, and the hybrid feature resulted in the best face detection rate among them. Second, the human detection rates of HOG, BHOG and the hybridization of LBP, LGP and BHOG features using INRIA database were 79, 80, and 86 %, respectively, which showed that BHOG feature had similar detection rate but 10 times faster than HOG feature and the hybrid feature resulted in the best human detection rate among them. From all the results, we can conclude that the proposed local transform features and its hybrid feature are very effective for the face and human detection rate in terms of the performance and operating speed.

Acknowledgements This work is supported by the Center for Integrated Smart Sensors funded by the Ministry of Science, ICT & Future Planning as the Global Frontier Project.

References

1. Ahonen T, Hadid A, Pietikainen M (2006) Face description with local binary patterns: application to face recognition. IEEE Trans Pattern Anal Mach Intell 28(12):2037–2041
2. Bay H, Ess A, Tuytelaars T, Gool LV (2008) SURF: speeded up robust features. Comput Vis Image Underst 110(3):346–359
3. Dahmane M, Meunier J (2011) Emotion recognition using dynamic grid-based HoG features. In: Proceedings of IEEE international conference on automatic face and gesture recognition, pp 884–888
4. Dalal N, Triggs B (2005) Histograms of oriented gradients for human detection. In: Proceedings of IEEE conference on computer vision and pattern recognition, pp 886–893
5. Deniza O, Buenoa G, Salido J (2011) Face recognition using histograms of oriented gradients. Pattern Recogn Lett 32(12):1598–1603
6. Dollar P, Belongie S, Perona P (2010) The fastest pedestrian detector in the west. In: Proceedings of the British machine vision conference, pp 1–11
7. Dollar P, Wojek C, Schiele B, Perona P (2009) Pedestrian detection: a benchmark. In: Proceedings of IEEE conference on computer vision and pattern recognition, pp 304–311
8. Dollar P, Wojek C, Schiele B, Perona P (2011) Pedestrian detection: an evaluation of the state of the art. IEEE Trans Pattern Anal Mach Intell 34(4):743–761
9. Enzweiler M, Gavrila DM (2009) Monocular pedestrian detection: survey and experiments. IEEE Trans Pattern Anal Mach Intell 31(12):2179–2195

10. Felzenszwalb P, Girshick R, McAllester D (2010) Cascade object detection with deformable part models. In: Proceedings of IEEE conference on computer vision and pattern recognition, pp 2241–2248
11. Felzenszwalb P, Girshick R, McAllester D, Ramanan D (2010) Object detection with discriminatively trained part based models. IEEE Trans Pattern Anal Mach Intell 32(9):1627–1645
12. Feng X, Pietikainen M, Hadid A (2005) Facial expression recognition with local binary patterns and linear programming. Pattern Recognit Image Anal 15(2):546–548
13. Froba B, Ernst A (2004) Face detection with the modified census transform. In: Proceedings of IEEE international conference on automatic face and gesture recognition, pp 91–96
14. Grimes DB, Rao RPN (2003) A bilinear model for sparse coding. Neural Inf Process Syst 15:1287–1294
15. Heikkila M, Pietikainen M, Heikkila J (2004) A texture-based method for detecting moving objects. In: Proceedings of British machine vision conference, pp 187–196
16. Heusch G, Rodriguez Y, Marcel S (2006) Local binary patterns as an image preprocessing for face authentication. In: Proceedings of international conference on automatic face and gesture recognition, pp 9–14
17. Huang X, Li SZ, Wang Y (2004) Shape localization based on statistical method using extended local binary pattern. In Proceedings of international conference on image and graphics, pp 184–187
18. Jain V, Miller EL (2010) FDDB: a benchmark for face detection in unconstrained settings. University of Massachusetts, Amherst
19. Jin H, Liu Q, Lu H, Tong X (2004) Face detection using improved LBP under Bayesian framework. In: Proceedings of international conference on image and graphics, pp 306–309
20. Ke Y, Sukthankar R (2004) PCA-SIFT: a more distinctive representation for local image descriptors. In: Proceedings of IEEE conference on computer vision and pattern recognition, pp 511–517
21. Kellokumpu V, Zhao G, Li S, Pietikainen M (2009) Dynamic texture based gait recognition. In: Proceedings of international conference on biometrics, pp 1000–1009
22. Lowe DG (2004) Distinctive image features from scale invariant keypoints. Int J Comput Vision 60(2):91–110
23. Mikolajczyk K, Schmid C, Zisserman A (2004) Human detection based on a probabilistic assembly of robust part detectors. In: Proceedings of the European conference on computer vision, pp 69–82
24. Ojala T, Pietikainen M, Harwood D (1996) A comparative study of texture measures with classification based on feature distributions. Pattern Recogn 29(1):51–59
25. Ojala T, Pietikainen M, Maenpaa T (2002) Multiresolution grayscale and rotation invariant texture classification with local binary patterns. IEEE Trans Pattern Anal Mach Intell 24 (7):971–987
26. Papageorgiou C, Poggio T (2000) CA trainable system for object detection. Int J Comput Vision 38(1):15–33
27. Porkili F (2005) Integral histogram: a fast way to extract histograms in cartesian spaces. In: Proceedings of IEEE conference on computer vision and pattern recognition, pp 829–836
28. Prisacariu V, Reid I (2009) FastHOG—a real-time GPU implementation of HOG. Department of Engineering Science, Oxford University
29. Randen T, Husoy JH (1999) Filtering for texture classification: a comparative study. IEEE Trans Pattern Anal Mach Intell 21(4):291–310
30. Rowley HA (1999) Neural network-based face detection. Ph.D. thesis, Carnegie Mellon University, Pitsburgh
31. Rowley H, Baluja S, Kanade T (1998) Neural network-based face detection. IEEE Trans Pattern Anal Mach Intell 20(1):23–38
32. Schneiderman H, Kanade T (2000) A statistical method for 3D object detection applied to faces and cars. In: Proceedings of IEEE conference on computer vision and pattern recognition, pp 746–751

33. Shan C, Gong S, McOwan P (2009) Facial expression recognition based on local binary patterns: a comprehensive study. Image Vis Comput 27:803–816
34. Shet VD, Neumann J, Ramesh V, Davis LS (2007) Bilattice-based logical reasoning for human detection. In: Proceedings of IEEE Conference on Computer Vision and Pattern Recognition, pp. 1–8
35. Subburaman V, Marcel S (2010) Fast bounding box estimation based face detection. In: Proceedings of ECCV workshop on face detection: where we are and what next?
36. Sun N, Zheng W, Sun C, Zou C, Zhao L (2006) Gender classification based on boosting local binary pattern. In: Proceedings of international symposium on neural networks, pp 194–201
37. Swain M, Ballard D (1991) Color indexing. Int J Comput Vision 7(1):11–32
38. Takala V, Ahonen T, Pietikainen M (2005) Block-based methods for image retrieval using local binary patterns. In: Proceedings of Scandinavian conference on image analysis, pp 882–891
39. Viola P, Jones M (2004) Robust real-time face detection. Int J Comput Vision 57(2):137–154
40. Viola P, Jones M, Snow D (2005) Detecting pedestrians using patterns of motion and appearance. Int J Comput Vision 63(2):153–161
41. Yan S, Shan S, Chen X, Gao W (2008) Locally assembled binary (LAB) feature with feature-centric cascade for fast and accurate face detection. In: Proceedings of IEEE conference on computer vision and pattern recognition, pp 1–7
42. Zabih R, Woodfill J (1994) Non-parametric local transforms for computing visual correspondence. In: Proceedings of European conference on computer vision, pp 151–158
43. Zhang L, Chu R, Xiang S, Liao S, Li S (2007) Face detection based on multi-block LBP representation. In: Proceedings of international conference on biometrics, pp 11–18
44. Zhang W, Shan S, Gao W, Chen X, Zhang H (2005) Local gabor binary pattern histogram sequence (LGBPHS): a novel non-statistical model for face representation and recognition. In Proceedings of IEEE international conference on computer vision, pp 786–791
45. Zhang L, Wu B, Nevatia R (2007) Detection and tracking of multiple humans with extensive pose articulation. In: Proceedings of IEEE international conference on computer vision, pp 1–8
46. Zhu Q, Avidan S, Yeh M, Cheng K (2006) Fast human detection using a cascade of histograms of oriented gradients. In: Proceedings of IEEE conference on computer vision and pattern recognition, pp 1491–1498

Adaptive Resource Management for Sensor Fusion in Visual Tracking

Bohyung Han, Seong-Wook Joo and Larry S. Davis

Abstract Sensor fusion for visual tracking is attractive since the integration of multiple sensors and/or features with different characteristics has potential to improve tracking performance. However, there exist several critical limitations to sensor fusion techniques: (1) the measurement cost increases typically as many times as the number of sensors, (2) it is not straightforward to quantify the confidence of each source and give each sensor a proper weight for state estimation, and (3) there is no principled algorithm for dynamic resource allocation to achieve better performance. We describe a method to combine information from multiple sensors and estimate the current tracker state by using a mixture of sequential Bayesian filters (e.g., particle filter)—one filter for each sensor, where each filter makes a different level of contribution to estimate the combined posterior in a reliable manner. In this framework, multiple sensors interact to determine an appropriate sensor for each particle dynamically; each particle is allocated to only one of the sensors for measurement and a different number of particles may be assigned to each sensor as a result. The level of the contribution of each sensor changes dynamically based on its prior information and relative measurement confidence. We apply this technique to visual tracking problems with multiple cameras or multiple features, and demonstrate its effectiveness through tracking results in real videos.

Keywords Visual tracking · Resource allocation · Sensor fusion · Multiple cameras · Multiple features · Kernel-based bayesian filtering · Mixture model

B. Han (✉)
Department of Computer Science and Engineering, POSTECH, Pohang, Korea
e-mail: bhhan@postech.ac.kr

S.-W. Joo
Google Inc., Mountain View, CA, USA
e-mail: swjoo@google.com

L.S. Davis
Department of Computer Science, University of Maryland, College Park MD 20742, USA
e-mail: lsd@umiacs.umd.edu

C. Kyung (ed.), *Theory and Applications of Smart Cameras*,
KAIST Research Series, DOI 10.1007/978-94-017-9987-4_9

1 Introduction

Rapid progress of video processing algorithms and the reduction of sensor prices make it possible that many computer vision systems, such as autonomous driving, visual surveillance, video conferencing, virtual/augmented reality, natural user interface system, etc., employ multiple cameras or sensors to develop new functions and improve system performance. For visual tracking, the combination of multiple sensors and/or tracking algorithms has the potential benefit by fusing complementary properties of heterogeneous sensors and algorithms. However, the integration process may not be straightforward, and typically requires additional cost for measurement and subsequent processing. Moreover, it is even more difficult how to allocate finite amount of resources to each sensor or algorithm and how much each source should be trusted to obtain the final solution, especially in large-scale systems.

Dynamic resource allocation is not a completely new problem in visual tracking. Kembhavi et al. [18] define the interaction groups of multiple targets using the Similarity Graph (SG) and allocate adaptive amount of resources to each group depending on the status of the groups and the associated targets. Tran and Davis [38] introduce a probabilistic framework of multiple resolution trackers in both spatial and temporal domain to achieve robustness and efficiency of trackers. An articulated object tracking algorithm is discussed in [30], where the amount of measurement for individual body parts and image frames are adjusted within a particle filter framework. Song et al. [36] present an algorithm to minimize the number of particles by monitoring the quality of the particles based on ranking SVM. However, these algorithms have not been investigated within a sensor fusion framework.

There are various types of sensor fusion algorithms for visual tracking. Fusion in the measurement step is the most typical one, where a single posterior is estimated by integrating multiple cues. Most tracking algorithms based on sensor fusion employ heuristic merge processes. For example, edge and color features are integrated to track elliptical objects in an ad hoc manner [3]. Slightly more advanced algorithms [37, 39] are proposed to combine multiple cues—motion, color, shape, etc., and they reduce the limitations of the individual modalities in practice. Simple sequential Bayesian filtering is adopted for tracking by fusion [2], where color, motion, and shape features are integrated using a variation of the Extended Kalman Filter (EKF).

The particle filter is an effective tool for fusion-based tracking, where a number of samples are drawn and the likelihood of each sample is typically computed based on the observations from all sensors. To improve observation quality, [44] proposes a straightforward method to combine color rectangle and edge features and [32, 41] describe fusion techniques of video and audio sensors for object tracking. Isard and Blake [16] employ skin color detection results to obtain better proposal distribution for contour tracking. A generic importance sampling mechanism for data fusion is discussed in [31], and a combination of top-down and bottom-up approaches is proposed in [8] to fuse multiple sensing modalities such as color, sound, and contour.

Although particle filter has been usefully applied to tracking by sensor fusion, their implementations have been mostly limited to

- combining observations of a particle from multiple sensors using the simple product of likelihoods based on independence assumption, and/or
- allocating the predefined number of particles to each sensor regardless of its reliability and usefulness.

More robust observations would be expected by such integration strategy of multiple sensors, but it is obvious that the cost of the measurement increases in proportion to the number of sensors. More importantly, assigning a fixed number of particles to each sensor, regardless of its reliability and usefulness, results in a potential waste or shortage of samples. This problem would be critical in large-scale systems that involve many sensors, and an intelligent resource allocation algorithm would be required. Note that the overall likelihood can be corrupted by the blind integration of the observations from multiple sensors if measurement process involves some noisy and/or non-discriminative sensors.

Graphical models have also been adopted to perform more sophisticated inferences for tracking by sensor fusion. Cue dependency is defined using a graphical model and Bayesian inference is employed for cue integration in [34, 45]. Tracker states with respect to shape and color are jointly optimized through co-inference technique in [43]. The relation between multiple modalities is used as a heuristic to estimate the reliability of each one in [34]. However, the relations are subjective and difficult to be generalized, and there is no discussion of their performance. The graphical model used in [45] might not be practical in systems with many sensors due to its complexity.

Another class of tracking by sensor fusion methods are in algorithm level ones, where trackers run independently and the final target state is estimated by merging their results through a post-processing step. People tracking results from multiple algorithms are combined using a heuristic in [35], and feature motions observed independently are merged by classification between inliers and outliers and cross-validation between trackers in [24]. Also, [22] proposed a fusion technique of multiple tracking algorithms within a Bayesian framework.

Although existing fusion-based tracking algorithms have been proposed to integrate multiple cues robustly, it is still not straightforward to handle how to measure the reliability of each cue and how to estimate target state and allocate resources based on the measured reliability. To address these problems, we introduce a mixture kernel-based Bayesian filter, where a mixture of the posteriors is propagated in a sequential Bayesian filtering framework and a useful sensor is selected for measurement probabilistically. Mixture models for posterior estimation in sequential Bayesian filtering is not new. The mixture particle filter is employed to maintain multi-modality in particle filters by modeling the posterior density as a nonparametric mixture model [40]. This technique is applied to multi-object tracking in a single camera setting [29, 40]. The mixture Kalman filter [7] and the

Interacting Multiple Model (IMM) algorithms [27] have similar ideas but continuous density functions such as Gaussian mixture are integrated instead of discrete ones. Unfortunately, these methods have not been discussed in the context of sensor fusion.

The mixture kernel-based Bayesian filtering (mKBF) was first proposed for sensor fusion in [12], which extended in [13] by presenting a new analysis of the update step for the fusion process and resource allocation. This chapter is based on [13]. The important features of our technique are discussed below.

- Our algorithm can be regarded as algorithm level fusion since a mixture of individual posteriors, which are continuous density functions, constructs the combined posterior for fusion in the update step.
- The individual posteriors have the different levels of contribution to the combined posterior; the weight of each posterior is determined by the prior and measurement confidence. The posterior estimation is expected to be more accurate by adopting a weighted mixture model instead of a single probability density function. This is because this method is effective to represent multi-modal density by giving more weight to reliable sensors for robust state estimation. Therefore, the performance of tracking algorithm can be improved, especially in the presence of clutter and occlusion.
- In our algorithm, significant interactions among sensors in the measurement step happen, and not all sensors are necessarily involved in the measurement of each sample. Instead, the sensor for the actual observation is determined probabilistically based on the expected likelihood of each sample. The proposal distribution is constructed from prediction as well as partial observations in each sensor. The sensor selection provides a framework to allocate an adaptive number of particles to each sensor based on its reliability.[1] It is not straightforward to implement this in conventional particle filters based on discrete distributions, since the density at an arbitrary location in the state space may not be available, so the expected likelihoods cannot be obtained; it is possible in our kernel-based Bayesian filtering, where all the relevant density functions are represented with a mixture of Gaussians.

Our approach is applied to a visual tracking problem with multiple cameras or multiple features. In addition, tracking in the presence of sensor failures is tested, where we assume that one of the cameras or features sometimes sends completely noisy signals and we expect that dynamic sensor weighting and adaptive particle allocation within mixture kernel-based Bayesian filtering framework handle the problem naturally.

The remaining sections of this chapter are organized as follows. Section 2 reviews kernel-based Bayesian filtering [14], and we discuss our sensor fusion

---

[1]The sample depletion in a sensor does not happen since a minimum number of particles is always allocated to each sensor and the sensor reliability can be obtained effectively with the minimum number of particles in our framework.

technique based on the mixture kernel-based Bayesian filtering in Sect. 3. The application of the proposed algorithm to visual tracking problem is illustrated in Sect. 4.

2 Background

In this section, we summarize Kernel-based Bayesian Filtering (KBF), which was originally introduced in [14]. The kernel-based Bayesian filtering is a state estimation technique, where a Gaussian mixture density function is propagated over time in the sequential Bayesian filtering framework. This framework is different from Kalman filter or extended Kalman filter in the sense that the posterior is not a Gaussian distribution any more and the multi-modal state estimation is achieved using a Gaussian mixture density function.

2.1 Overview

Let $\mathbf{x}_t$ and $\mathbf{z}_t$ $(t = 0, \ldots, T)$ be the state and measurement variables. In the sequential Bayesian filtering, the conditional density function of $\mathbf{x}_t$ given the history of measurement $\mathbf{z}_{1:t}$ is propagated through two steps—prediction and update, which are given by

$$\text{Prediction: } p(\mathbf{x}_t|\mathbf{z}_{1:t-1}) = \int p(\mathbf{x}_t|\mathbf{x}_{t-1})p(\mathbf{x}_{t-1}|\mathbf{z}_{1:t-1})\mathrm{d}\mathbf{x}_{t-1} \tag{1}$$

$$\text{Update: } p(\mathbf{x}_t|\mathbf{z}_{1:t}) = \frac{1}{\mathscr{C}}p(\mathbf{z}_t|\mathbf{x}_t)p(\mathbf{x}_t|\mathbf{z}_{1:t-1}), \tag{2}$$

where $\mathscr{C} = \int p(\mathbf{z}_t|\mathbf{x}_t)p(\mathbf{x}_t|\mathbf{z}_{1:t-1})d\mathbf{x}_t$ is a normalization constant that does not depend on $\mathbf{x}_t$. The posterior density at time step t denoted by $p(\mathbf{x}_t|\mathbf{z}_{1:t})$, is used to estimate the prior term in the next time step.

In kernel-based Bayesian filtering, when the posterior density function at the previous time step is given by a weighted mixture of Gaussians, the same mixture representation is obtained in the posterior at the current time step through the prediction and update steps as discussed in [14].

2.2 Kernel-Based Bayesian Filtering

Denote by $\mathbf{x}_t^i \in \mathbb{R}^d$ and $\mathbf{P}_t^i \in \mathbb{R}^{d\times d}$ $(i = 1, \ldots, n_t)$ a set of mean vectors and their corresponding covariance matrices, respectively, at time step t. Let each Gaussian have a weight ω_t^i with $\sum_{i=1}^{n_t} \omega_t^i = 1$, and let the posterior density function at time step t–1 be given by

$$p(\mathbf{x}_{t-1}|\mathbf{z}_{1:t-1}) = \sum_{i=1}^{n_{t-1}} \omega_{t-1}^i, N\left(\mathbf{x}_{t-1}^i, \mathbf{P}_{t-1}^i\right), \tag{3}$$

where $N(\mathbf{m}, \mathbf{C})$ represents a normal distribution with mean m and covariance $\mathbf{C}$.

In the prediction step, we employ the Unscented Transformation (UT) [17, 25] to each mode of the density function in Eq. (3) to make a prediction based on non-linear process models accurately. After applying the UT, we obtain the prior density function, which is also a mixture of Gaussians, as

$$p(\mathbf{x}_t|\mathbf{z}_{1:t-1}) = \sum_{i=1}^{n_{t-1}} \hat{\omega}_t^i, N\left(\hat{\mathbf{x}}_t^i, \hat{\mathbf{P}}_t^i\right), \tag{4}$$

where $\hat{\omega}_t^i$ is identical to ω_{t-1}^i, and $\hat{\mathbf{x}}_t^i$ and $\hat{\mathbf{P}}_t^i$ are the transformed mean and covariance by the UT, respectively.

The measurement density function is parameterized with a Gaussian mixture, which is achieved by density interpolation based on the Nonnegative Least Square (NNLS) technique [1, 6, 21]. The estimated measurement density function is denoted by

$$p(\mathbf{z}_t|\mathbf{x}_t) = \sum_{i=1}^{m_t} \tau_t^i N\left(\mathbf{x}_t^i, \mathbf{R}_t^i\right), \tag{5}$$

where τ_t^i is the unnormalized weight of each Gaussian obtained from the NNLS and $\mathbf{R}_t^i \in \mathbb{R}^{d \times d}$ is the covariance matrix corresponding to the mean $\mathbf{x}_t^i$ $(i = 1, \ldots, m_t)$.

In the update step, the posterior is computed by the products of a pair of Gaussian mixtures, which correspond to prior and measurement density functions (Eqs. (4) and (5), respectively). It is true that the derived density function is also a weighted Gaussian mixture, but the number of components in the mixture density increases exponentially over time, which is prohibitive in sequential Bayesian filtering framework. To figure out this issue, we employ kernel density approximation technique [11], which enables us to maintain a compact and accurate representation of a Gaussian mixture density function even after many stages of density propagation. Note that, although there are a few alternative techniques to reduce the number of components in a Gaussian mixture [33, 42], kernel density approximation is more principled and effective conceptually. After the update step, the final posterior distribution is given by the following equation:

$$p(\mathbf{x}_t|\mathbf{z}_{1:t}) = \sum_{i=1}^{n_t} \omega_t^i N\left(\mathbf{x}_t^i, \mathbf{P}_t^i\right), \tag{6}$$

where n_t is the number of Gaussian components at time step t and the sum of ω_t^i is equal to 1.

2.3 *Discussion of Kernel-Based Bayesian Filtering*

Kernel-based Bayesian filtering is advantageous compared to conventional methods based on discrete density functions such as particle filtering. It is generally known that a continuous proposal distribution would be helpful to improve sampling quality [10], so the natural filtering algorithm based on continuous density functions may reduce inherent limitation of particle filter—*degeneracy* or *loss of diversity* problem. In practice, the kernel-based Bayesian filter demonstrates equivalent accuracy with a smaller number of samples compared to conventional particle filters [14].

Another important characteristic of kernel-based Bayesian filter is that, unlike the particle filters based on a discrete representation of probability density functions, the probability at an arbitrary location in the state space can be computed straightforwardly regardless of sample locations. This property plays a crucial role in our sensor fusion framework, where the expected likelihood of each sample is supposed to be estimated even before the "real" observation. In the next section, we describe how the kernel-based Bayesian filtering is utilized in sensor fusion for visual tracking.

3 Fusion Tracking by Mixture KBF

Suppose that we have K sensors and hope to fuse data from those sensors. If the mixture weights of the sensors are given by π_{t-1}^i $(i = 1, \ldots, K)$ at time $t - 1$, the posterior at time step $t - 1$ is given by

$$p(\mathbf{x}_{t-1}|\mathbf{z}_{1:t-1}) = \sum_{k=1}^{K} \pi_{t-1}^k p_k(\mathbf{x}_{t-1}|\mathbf{z}_{1:t-1}), \tag{7}$$

where $p_k(\mathbf{x}_{t-1}|\mathbf{z}_{1:t-1})$ is the posterior of an individual sensor at time t–1, which is represented by a mixture of Gaussians.

Mixture density propagation in the sequential Bayesian filtering framework has been employed to maintain multi-modality. Interacting Multiple Model (IMM) filters are used to handle multiple process models [5, 23, 27, 28] or multiple measurements [15] effectively. The mixture particle filter presents reasonable performance in preserving and maintain multiple modes in the posterior [40]. Explicit mixture modeling is often useful to maneuver multi-modal characteristics observed in multi-sensor tracking. In our framework, each mixture component in Eq. (7),

$p_k(\mathbf{x}_{t-1}|\mathbf{z}_{1:t-1})$, belongs to an independent dynamic system[2] that has a separate measurement model. Note that the fusion of the posteriors by the weighted sum is a reasonable choice to estimate target state with multiple sensors as discussed in [5, 15, 23, 27, 28] although it is not a standard method to combine the measurements from multiple sensors by Bayesian way using an independent assumption—product of likelihood densities.

Our goal is to preserve the mixture representation through the iterations of sequential Bayesian filtering. The procedure for an individual Bayesian filter is similar to the description in Sect. 2, and we next explain how to combine the information from multiple sensors and how sensors interact with each other for resource allocation.

3.1 Prediction Step and Proposal Distribution

We make an independent prediction for an individual Bayesian filter using the unscented transformation as described in 2.2, and obtain the prior density function, which is given by

$$\begin{aligned} p(\mathbf{x}_t|\mathbf{z}_{1:t-1}) &= \sum_{k=1}^{K} \pi_{t-1}^k \int p_k(\mathbf{x}_t|\mathbf{x}_{t-1}) p_k(\mathbf{x}_{t-1}|\mathbf{z}_{1:t-1}) d\mathbf{x}_{t-1} \\ &= \sum_{k=1}^{K} \pi_{t-1}^k p_k(\mathbf{x}_t|\mathbf{z}_{1:t-1}). \end{aligned} \tag{8}$$

The proposal distribution is a critical factor to overall performance of our algorithm since it probabilistically selects a sensor for observation of each sample. There are several techniques to improve the proposal distribution in particle filter, which include use of an auxiliary tracker with different features [16], unscented particle filter [25, 32], and multi-stage sampling [14, 29].

We employ a 2-stage sampling technique to improve the effectiveness of particles, which combines the prior and partial observation distributions from each individual filter to construct the proposal distribution. The first proposal distribution denoted by $q^1(\mathbf{x}_t|\mathbf{x}_{t-1}, \mathbf{z}_{1:t})$ is common for every sensor and is equal to the prior density in Eq. (8):

---

[2]It is not completely independent since there are substantial interactions in sampling and measurement steps, but the posterior density function is propagated independently.

$$q^1(\mathbf{x}_t|\mathbf{x}_{t-1}, \mathbf{z}_{1:t}) = p_k(\mathbf{x}_t|\mathbf{z}_{1:t-1}). \tag{9}$$

The main idea behind this strategy is that, since the posterior in the previous step in Eq. (7) is based on the information from all sensors, it should be more reliable than the individual posteriors. In the second stage, the proposal distribution for each sensor, $q_k^2(\mathbf{x}_t|\mathbf{x}_{t-1}, \mathbf{z}_{1:t})$, is based on the combination of the initial proposal distribution and the partial observation in each sensor, which is formally given by

$$q_k^2(\mathbf{x}_t|\mathbf{x}_{t-1}, \mathbf{z}_{1:t}) = (1-\alpha)q^1(\mathbf{x}_t|\mathbf{x}_{t-1}, \mathbf{z}_{1:t}) + \alpha p_k^1(\mathbf{z}_t|\mathbf{x}_t), \tag{10}$$

where $p_k^1(\mathbf{z}_t|\mathbf{x}_t)$ denotes the initial normalized measurement density and α is a constant in $[0, 1]$. Then, the combined proposal distribution is given by

$$\begin{aligned} q^2(\mathbf{x}_t|\mathbf{x}_{t-1}, \mathbf{z}_{1:t}) &= \sum_{k=1}^{K} \pi_{t-1}^k q_k^2(\mathbf{x}_t|\mathbf{x}_{t-1}, \mathbf{z}_{1:t}) \\ &= \sum_{k=1}^{K} \pi_{t-1}^k \left((1-\alpha)q^1(\mathbf{x}_t|\mathbf{x}_{t-1}, \mathbf{z}_{1:t}) + \alpha p_k^1(\mathbf{z}_t|\mathbf{x}_t)\right). \end{aligned} \tag{11}$$

This 2-stage sampling strategy typically improves the sampling quality and reduces the number of samples required for robust observation since the proposal distribution uses the information from the priors of all sensors and the partial observations in the current step.

3.2 Measurement Step

The measurement step also has two stages in accordance with the 2-stage sampling strategy. The main purpose of the 2-stage sampling is to improve the proposal distribution and observation quality in a progressive manner. By assigning a fixed number of particles to all sensors in the first stage, the degeneracy problem—the situation that no particle is assigned to one or more sensors for observation and the measurement densities corresponding to the sensors do not become available—can be avoided.

In the first stage, the samples are drawn from the common proposal distribution in Eq. (9), and the observations are performed at the same locations in all sensors. Then, the initial observation in each sensor $p_k^1(\mathbf{z}_t|\mathbf{x}_t)$ $(k = 1, \ldots, K)$ is reflected in the second proposal distribution, as shown in Eq. (11), from which additional samples are drawn. In the second stage, each sample is used for observation in only one sensor, which is determined probabilistically by considering the prior and likelihood expectation. The probability that the kth sensor is selected is given by

$$p(sel(i)=k)=\frac{\pi_{t-1}^{k}\left(\beta p_k\left(\mathbf{x}_t^{(i)}|z_{1:t-1}\right)+(1-\beta)p_k^1\left(\mathbf{z}_t|\mathbf{x}_t^{(i)}\right)\right)}{\sum_{j=1}^{K}\pi_{t-1}^{j}\left(\beta p_j\left(\mathbf{x}_t^{(i)}|\mathbf{z}_{1:t-1}\right)+(1-\beta)p_j^1\left(\mathbf{z}_t|\mathbf{x}_t^{(i)}\right)\right)}, \tag{12}$$

where $sel(i)$ is the selected sensor index for the ith sample and $\beta \in [0,1]$ is a constant. Note that $p_k\left(\mathbf{x}_t^{(i)}|\mathbf{z}_{1:t-1}\right)$ is the prior density of the ith sample in the kth sensor, and $p_k^1\left(\mathbf{z}_t|\mathbf{x}_t^{(i)}\right)$ is the likelihood of the ith sample given the initial measurement density. The sensor selection for the ith sample is performed by the following equation:

$$sel(i)=\arg\min_{s}\left(\sum_{k=1}^{s}p(sel(i)=k)>r_i\right) \tag{13}$$

where r_i is a random number from a uniform distribution on $[0,1)$.

It is not straightforward to perform this procedure in conventional particle filters since it is difficult to obtain probabilities at arbitrary locations from discrete representation of density. The sensor that is likely to produce the highest likelihood is prioritized for observation, and receives more particles to improve the quality of the measurement density. The sampling and measurement procedure in the second stage is illustrated in Fig. 1.

The multi-stage measurements in the individual filter is identical to the kernel-based Bayesian filter [14], where the nonnegative least square method is used to approximate measurement density functions. The un-normalized measurement density function of the k-th sensor at time step t, denoted by $\tilde{p}_k(\mathbf{z}_t|\mathbf{x}_t)$, is given by

$$\tilde{p}_k(\mathbf{z}_t|\mathbf{x}_t)=\sum_{i=1}^{m_{t,k}}\kappa_{t,k}^{i}N\left(\mathbf{x}_{t,k}^{i},\mathbf{R}_{t,k}^{i}\right), \tag{14}$$

where $m_{t,k}$ is the number of components, $\kappa_{t,k}^{i}$ is an un-normalized weight of each Gaussian component, and $\mathbf{x}_{t,k}^{i}$ and $\mathbf{R}_{t,k}^{i}$ are the mean and covariance in the k-th measurement density, respectively.

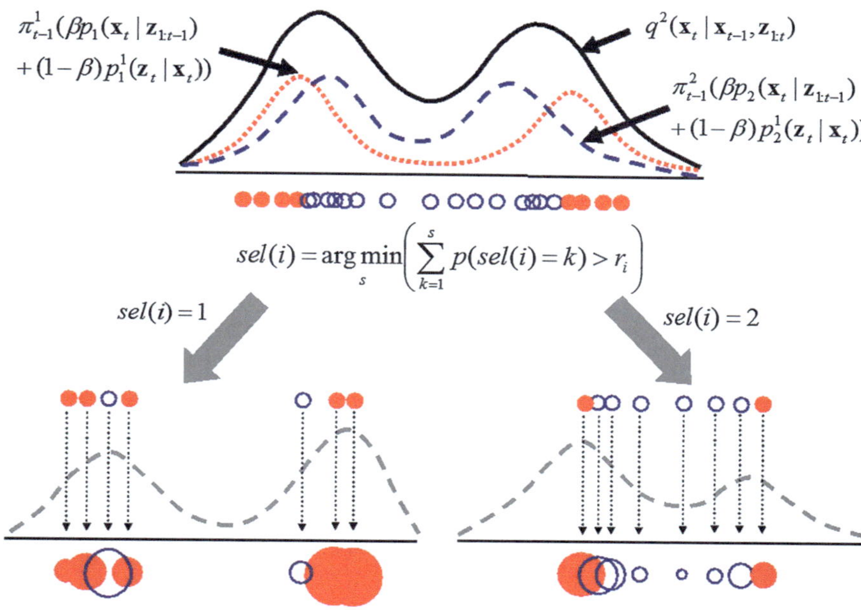

Fig. 1 An example of sampling and measurement procedure in the second stage. The proposal distribution $q_k^2(\ x_t|\ x_{t-1},\ z_{1:t})$ is constructed based on the prior and the partial measurement density function of the kth sensor, and $q^2(\ x_t|\ x_{t-1},\ z_{1:t})$ is the mixture of q_k^2, $(k = 1, 2)$. *(Top)* The samples such that $p(sel(i) = 1) \geq p(sel(i) = 2)$ are represented with *red (shaded) circles*, and the rest are represented with *blue (hollow) circles*. The sensor selection for each sample is performed by Eq. (13). *(Bottom)* Because the sensor selection for each particle is probabilistic, *red* and *blue* particles are mixed in each sensor. Based on the measurements of each sensor, the final measurement density functions are constructed by the nonnegative least square method

3.3 Update Step

The update step combines the prior and the measurement information to construct the individual posteriors, which are integrated to derive the overall posterior probability density function. Recall Eq. (7), where the overall posterior is given by a mixture of normalized individual posteriors. After one more step of sequential Bayesian filtering, we obtain the un-normalized posterior of each sensor, which models the relative confidence for target state estimation induced only from the current time step. Therefore, the fusion-based posterior is estimated by the sum of the product of mixture weight at the previous time step and un-normalized posterior, and is converted to the same kind of representation of Eq. (7).

Suppose that $\tilde{p}_k(\mathbf{x}_t|\mathbf{z}_{1:t})$ and $\tilde{p}_k(\mathbf{z}_t|\mathbf{x}_t)$ are un-normalized posterior and measurement density for the k-th sensor at time step t, respectively; then the overall posterior is given by

$$
\begin{aligned}
p(\mathbf{x}_t|\mathbf{z}_{1:t}) &\equiv \frac{1}{C}\sum_{k=1}^{K}\pi_{t-1}^{k}\tilde{p}_k(\mathbf{x}_t|\mathbf{z}_{1:t}) \\
&= \frac{1}{C}\sum_{k=1}^{K}\pi_{t-1}^{k}\tilde{p}_k(\mathbf{z}_t|\mathbf{x}_t)p_k(\mathbf{x}_t|\mathbf{z}_{1:t-1}) \\
&= \frac{1}{C}\sum_{k=1}^{K}\pi_{t-1}^{k}\psi_t^k p_k(\mathbf{z}_t|\mathbf{x}_t)p_k(\mathbf{x}_t|\mathbf{z}_{1:t-1}) \\
&= \frac{1}{C}\sum_{k=1}^{K}\pi_{t-1}^{k}\psi_t^k \int p_k(\mathbf{z}_t|\mathbf{x}_t)p_k(\mathbf{x}_t|\mathbf{z}_{1:t-1})\mathrm{d}\mathbf{x}_t \frac{p_k(\mathbf{z}_t|\mathbf{x}_t)p_k(\mathbf{x}_t|\mathbf{z}_{1:t-1})}{\int p_k(\mathbf{z}_t|\mathbf{x}_t)p_k(\mathbf{x}_t|\mathbf{z}_{1:t-1})\mathrm{d}x_t} \\
&= \frac{1}{C}\sum_{k=1}^{K}\pi_{t-1}^{k}\psi_t^k \int p_k(\mathbf{z}_t|\mathbf{x}_t)p_k(\mathbf{x}_t|\mathbf{z}_{1:t-1})\mathrm{d}\mathbf{x}_t p_k(\mathbf{x}_t|\mathbf{z}_{1:t}) \\
&= \sum_{k=1}^{K}\pi_t^k p_k(\mathbf{x}_t|\mathbf{z}_{1:t})
\end{aligned}
\tag{15}
$$

where

$$
C = \int \sum_{k=1}^{K}\pi_{t-1}^{k}\tilde{p}_k(\mathbf{x}_t|\mathbf{z}_{1:t})\mathrm{d}\mathbf{x}_t \tag{16}
$$

is the normalization constant,

$$
\psi_t^k = \sum_{i=1}^{m_{t,k}}\kappa_{t,k}^{i} = \int \tilde{p}_k(\mathbf{z}_t|\mathbf{x}_t)\mathrm{d}\mathbf{x}_t \tag{17}
$$

is the measurement confidence for each sensor, and

$$
\pi_t^k = \frac{1}{C}\pi_{t-1}^{k}\psi_t^k \int p_k(\mathbf{z}_t|\mathbf{x}_t)p_k(\mathbf{x}_t|\mathbf{z}_{1:t-1})\mathrm{d}\mathbf{x}_t \tag{18}
$$

is the new mixture weight for the kth component at time t. Note that, as illustrated in Fig. 2, the measurement density function and the measurement confidence denoted by $\tilde{p}_k(\mathbf{z}_t|\mathbf{x}_t)$ and ψ_t^k, respectively, are hardly affected by the number of particles. This is because the measurement density function can be reconstructed using a small number of control points based on the nonnegative least square technique; the extra samples typically improve the details of the density function, but its overall shape is modeled effectively by the small number of samples with high likelihoods. This is a very important property in our algorithm since it allocates a different number of samples to each sensor, and the sensor confidence computed by integration of measurement density function should be invariant to the number of

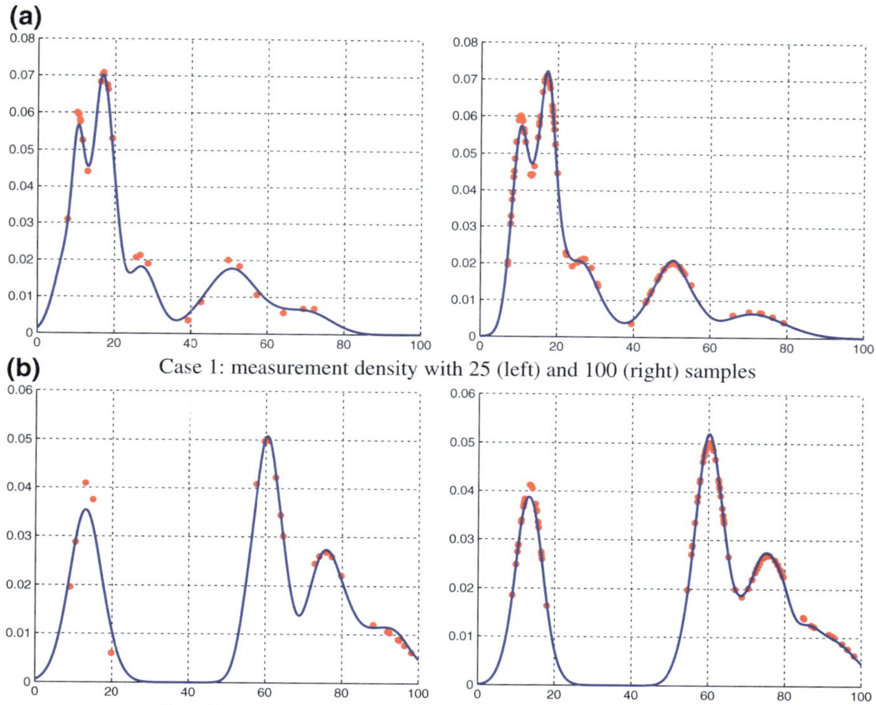

Case 2: measurement density with 25 (left) and 100 (right) samples

Fig. 2 Comparison of measurement density functions with different number of samples. As illustrated, the measurement density functions based on the nonnegative least square are almost invariant to the number of samples. Note that the measurement confidences, ψ, are approximately 1.4 with 100 samples and 1.3 with 25 samples for both cases in average. **a** Case 1: measurement density with 25 (*left*) and 100 (*right*) samples. **b** Case 2: measurement density with 25 (*left*) and 100 (*right*) samples

samples for accurate weight estimation of the sensor; otherwise, the sensors with more samples are likely to have more weight consistently or we need to normalize the confidence of each sensor based on the number of samples, which is not stable enough according to our simulation.

The new mixture weight π_t^k is proportional to three terms—previous mixture weight, measurement confidence, and an integration term. Note that the integration term, $\int p_k(\mathbf{z}_t|\mathbf{x}_t)p_k(\mathbf{x}_t|\mathbf{z}_{1:t-1})d\mathbf{x}_t$, reveals the coherency between prior and measurement density functions. Since both densities are Gaussian mixtures, their product is also a mixture of Gaussians and the integration is equal to the sum of the weights of Gaussians in the new mixture. Let $p_k(\mathbf{x}_t|\mathbf{z}_{1:t-1}) = \sum_{i=1}^{n_{t-1}} \omega_{t,k}^i N(\mathbf{x}_{t,k}^i, \mathbf{P}_{t,k}^i)$ and $p_k(\mathbf{z}_t|\mathbf{x}_t) = \sum_{i=1}^{m_t} \tau_t^i N(\mathbf{x}_{t,k}^i, \mathbf{R}_{t,k}^i)$ be the prior and measurement density function, respectively. Then, the product of two density functions is given by

$$\sum_{i=1}^{n_{t-1}} \sum_{j=1}^{m_t} \omega_t^{ij} N(\mathbf{m}_t^{ij}, \mathbf{C}_t^{ij}), \tag{19}$$

where

$$\omega_t^{ij} = \kappa_t^i \tau_t^j N(\mathbf{x}_t^i, \mathbf{P}_{t,k}^i + \mathbf{R}_{t,k}^j) \tag{20}$$

$$\mathbf{m}_t^{ij} = \mathbf{C}_t^{ij} \left((\mathbf{P}_{t,k}^i)^{-1} \mathbf{x}_t^i + (\mathbf{R}_{t,k}^j)^{-1} \mathbf{x}_t^j \right) \tag{21}$$

$$\mathbf{C}_t^{ij} = \left((\mathbf{P}_{t,k}^i)^{-1} + (\mathbf{R}_{t,k}^j)^{-1} \right)^{-1}. \tag{22}$$

Therefore, the integration term is given by

$$\int p_k(\mathbf{z}_t | \mathbf{x}_t) p_k(\mathbf{x}_t | \mathbf{z}_{1:t-1}) d\mathbf{x}_t = \sum_i \sum_j \kappa_t^i \tau_t^j N(\mathbf{x}_t^i, \mathbf{P}_{t,k}^i + \mathbf{R}_{t,k}^j), \tag{23}$$

where it will be larger when two density functions are similar to each other.

4 Experiments

We apply the proposed sensor fusion technique to visual tracking problem with multiple cameras and multiple features, where the weighted mixture density function is propagated in the framework of mixture Kernel-based Bayesian Filtering (KBF).

4.1 Sensor Fusion with Multiple Cameras

4.1.1 Implementation Issues

Multi-camera tracking is useful compared to single camera tracking especially for handling occlusion. There have been a large number of prior studies on tracking using multiple cameras [4, 9, 19, 20, 26], but little efforts have been made to control the degree of contribution from each cameras. Our sensor fusion technique adjusts the degree of contribution of each camera dynamically based on observation history and improves tracking performance by adaptive resource allocation. We describe implementation details about our probabilistic sensor fusion framework based on multiple cameras, and demonstrate tracking results compared to the conventional algorithm.

We assume that objects are moving on a ground plane and all cameras share some field of view of those objects. The common state space is defined as the 2D location (x, y) in the canonical top view, and the state vector can be transformed into each view for observation using the ground plane homography. Even though the cameras are static, no background subtraction is performed for tracking.

In our algorithm, the process model is the random walk, and the likelihood of each sample is computed by the similarity of the RGB color histograms between the target and the candidates. The distance measure of two histograms is Bhattacharyya distance. The measurement process is performed in each camera independently, so the measurement density of camera k is given by

$$p_k(\mathbf{z}_t|\mathbf{x}_t) = p_k(\mathbf{z}_t^k|T_k(\mathbf{x}_t)), \tag{24}$$

where $\mathbf{z}_t^k$ represents the observation data in camera k and $T_k(\cdot)$ denotes the homography transformation of the common state into the corresponding view.

A tricky problem in the measurement is that the absolute values of likelihoods are not normalized properly across cameras. Therefore, the measurement confidence ψ_t^k may have a significantly different order of magnitude in each camera due to its characteristics, and it may not be appropriate to use the likelihoods directly. So, instead of simply computing distances between target and candidate histograms, the likelihood of each sample is obtained by computing the ratio of candidate-target distance to candidate-uniform distribution distance. Then, the likelihood of the i-th sample, $p_k(\mathbf{z}_t^{(i)}|\mathbf{x}_t)$ is given by

$$p_k(\mathbf{z}_t|\mathbf{x}_t^{(i)}) \propto \exp\left(-\lambda \frac{D^2(\mathbf{q}, \mathbf{p}_i)}{D^2(\mathbf{q}, \mathbf{u})}\right), \tag{25}$$

where $D^2(\cdot, \cdot)$ is squared Bhattacharyya distance between two histograms, and λ is a constant. Also, $\mathbf{p}$, $\mathbf{q}$, and $\mathbf{u}$ are normalized target, candidate, and uniform histograms, respectively. Note that the denominator in Eq. (25) can be pretty different across each camera, especially when the color characteristics of camera sensors are different. This method provides a practically reasonable solution for normalizing the likelihoods from different cameras.

Throughout our experiments, the RGB color histograms are constructed based on $16 \times 16 \times 16$ bins and we used fixed parameter values : $\alpha = \beta = 0.5$ and $\lambda = 30$. According to our experiments, small changes to these parameters have negligible effects on tracking results; the performance is slightly worse with $\alpha = \beta = 0.3$, but is essentially unaffected over variations when $15 \leq \lambda \leq 30$.

4.1.2 Results

We tested our algorithm on an indoor sequence captured by two cameras, where walking people are tracked. The appearance model of a target is constructed based on two separate histograms—one for the upper and the other for the lower body, and we compute the joint likelihood.

Figure 3 illustrates the result of tracking two persons using two cameras in an indoor environment. Note that we employed the approximate height information of each person for measurement and visualization. In this example, we constructed a measurement density from 50 observations—5 samples are given to each camera in the first stage ($5 \times 2 = 10$ observations) and 40 samples are dynamically allocated to the two cameras in the second stage (40 observations). Despite frequent occlusion and clutter, the persons are successfully tracked throughout the sequence by the active collaboration of two cameras. Obviously, the mixture weights and the number of particles assigned to each camera are updated at each frame depending on visibility and the discriminativeness of the target in each view, which are illustrated in Fig. 4.

We also compared tracking performance of our method with a conventional product-of-likelihood fusion algorithm by particle filter, which is presented in Fig. 5. The sequence for this experiment is similar to that used for Fig. 3, but multiple severe dynamic occlusions occur between two people whose appearances are pretty similar because both are wearing white T-shirts. The same number of measurements (50 altogether) are performed for both methods; in the case of our method, 5 samples are drawn at the first stage of measurement step, and 40 samples are then dynamically allocated to both cameras at the second stage. After the first occlusion, both tracking algorithms recovered from short-term failures successfully but the conventional fusion method based on particle filtering lost the target after the second occlusion. On the other hand, our algorithm succeeded in tracking the target even after the second occlusion.

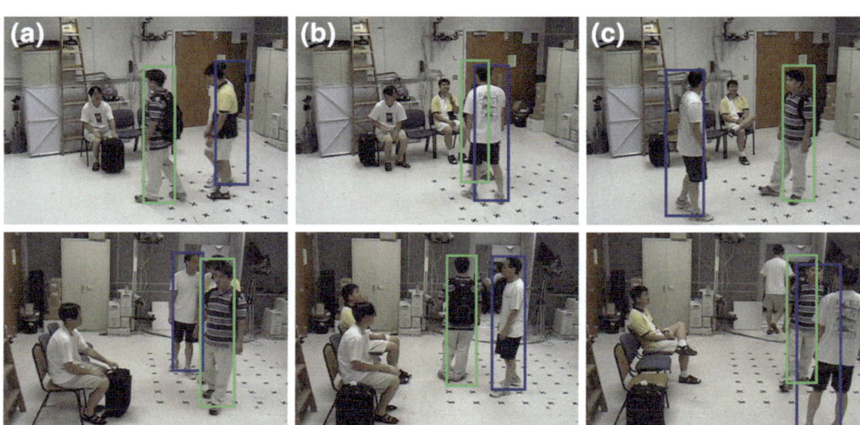

Fig. 3 A tracking example. Results in camera 1 (*top*) and 2 (*bottom*) are presented at 3 different time steps. Person 1 and 2—*blue* and *green* bounding box, respectively—are successfully tracked in the presence of multiple dynamic occlusions. **a** $t = 158$. **b** $t = 215$. **c** $t = 284$

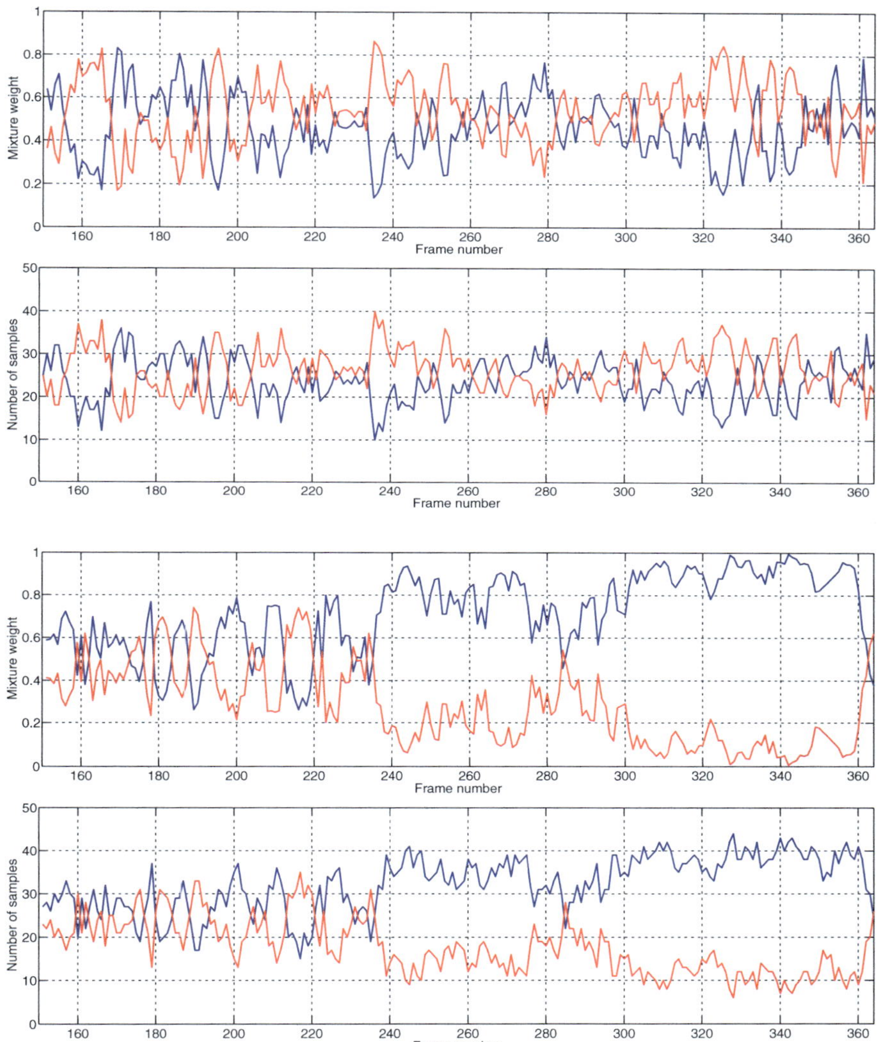

Fig. 4 Mixture weights and particle allocations for each person and each sensor in each frame. *Blue* and *red lines* represent camera 1 and camera 2, respectively. Note that the mixture weights and particle assignments mostly correspond to visibility of targets. **a** Mixture weights and particle assignment for person 1. **b** Mixture weights and particle assignment for person 2

In our method, the mixture weight and the number of observations in camera 2 during occlusion (around $t = 60$) are consistently higher than those in camera 1 as illustrated in Fig. 6. It suggests that tracking by mixture KBF is successful because the the more reliable sensor (camera 2) is prioritized for observation by our dynamic sample allocation technique.

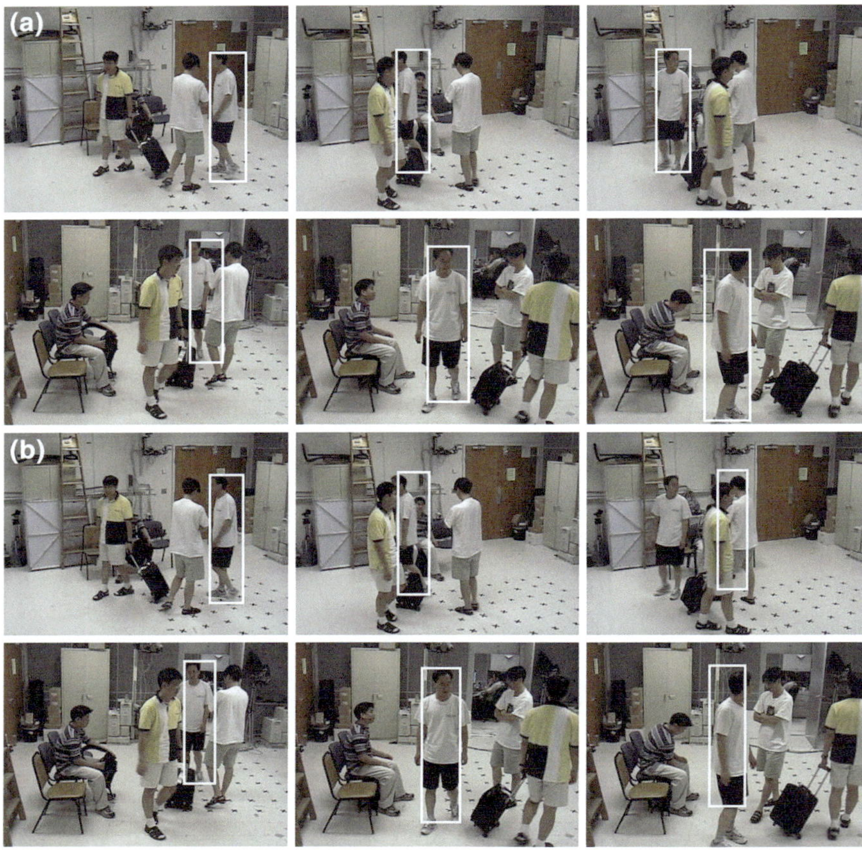

Fig. 5 Comparison between mixture KBF and conventional fusion by particle filter. The results at time $t = 18, 54, 67$ are presented for each algorithm in (**a**) and (**b**), where the first and second rows represent results in camera 1 and camera 2, respectively. Note that the target is lost after the second occlusions around $t = 60$ in (**b**). **a** Tracking by mixture KBF. **b** Tracking by conventional fusion method

In many computer vision systems, it is fairly common that some sensor data are temporarily missing or become totally unreliable due to sensor noises, occlusion, hardware/software failures, etc. To investigate the performance of our algorithm in this challenge, we used another video sequence captured by three outdoor cameras. The temporary failures in one of the cameras are simulated by replacing the original image with completely noisy signals in some frames shown in Table 1.

The performance of our sensor fusion tracking algorithm, denoted by mKBF, is also tested in the presence of sensor failures and compared with the following other fusion-based tracking algorithms:

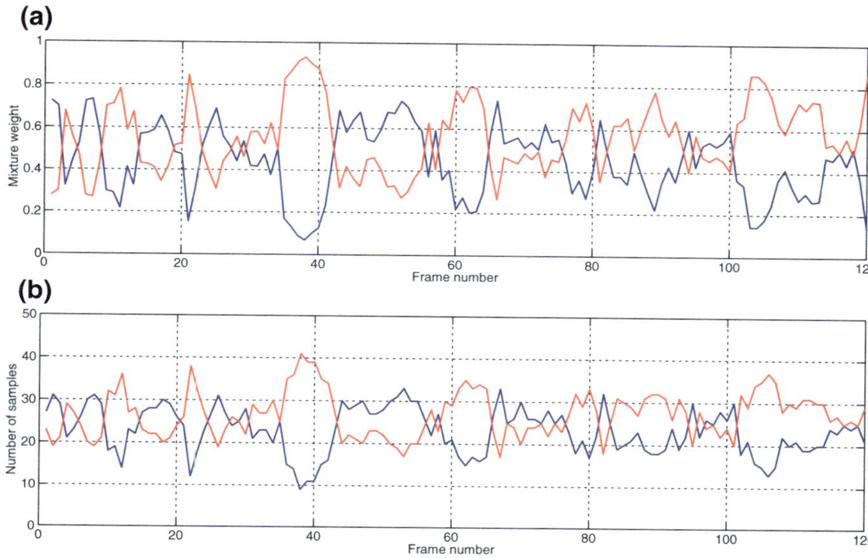

Fig. 6 Mixture weights and particle assignments for each sensor in each frame. *Blue* and *red lines* are for camera 1 and camera 2, respectively. Note that the mixture weight and the number of particles are significantly larger in camera 2 during the second occlusions around $t = 60$. **a** Mixture weights. **b** Particle assignment

Table 1 Frames with camera failures

Camera	Frames with camera failures
1	From 581 to 670
2	From 501 to 575
3	From 366 to 415

- **KBF**: Tracking by KBF based on the standard sensor fusion technique (same number and locations of samples in each sensor, product-of-likelihood fusion)
- **PF**: Tracking by Particle Filter (PF) based on the standard sensor fusion technique (same number and locations of samples in each sensor, product-of-likelihood fusion)
- **mKBFe**: Tracking by mKBF without adaptive resource allocation (same number of samples in each sensor, but different locations, sum-of-posterior fusion)

To track people in our algorithm, 10 particles are used at the first stage of measurement and 30 particles are distributed over three cameras dynamically, resulting in a total of 60 observations. On the other hand, 20 particles are distributed evenly to each camera in the other three algorithms, and the total number of observations is the same as our method.

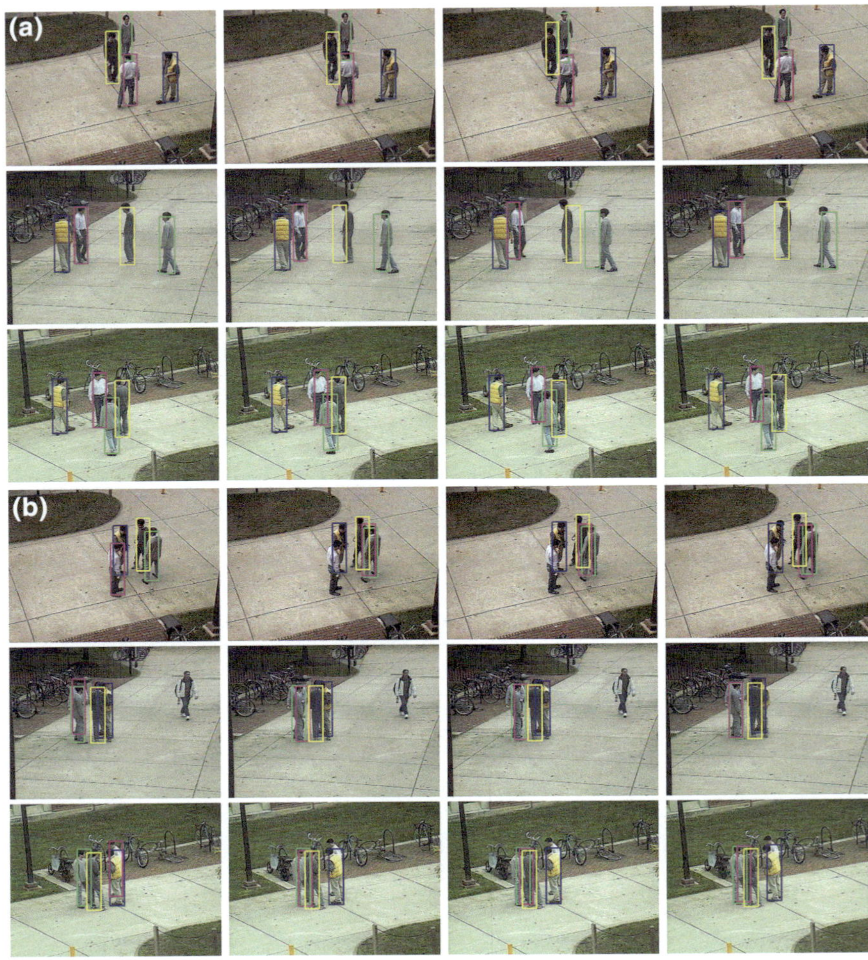

Fig. 7 Comparison of people tracking in presence of temporal sensor failures. (*Col1*) mKBF (*Col2*) KBF (*Col3*) PF (*Col4*) mKBF with even sample distribution. (*Top*) camera1 (*Middle*) camera2 (*Bottom*) camera 3. The errors are significant in person 2 (*green*) and 4 (*yellow*) at $t = 529$ and person 3 (*magenta*) at $t = 685$. Note that the signal from camera 2 is completely noisy $t = 529$ but we presented a normal image to show tracking performance effectively. **a** Tracking results at $t = 529$. **b** Tracking results at $t = 685$

Figure 7 illustrates the result of multi-object tracking using three cameras in the outdoor scene. Even with frequent occlusions amongst the group of people and temporary simulated sensor failures, tracking by mKBF with adaptive resource allocation is successful for the entire 900 frames. Note that results by three compared algorithms are less stable than the proposed technique.

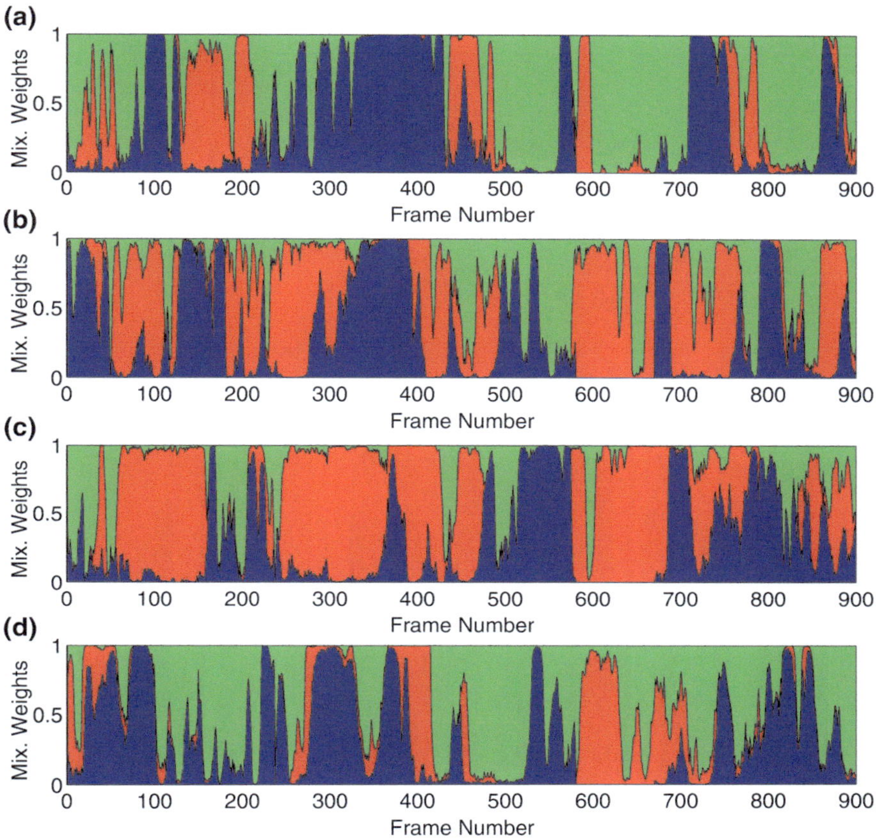

Fig. 8 Mixture weight for each sensor in people tracking sequence. *Blue*, *red* and *green* area denote camera 1, 2, and 3, respectively. The mixture weights changes dynamically due to various reasons such as target visibility, appearance changes, and so on. **a** Person 1 (*blue bounding box*). **b** Person 2 (*green bounding box*). **c** Person 3 (*magenta bounding box*). **d** Person 4 (*yellow bounding box*)

The mixture weight for each camera is illustrated in Fig. 8. As observed, the mixture weight of the failed sensor was negligible so that only the minimum number of particles was allocated.

Figure 9 illustrates the trajectories of four tracked people, where numerous dynamic occlusions are observed.

We also performed a quantitative performance evaluation for the four different algorithms, where the ground-truths are created manually and the error is measured by the Euclidean distance between the ground-truth and tracking results computed in the canonical top-view plane. The quantitative comparison results are illustrated in Fig. 10, which are the average of 10 independent runs for each algorithm due to randomness of particle filtering. As a result, the performance of our method turns

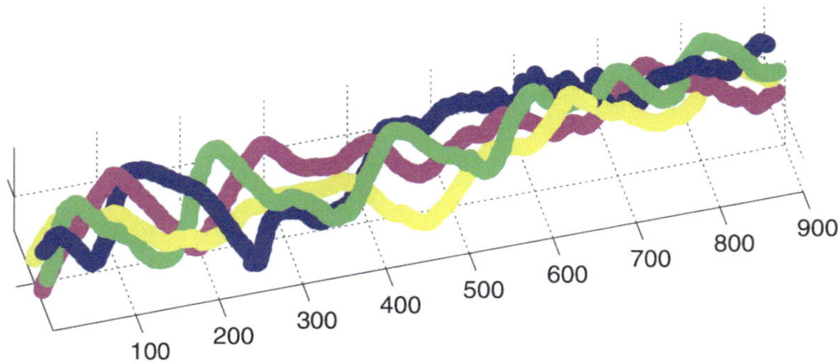

Fig. 9 Space-time trajectories for four people in the outdoor sequence

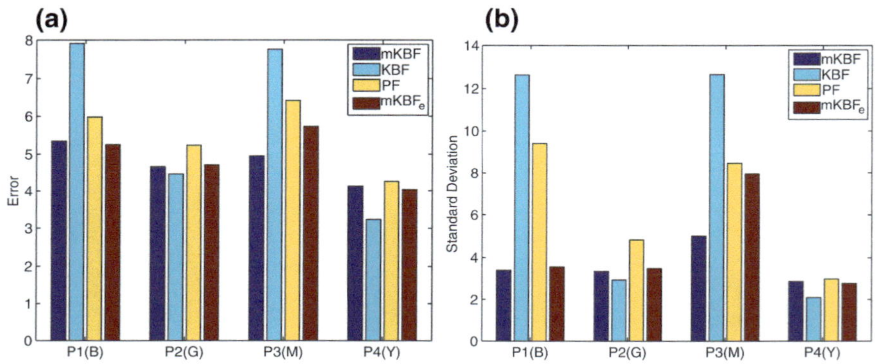

Fig. 10 Quantitative comparison of the four different sensor fusion algorithms in terms of errors and error variations for each person. The labels in *x*-axis denote person IDs (the colors of bounding boxes in Fig. 7). **a** Error for each person in the four sensor fusion methods. **b** Error variation for each person in four sensor fusion methods

out to be better than other algorithms with the same number of measurements in general. The variance of tracking errors in KBF is high although tracking results are sometimes very accurate, and the errors and their variations of PF are consistently higher than our method. The performance of mKBFe is close to our algorithm, but it exhibits noticeably higher errors in tracking person 3.

Now, we present the benefit of mKBF by comparing the tracking errors of mKBF to the errors of KBF and PF with more observations; tracking with 60, 90, and 120 measurements are performed 10 times and averaged, where 20, 30, and 40 samples are given to each sensor, respectively. The accuracy of mKBF with 60 measurements is almost equivalent with that of KBF and PF with 120 measurements although KBF is slightly better than PF; the error variance of mKBF with 60 measurements is lower than that of KBF and PF with 120 observations by more than 30 %. These results are presented in Fig. 11.

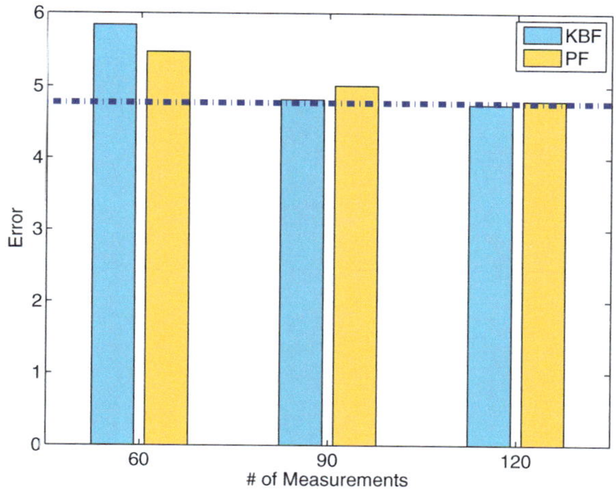

Fig. 11 Error comparison between mKBF, KBF, and PF by varying the number of measurements. The error bars of KBF and PF are obtained for the three different numbers of measurements, and the error for mKBF with 60 measurements is illustrated with a *blue dotted horizontal line*

4.2 Sensor Fusion with Multiple Features

We also applied our sensor fusion technique to object tracking using multiple logical sensors, i.e., features; the logical sensors employed in our experiment are color, gradient, template, and contour. For the color and the gradient sensor, the target appearances are modeled by histograms and the Bhattacharyya distance is used to compute likelihoods. The template sensor measures the mean squared differences of the color pixels between the smoothed image templates of target and candidates. Finally, the contour sensor computes the sum of the gradient magnitudes along the several normal direction around the perimeter of an ellipse. For each sensor, we performed tracking independently by KBF, but all the sensors interact actively and compete for particle allocation as discussed in Sect. 3. The object is tracked in a 4D state space consisting of image location (x, y), in-plane rotation, and scale, and the random walk is adopted as the process model.

Figure 12 presents the results of tracking with four different sensors by the mKBF. Our algorithm tracked a target successfully under significant pose variations and severe appearance changes throughout the entire 500 frames. The number of samples for observation is 90 altogether—10 in the first stage and 80 in the second stage, so the total number of observations in all sensors is $10 \times 4 + 80 = 120$. The mixture weight and sample allocations results for each sensor are presented in Fig. 13, where the dynamic changes in the mixture weights and the number of samples are observed. It is interesting that the mixture weight for contour feature is significantly high when the back of woman's head is shown and the gradient feature has high weights occasionally.

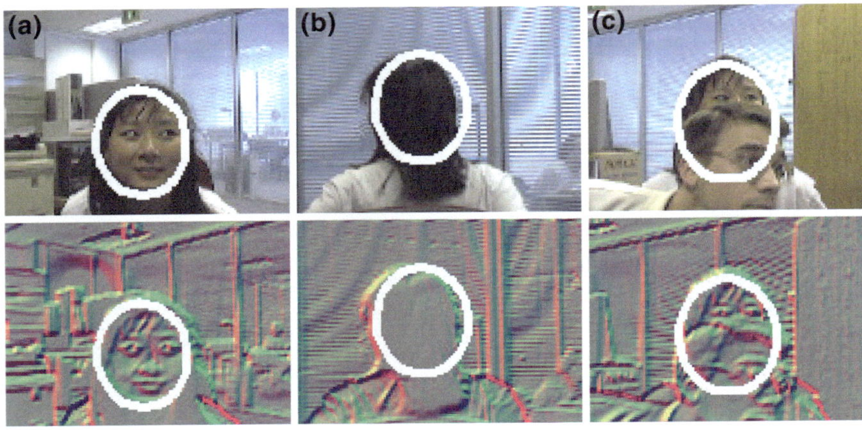

Fig. 12 Tracking results by logical sensor fusion. (*Top*) Results in the color images (*Bottom*) Results in the gradient images. Note that the gradients in the x and y direction are mapped to R and G space in the gradient images, respectively. **a** $t = 49$. **b** $t = 208$. **c** $t = 440$

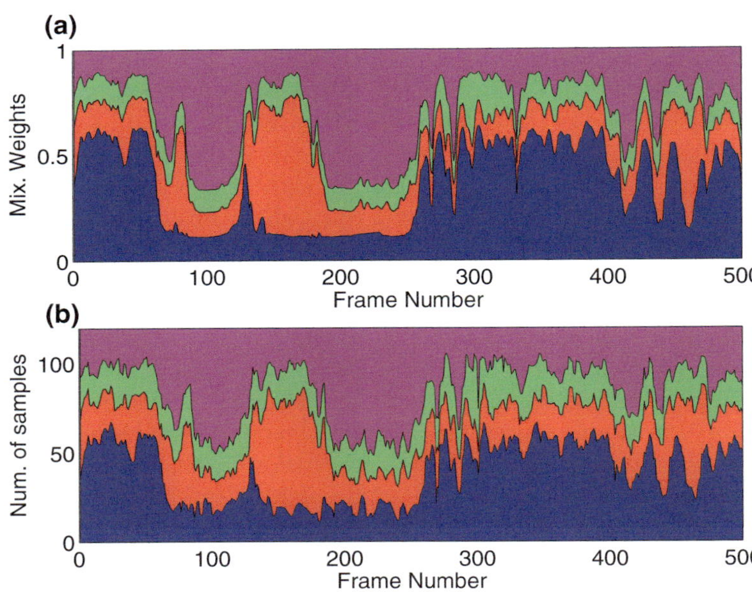

Fig. 13 Mixture weights and resource allocation results in each frame. (*Blue*) Color, (*Red*) Gradient (*Green*) Template (*Magenta*) Contour. The mixture weight of contour feature is significantly high when the back of woman's head is shown around at $t = 90 \sim 110$ and $t = 180 \sim 240$. **a** Mixture weights for each sensor in each frame. **b** Number of samples for each sensor in each frame

5 Conclusion

We presented a probabilistic framework of sensor fusion based on the mixture kernel-based Bayesian filtering. This framework provides a methodology to select effective sensors for measurements probabilistically and to maintain the multi-modality of the combined posterior density function effectively. By assigning particles to a sensor based on its reliability, we can expect more robust observations and improve the effectiveness of particles. We applied our algorithm to various sensor fusion scenarios in multi-camera and multi-feature tracking scenarios, and demonstrated tracking results in the presence of severe occlusions, clutter, and sensor failures. Our experiment shows that tracking by the proposed algorithm, mKBF, is advantageous over other sensor fusion techniques such as KBF, PF, and mKBFe, qualitatively and quantitatively.

Acknowledgment This work is supported by the Center for Integrated Smart Sensors funded by the Ministry of Science, ICT & Future Planning as the Global Frontier Project.

References

1. Adlers M (2000) Topics in sparse least squares problems. Ph.D. thesis, Linköpings Universitet, Sweden. http://www.math.liu.se/~milun/thesis
2. Azoz Y, Devi L, Sharma R (1998) Reliable tracking of human arm dynamics by multiple cue integration and constraint fusion. In: Proceedings IEEE conference on computer vision and pattern recognition. Santa Barbara, California
3. Birchfield S (1998) Elliptical head tracking using intensity gradients and color histograms. In: Proceedings IEEE conference on computer vision and pattern recognition. Santa Barbara, CA, pp 232–237
4. Black J, Ellis T (2006) Multi camera image tracking. Image Vision Comput J 24(11):1256–1267
5. Blom H, Bar-Shalom Y (1988) The interacting multiple model algorithm for systems with markovianswitching coefficients. IEEE Trans Autom Control 33(8):780–783
6. Cantarella J, Piatek M (2004) tsnnls: a solver for large sparse least squares problem with non-negative variables. http://www.cs.duq.edu/~piatek/tsnnls/ (preprint)
7. Chen R, Liu J (2000) Mixture kalman filters. J Roy Statist Soc B 62(3):493–508
8. Chen Y, Rui Y (2004) Real-time speaker tracking using particle filter sensor fusion. Proc IEEE 92(3):485–494
9. Dockstader S, Tekalp A (2001) Multiple camera tracking of interacting and occluded human motion. Proc IEEE 89(10):1441–1455
10. Doucet A, de Freitas N, Gordon N (2001) Sequential Monte Carlo methods in practice. Springer, Berlin
11. Han B, Comaniciu D, Zhu Y, Davis L (2008) Sequential kernel density approximation and its application to real-time visual tracking. IEEE Trans Pattern Anal Mach Intell 30(7):1186–1197
12. Han B, Joo SW, Davis L (2007) Probabilistic fusion tracking using mixture kernel-based bayesian filtering. In: Proceedings 11th international conference on computer vision. Rio de Janeiro, Brazil
13. Han B, Joo SW, Davis LS (2011) Multi-camera tracking with adaptive resource allocation. Int J Comput Vision 91(1):45–58

14. Han B, Zhu Y, Comaniciu D, Davis L (2009) Visual tracking by continuous density propagation in sequential bayesian filtering framework. IEEE Trans Pattern Anal Mach Intell 31(5):919–930
15. Hoffmann C, Dang T (2009) Cheap joint probabilistic data association filters in an interacting multiple model design. Rob Auton Syst 57(3):268–278
16. Isard M, Blake A (1998) ICONDENSATION: unifying low-level and high-level tracking in a stochastic framework. In: Proceedings European conference on computer vision. Freiburg, Germany, pp 893–908
17. Julier S, Uhlmann J (1997) A new extension of the Kalman filter to nonlinear systems. Proc SPIE 3068:182–193
18. Kembhavi A, Schwartz W, Davis L (2008) Resource allocation for tracking multiple targets using particle filters. In: Worshop on visual surveillance, in conjunction with ECCV
19. Khan S, Shah M (2006) A multiview approach to tracking people in crowded scenes using a planar homography constraint. In: Proceedings European conference on computer vision, part IV. Graz, Austria, pp 133–146
20. Kim K, Davis L (2006) Multi-camera tracking and segmentation of occluded people on ground plane using search-guided particle filtering. In: Proceedings European conference on computer vision, part III. Graz, Austria, pp 98–109
21. Lawson CL, Hanson BJ (1974) Solving least squares problems. Prentice-Hall, New Jersey
22. Leichter I, Lindenbaum M, Rivlin E (2004) A probabilistic framework for combining tracking algorithms. In: Proceedings IEEE conference on computer vision and pattern recognition. Washington DC, pp 445–451
23. Mazor E, Averbuch A, Bar-Shalom Y, Dayan J (1998) Interacting multiple model methods in target tracking: a survey. IEEE Trans Aerosp Electron Syst 34(1):103–123
24. Mccane B, Galvin B, Novins K (2002) Algorithmic fusion for more robust feature tracking. Int J Comput Vision 49(1):79–89
25. von der Merwe R, Doucet A, de Freitas N, Wan E (2000) The unscented particle filter. Technical report CUED/F-INFENG/TR 380. Cambridge University Engineering Department, Cambridge
26. Mittal A, Davis LS (2003) M2tracker: a multi-view approach to segmenting and tracking people in a cluttered scene. Int J Comput Vision 51(3):189–203
27. Musicki D (2008) Bearings only multi-sensor maneuvering target tracking. Syst Control Lett 57(3):216–221
28. Musicki D, Scala BL (2008) Multi-target tracking in clutter without measurement assignment. IEEE Trans Aerosp Electron Syst 44(3):877–896
29. Okuma K, Taleghani A, de Freitas N, Little J, Lowe D (2004) A boosted particle filter: multitarget detection and tracking. In: Proceedings European conference on computer vision. Prague, Czech Republic
30. Pan P, Schonfeld D (2008) Adaptive resource allocation in particle filtering for articulated object tracking. In: Proceedings of the IEEE international conference on acoustics, speech, and signal processing. Las Vegas, Nevada, USA, pp 729–732
31. Perez P, Vermaak J, Blake A (2004) Data fusion for visual tracking with particles. Proc IEEE 92(3):495–513
32. Rui Y, Chen Y (2001) Better proposal distributions: object tracking using unscented particle filter. In: Proceedings IEEE conference on computer vision and pattern recognition, vol. II. Kauai, Hawaii, pp 786–793
33. Salmond D (1990) Mixture reduction algorithms for target tracking in clutter. In: SPIE signal and data processing of small targets, vol 1305. Orlando, Florida pp 434–445
34. Sherrah J, Gong S (2001) Continuous global evidence-based bayesian modality fusion for simulataneous tracking of multiple objects. In: Proceedings 8th international conference on computer vision. Vancouver, Canada
35. Siebel NT, Maybank SJ (2002) Fusion of multiple tracking algorithms for robust people tracking. In: Proceedings European conference on computer vision, Copenhagen, vol IV. Denmark, pp 373–387

36. Song C, Son J, Kwak S, Han B (2011) Dynamic resource allocation by ranking SVM for particle filter tracking. In: British machine vision conference. Dundee, UK
37. Spengler M, Schiele B (2003) Towards robust multi-cue integration for visual tracking. Mach Vis Appl 14(1):50–58
38. Tran S, Davis L (2007) Object tracking at multiple levels of spatial resolutions. In: Proceedings of international conference on image analysis and processing, pp 149–154
39. Triesch J, von der Malsburg C (2001) Democratic integration: self-organized integration of adaptive cues. Neural Comput 13(9):2049–2074
40. Vermaak J, Doucet A, Perez P (2003) Maintaining multi-modality through mixture tracking. In: Proceedings 9th internatinal conference on computer vision, vol II. Nice, France
41. Vermaak J, Gangnet M, Blake A, Perez P (2001) Sequential monte carlo fusion of sound and vision for speaker tracking. In: Proceedings 8th international conference on computer vision, vol I. Vancouver, Canada, pp 741–746
42. Williams J, Maybeck P (2003) Cost-function-based gaussian mixture reduction for target tracking. In: International conference on information fusion, vol 2, pp 1047–1054
43. Wu Y, Huang T (2001) A co-inference approach to robust visual tracking. In: Proceedings 8th international conference on computer vision, vol II. Vancouver, Canada, pp 26–33
44. Yang C, Duraiswami R, Davis L (2005) Fast multiple object tracking via a hierarchical particle filter. In: Proceedings 10th international conference on computer vision, vol. I. Beijing, China, pp 212–219
45. Zhong X, Xue J, Zheng N (2006) Graphical model based cue integration strategy for head tracking. In: Proceedings British machine vision conference. Edinburgh, UK

Traffic Pattern Analysis and Anomaly Detection via Probabilistic Inference Model

Hawook Jeong, Youngjoon Yoo, Kwang Moo Yi and Jin Young Choi

Abstract In this chapter, we introduce a method for trajectory pattern analysis through the probabilistic inference model with both regional and velocity observations. By embedding Gaussian models into the discrete topic model framework, our method uses continuous velocity as well as regional observations unlike the existing approaches. In addition, the proposed framework combined with Hidden Markov Model can cover the temporal transition of the scene state, which is useful in checking violation of the rule that some conflict topics (e.g., two cross traffic patterns) should not occur at the same time. To achieve online learning even with the complexity of the proposed model, we suggest a novel learning scheme instead of collapsed Gibbs sampling. The proposed two-stage greedy learning scheme is not only efficient at reducing the search space but also accurate in a way that the accuracy of online learning becomes not worse than that of the batch learning. To validate the performance of our method, experiments were conducted on various datasets. Experimental results show that our model explains satisfactorily the trajectory patterns with respect to scene understanding, anomaly detection, and prediction.

This chapter is the reduced version of the authors paper [12] with Copyright ©Springer-Verlag Berlin Heidelberg, 2014.

H. Jeong (✉) · Y. Yoo · J.Y. Choi
Perception and Intelligence Laboratory, ASRI Room 413, Bldg 133, Seoul National University, 599 Gwanak-ro, Gwanak-gu, Seoul, Korea
e-mail: hwjeong@snu.ac.kr

Y. Yoo
e-mail: i0you200@snu.ac.kr

J.Y. Choi
e-mail: jychoi@snu.ac.kr

K.M. Yi
Ecole Polytechnique Federale de Lausanne, EPFL, CVLAB, BC 309 (Btiment BC), Station 14, 1015 Lausanne, Switzerland
e-mail: kwang.yi@epfl.ch

C. Kyung (ed.), *Theory and Applications of Smart Cameras*,
KAIST Research Series, DOI 10.1007/978-94-017-9987-4_10

Keywords Trajectory pattern analysis · Anomaly detection in traffic · Probabilistic inference model · Topic model · Online inference learning · Scene understanding

1 Introduction

1.1 Objective and Contribution

Analyzing motion patterns and detecting abnormal activities are essential functions for intelligent surveillance. In most cases, moving objects follow specific motion patterns, for example, most cars and pedestrians move according to specific traffic rules. Abnormal events are defined as outliers that are far from the typical patterns (e.g., go straight, U-turn, turn right, etc.) following traffic rules. Hence detection of anomalies in this case becomes a process of finding motions which do not obey these rules. To achieve automatic detection of this anomaly without human labor, the surveillance system should learn the normal patterns in an unsupervised way from a large amount of crude data as shown in Fig. 1. Many researchers have proposed various learning models to discover the typical normal motion patterns from raw data in video [1, 7, 10, 11, 14, 17, 19, 32, 33].

Through analyzing strength and weakness of the existing works on unsupervised learning of motion patterns, we establish the following five requirements that the learning model should satisfy to work well in actual environments. First, the model should recognize regions showing normal movement patterns. The regions should be categorized into semantic regions representing typical activities (e.g., go straight upward, turn right, walk across the street, etc.). This is important for explaining the activities in an intersection, detecting intrusions of restricted areas, and detecting illegal U-turns. Second, the model should include not only direction information but also speed information for each activity region. This would increase the

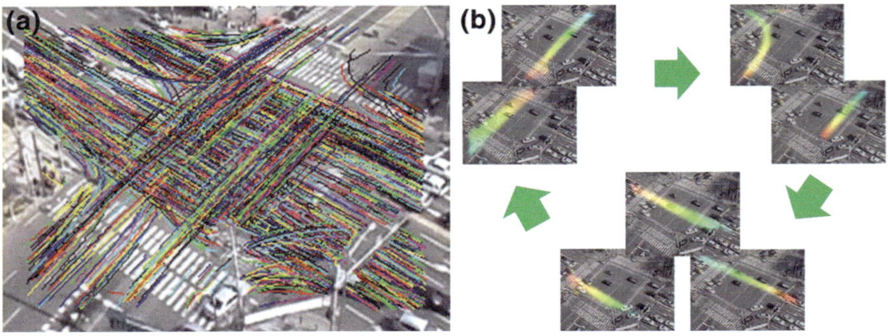

Fig. 1 An example of motion pattern analysis. **a** Crude motion data (unlabeled trajectories) in a surveillance scene. **b** Results of learning typical activities. The typical patterns are denoted with *red* and *blue* coloring, where objects move from *red* to *blue*. Some typical patterns occur at the same time, and their occurrences have temporal rules (best viewed in color)

discrimination ability of the model to detect abnormal patterns such as pedestrians walking along the path of vehicles, bikes running in pedestrian road, cars driving with over-speed, cars stopping in a railroad crossing, and so on. Third, spatio-temporal relationship between typical activity patterns needs to be considered. For instance, it is impossible for two straight movements, "moving from left to right" and "moving from top to bottom," to occur in an intersection at the same time. The model also needs to recognize the temporal order of activities such as governed by a traffic signal. Fourth, the algorithm should be robust to crowded scenes. In crowded scenes, it is hard to extract motions of individual objects. Even the current state-of-the-art methods for multi-object tracking [20, 29] are still limited for applying to the crowded scenes. Fifth, the model should be able to adapt itself to temporal changes of the scene (e.g., reversible lane, traffic volume changes). Online learning approach will not only enable the adaptation but also save memory and computational load because the model does not need to keep old data. A surveillance system running over months or even years, for example, would require an online model if it needs to keep running.

According to the authors' survey, there is no existing work satisfying all of the aforementioned requirements until now; the details on this issue will be described in related works of Sect. 2 and here we would give a brief mention. Object tracking based approach [1, 11, 17, 19, 33], whose observations are actual velocity from trajectories, can satisfy the first and second requirements but hardly fulfill the third and fourth requirements. On the other hand, the topic model based approach [7, 10, 14, 32], whose observations are quantized directions in a local region, are particularly useful for the first, third and fourth requirements. These kinds of observations, however, cannot deal with precise velocities (second requirement). Furthermore, most of the motion learning methods are restricted to offline learning not allowing to adapt to the changing situations (fifth requirement). The crowd motion approach [13, 21, 30] does not fulfill the first and third requirements since it is designed to understand only the crowd motion rather than typical motion paths.

In this article, we propose an approach to meet all of the aforementioned requirements for motion pattern analysis. This purpose is achieved through embedding the precise velocity pattern model, spatiotemporal pattern transition model, and the topic model into a probabilistic graphical framework. In particular, the newly defined continuous velocity model is distinctive from the existing models [7, 10, 14, 28, 31, 32], which do not provide satisfactory performance on the second requirement. In addition, to achieve online learning even with the enormous complexity of the proposed model, we suggest an efficient two-stage greedy learning method. The learning method of collapsed Gibbs sampling [8] restricts the existing models to offline learning. On the other hand, our learning method is designed to infer latent variables step by step in a greedy manner to reduce the search space. Moreover, the sub-model in each step is learned in a way that the online learning should not lose the learning capabilities shown in the offline learning. The whole learning process allows online adaptation of the model quickly and accurately. We evaluate our method on six datasets for activity pattern modeling and anomaly detection, showing that our method outperforms the state-of-the-art methods.

1.2 Related Works

One of the conventional approaches used for unsupervised activity analysis is to learn trajectory patterns through clustering. This approach defines distance measure between two different trajectories and groups similar ones together [11, 17–19, 33]. These methods use a similarity measure to group similar trajectories and can model trajectories in a whole path. Therefore, they can deal with the long-term characteristics of trajectories. However, these methods suffer from errors due to projective distortions and fragmentation of trajectories. In addition, the computation to obtain the distance for every pair of trajectories is heavy.

Another trajectory based approach learns the transition probabilities of each pixel to its nearby pixels using Gaussian mixture models (GMM) [1] or kernel density estimation (KDE) [23]. This approach statistically learns the velocities and the sizes of moving objects at each position. It is more invariant to scene variation and more robust to trajectory fragmentation than similarity based approach. However, these methods may fail to detect anomalies in regions where movements are diverse, such as the center of an intersection. In such situations, the trained model would count all patterns as normal because it is not fully aware of mutual dependence among trajectories; that is, it does not handle spatiotemporal relationship among typical activity patterns (third requirement). Moreover, object tracking based methods have difficulty in extracting individual object trajectories in a crowded scene due to the inevitable tracking failures.

On the other hand, optical flow based methods have been proposed recently to overcome the problem of object tracking failure in a crowded scene. These methods adopt mixture of Gaussians [22], sparse coding [35], Markov random field [2], dynamic textures [16], probabilistic topic models [7, 10, 14, 28, 32], and so on. In particular, the topic models have been prevalently employed to learn motion patterns because they can well discover typical activities using co-occurrence property. The Dual Hierarchical Dirichlet Process (Dual-HDP) [32] discovers typical activity patterns by modeling spatial relation of activities. Markov Clustering Topic Model (MCTM) [10] additionally considers temporal relationships between activities, and Dependent Dirichlet Process Hidden Markov Model (DDP-HMM) [14] solves the same problem in a non-parametric manner. However, the above methods ignore the temporal order of low-level motion features, which leads to incomplete modeling of long-term path. This approach has been extended by considering the temporal information inside the topic [7, 28]. Nevertheless, all of these topic model based approaches cannot completely address the precise velocity of a whole trajectory since they only use quantized directions obtained from optical flows in a local cell (i.e., it does not fulfill the second requirements). Moreover, the collapsed Gibbs sampling, which is commonly utilized for learning of the topic models, is not only ineffective in dealing with a large solution space of a complex model but also restricted to offline learning making it unable to adapt to a changing situation (i.e., it does not fulfill the fifth requirements).

Crowd motion analysis [13, 21, 30] has also been conducted to detect strange motion patterns in an extreamly crowded scene. Probabilistic Crowd Model [21] allows objects to be tracked even in extremely crowded scenes, and local spatio-temporal motion pattern [13, 30] is modeled in the dense crowded scenes. These methods, however, allow their model to understand only the crowd motion rather than typical motion paths (i.e., it does not fulfill the first and the third requirements). Hence, this approach is not suitable for the task of deducing traffic rules though it gives good performance on anomaly detection in the crowded scene.

2 Proposed Approach

Figure 2 shows the schematic diagram of the proposed framework. We first apply a simple background subtraction [24] to extract foreground map and detect corner points on the foreground pixels. We perform KLT [25] on these corner points to extract trajectories. By using the KLT trajectories, we can reduce the tracking error in a crowded scene because KLT tracks corner points, which are relatively easier to track than each object in a crowded scene. Of course, the tracking of corner points under the far-field view may generate broken trajectories. Despite the broken trajectories, our framework can cope with this problem by considering co-occurrence property of many corner point trajectories. After KLT tracking, consequent trajectories are collected during a time interval. The trajectories in the same time interval compose a *collection* that is a mixture of diverse activities. The dozens of trajectory collections are piled as in Fig. 2, and a recent set of collections is used as an input to the proposed inference model for online update.

The proposed inference model is formulated in a probabilistic graphical framework including trajectory pattern model, spatiotemporal relation of trajectories, and velocity model of each trajectory. To infer this model in online manner, instead of exact inference, an approximate method is proposed by two-stage greedy inference with three sub-models of trajectory clustering, spatiotemporal dependency modeling, and velocity learning. Lastly, the recently observed scene is tested by the trained model to detect anomalies in the current scene.

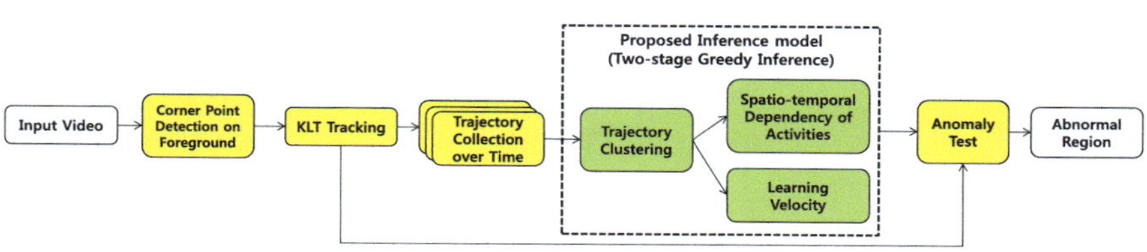

Fig. 2 Overall scheme of the proposed method

2.1 Probabilistic Inference Model

In this section, we describe the proposed model denoted with green in Fig. 2. The main frame of our approach is topic model such as Latent Dirichlet Allocation (LDA) [4], which is proposed for analysis of relationships between a set of documents and words in the documents. In this approach, the frequency of occurrence of each word in a document is used as a feature to train the model. For example, a word "relativity" tends to *co-occur* with words such as "Einstein", "energy", "gravity", "universe" in each document, so a set of the words is interpreted by the viewer as the physics-related topic. Because of the ability of co-occurrence modeling, LDA is adopted as a baseline of many motion pattern learning frameworks [7, 10, 14, 28, 32]. In these works, quantized local motions are treated as words, a set of the local motions in a video clip is treated as a document, and the topic can be treated as typical motion patterns.

In our approach, we also have to define variables corresponding to "word", "document", and "topic" in the topic model literatures. We define "words" as grid cells dividing a scene, where all of the cells in a scene have the same height and width. Instead, we newly define the velocity of trajectory (details are defined in the following), which can handle not only quantized direction inside a cell but also long-term actual velocity over dozens of frames. The trajectory is denoted by a set of grid cells as in Fig. 3 and velocity vector defined as in Fig. 4. A "document" in the topic model corresponds to a collection of trajectories defined by a set of trajectories collected in a time interval. The trajectories are categorized into multiple typical patterns (topics), referred to as trajectory patterns (e.g., turn left from south to west, gostraight downwards, etc.).

The indexed variables for the proposed model are defined as following. The index of i-th cell of j-th trajectory in t-th collection of trajectories is denoted by $c_{tji} \in \{1, 2, \ldots, C\}$, where C is the number of grid cells in a scene. As depicted in Fig. 4, the velocity vector $v_{tjif} \in \mathbb{R}^2$ is defined as a relative vector from a point in the i-th cell on the j-th trajectory to the point at the frame of f-steps ahead.

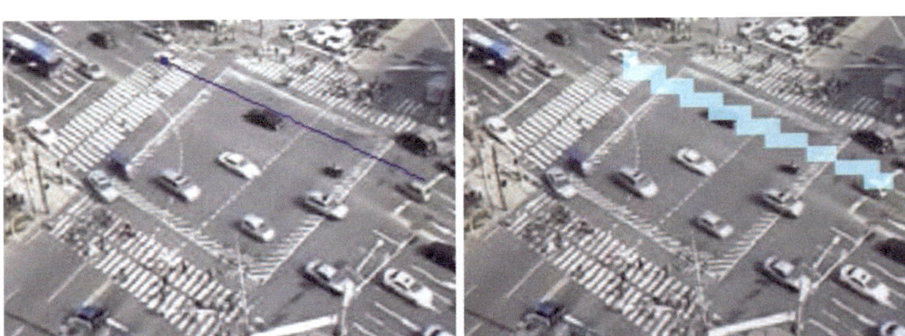

Fig. 3 Example of a single trajectory corresponding with a set of cells

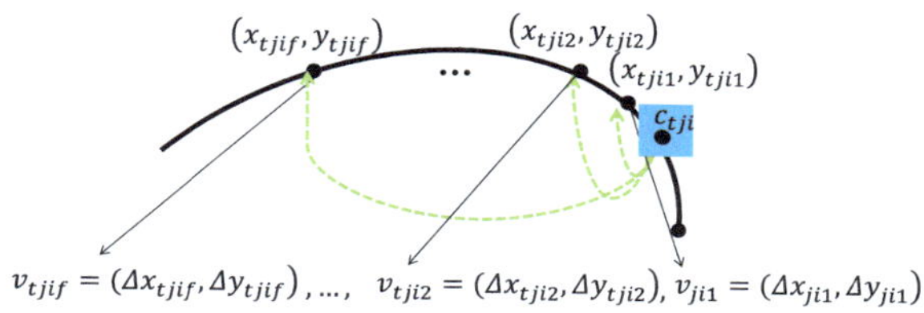

Fig. 4 Synthetic trajectory with marked points and relative vectors from origin coordinate in cell c_{tji}

Following the above definition of variables, observed trajectories in the collection of the t-th time interval can be expressed by a set of cells $\{c_{tji}\}_{i=1,j=1}^{N_{tj},M}$ and a set of velocity vectors $\{v_{tjif}\}_{f=1,i=1,j=1}^{F_{tji},N_{tj},M}$, where M is the number of trajectories in the collection, N_{tj} is the number of cells where the j-th trajectory passes, and F_{tji} is the maximum value of f according to the length of the observed trajectory. We also define a design parameter F, acting as the maximum possible value for F_{tji}.

The state of t-th collection $s_t \in \{1, 2, \ldots, S\}$ is a set of trajectory patterns that can occur at the same time, such as a vertical moving state (a mixture of go straight upwards and downwards) governed by a traffic light. The sequence of the state s_t is modeled so that it transits from one state to another over time, according to multinomial distribution with transition probability matrix π as follows:

$$s_t | s_{t-1} \sim \text{Multi}(\pi_{s_{t-1}}), \tag{1}$$

For this example, the sequence of states $\{s_t\}$ is formed according to the change of a traffic signal as time passes. The constant S is a design parameter determining the number of states, usually selected to 2 or 3 according to the traffic changes in an intersection case. If the state s_t is given, the distribution of topic occurrence in the state can be determined. The topic occurrence probability vector for t-th collection is defined by $\theta_t \in \mathbb{R}^K$, where K is a design parameter that stands for the number of typical trajectory patterns in a scene. The θ_t is represented with a histogram that must sum to 1, and the distribution of θ_t is assumed to be Dirichlet distribution with given parameter α, i.e.,

$$\theta_t | s_t, \alpha \sim \text{Dir}(\alpha_{s_t}). \tag{2}$$

The θ_t is used as the parameter of multinomial distribution over the K trajectory patterns (topics) for the t-th collection. The trajectory pattern of the j-th trajectory in the t-th collection is denoted with $z_{tj} \in \{1, 2, \ldots, K\}$, which is assumed to follow a multinomial distribution with the parameter θ_t, i.e.,

$$z_{tj} | \theta_t \sim \text{Multi}(\theta_t). \tag{3}$$

We design the cell c_{tji} to be generated by a multinomial distribution with the parameters $\phi_{z_{tj}} \in \mathbb{R}^C$ being related to the pattern z_{tj}, given by:

$$c_{tji}|z_{tj}, \phi \sim \mathrm{Multi}(\phi_{z_{tj}}). \tag{4}$$

The multinomial parameter $\phi_k \in \mathbb{R}^C$ in Eq. (4) holds information about which cell has high probability to appear in the k-th trajectory pattern. The distribution of ϕ_{ztj} is assumed to be Dirichlet distribution with parameter β, i.e.,

$$\phi_k|\beta \sim \mathrm{Dir}(\beta). \tag{5}$$

Also, the velocity vector v_{tjif} is modeled to be drawn from a Gaussian distribution with its mean $\mu_{c_{tji}z_{tjf}}$ and variance $\Sigma_{c_{tji}z_{tjf}}$ as follows:

$$v_{tjif}|z_{tj}, c_{tji}, \mu, \Sigma \sim \mathcal{N}(\mu_{c_{tji}z_{tjf}}, \Sigma_{c_{tji}z_{tjf}}). \tag{6}$$

Using variables and their dependence defined in the above, the overall model to consider trajectory patterns (topics), velocity patterns of the trajectories, and spatiotemporal transition patterns of the states is graphically represented as shown in Fig. 5. The figure can be interpreted in a top-down order through the generative process [4], where the nodes denote random variables, and the edges denote possible dependence between random variables.

The primary goal of our framework is to infer the latent variables and parameters from the given observations $\{c_{tji}\}$ and $\{v_{tjif}\}$ in a surveillance video through an online unsupervised learning scheme.[1] This task can be done by posterior inference, which can be regarded as a reversal of the generative process that the graphical model illustrates. The posterior inference for all latent variables $s, \phi, \theta, z, \mu, \Sigma$ given the observations c, v and hyper-parameters α, β is as follows:

$$s^*, \phi^*, \theta^*, z^*, \mu^*, \Sigma^* = \underset{s,\phi,\theta,z,\mu,\Sigma}{\arg\max}\, p(s, \phi, \theta, z, \mu, \Sigma|c, v, \alpha, \beta), \tag{7}$$

where,

$$p(s, \phi, \theta, z, \mu, \Sigma|c, v, \alpha, \beta) = \frac{p(s, \phi, \theta, z, \mu, \Sigma, c, v|\alpha, \beta)}{p(c, v|\alpha, \beta)}. \tag{8}$$

The numerator on the right-hand side in Eq. (8) corresponds to a joint probability distribution represented by the proposed model. Also, using the chain rule and

[1]To concisely represent notations, the set notation $\{\cdot\}$ without the range of index is defined as a set of variables containing all possible indices. Also, the variables without indices imply that they deal with all possible indices, such as,

$$c = \{c_{tji}\} = \{c_{tji}\}_{t=1,j=1,i=1}^{T,M,N_j}, p(s) = p(\{s_t\}_{t=1}^{T}) = \prod_{t=1}^{T} p(s_t)..$$

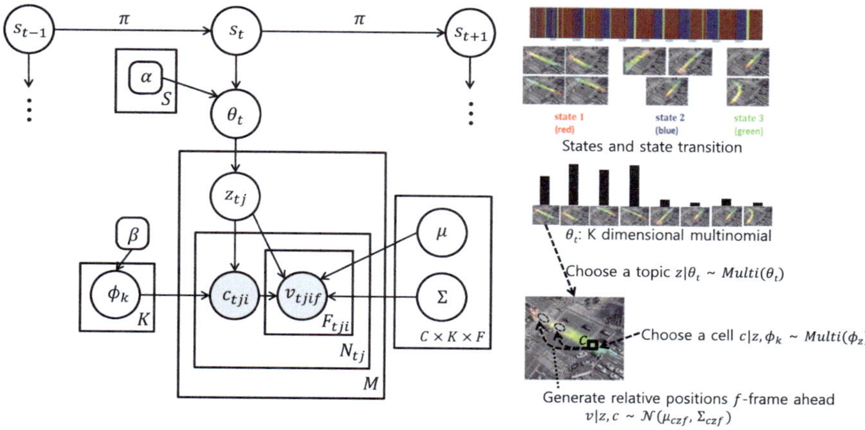

Fig. 5 Graphical representation of the proposed model. The hidden variables are *unshaded* and the observed variables are *shaded*. The *rectangles* are plate notation which denotes replication

assumptions of independence among variables, the joint probability can be factorized into Eq. (9), which consists of the probability distributions defined in Eq. (1)–(6).

$$p(s, \phi, \theta, z, \mu, \Sigma, c, v|\alpha, \beta) = \left(\prod_{k=1}^{K} p(\phi_k|\beta)\right)$$
$$\prod_{t=1}^{T} p(s_t|s_{t-1})p(\theta_t|s_t, \alpha) \prod_{j=1}^{M} p(z_{tj}|\theta_t) \tag{9}$$
$$\prod_{i=1}^{N_{tj}} p(c_{tji}|z_{tj}, \phi) \prod_{f=1}^{F_{tji}} p(v_{tjif}|z_{tj}, c_{tji}, \mu, \Sigma).$$

The learning of distribution parameters $(\phi, \theta, \mu, \Sigma)$ for the proposed model can be achieved by maximizing the probability $p(s, \phi, \theta, z, \mu, \Sigma, c, v|\alpha, \beta)$ with latent variables s, ϕ, θ, z, μ, Σ to be inferred under the given observations c, v and the hyper-parameters α, β. However, the exact inference is intractable due to non-convexity of the joint probability function and a tremendous search space caused by calculating the joint probability for all possible configurations of the latent variables to find the best case. Instead of exact inference, we propose an approximate inference method that will be presented in the Sect. 2.2.1.

As for an application of inference results of the proposed model, anomaly detection can be performed. Using the distribution parameters $\mu, \Sigma, \phi, \theta$ inferred from the learning phase and the current observations $\{c_{t'ji}\}$, $\{v_{t'jif}\}$ at the current time t',[2] the a state $s^*_{t'}$ and a topic assignment $z^*_{t'j}$ for each trajectory j are estimated by maximizing a posterior:

$$s^*_{t'}, \{z^*_{t'1}, z^*_{t'2}, .., z^*_{t'M}\} = \arg\max_{s_{t'}, \{z_{t'j}\}} \left[p(s_{t'}, \{z_{t'j}\} | \{c_{t'ji}\}, \{v_{t'jif}\}, \mu, \Sigma, \phi, \theta, \alpha, \beta)\right]. \tag{10}$$

Here,

$$p(s_{t'}, \{z_{t'j}\} | \{c_{t'ji}\}, \{v_{t'jif}\}, \mu, \Sigma, \phi, \theta, \alpha, \beta) = \frac{p(s_{t'}, \{z_{t'j}\}, \{c_{t'ji}\}, \{v_{t'jif}\}, \mu, \Sigma, \phi, \theta | \alpha, \beta)}{p(\{c_{t'ji}\}, \{v_{t'jif}\}, \mu, \Sigma, \phi, \theta, \alpha, \beta)}. \tag{11}$$

The denominator of Eq. (11) is constant to the variation of optimization variables s, z, so it is enough to maximize the numerator (joint probability) of Eq. (11) to achieve Eq. (10). Therefore, the joint probability in Eq. (9) can substitute for the posterior in Eq. (10) by fixing $t = t'$ and removing $\prod_{t=1}^{T}$. The observations are extracted from trajectories of the current frame and $j \in [1, M], i \in [1, N_j]$, $f \in [1, F_{ji}]$. Indeed, if the joint probability $p(s^*_{t'}, \{z^*_{t'j}\}, \{c_{t'ji}\}, \{v_{t'jif}\}, \mu, \Sigma, \phi, \theta | \alpha, \beta)$ in Eq. (9) has low value even with the optimal $s^*_{t'}, \{z^*_{t'j}\}$, the current scene is decided to be abnormal. However, as in case of model learning, exact inference of Eq. (10) is intractable. The details for anomaly detection with approximate method are described in Sect. 2.2.2.

2.2 *Two-Stage Greedy Inference*

2.2.1 Model Learning

An exact learning of the proposed model by maximizing the joint probability Eq. (9) is intractable because of the aforementioned reasons in the previous section. Hence, many conventional methods using various topic models [7, 10, 14, 31, 32] commonly employ collapsed Gibbs sampling (CGS) for an approximate learning of the models. CGS is a popular Markov Chain Monte Carlo (MCMC) approach for topic model learning. However, on the results of online MCMC learning for topic models [5], the results have shown that online MCMC learning is inferior to the offline learning. According to [34], in case of distributed processing for the learning of the topic models, variational inference (VI) [3, 4] gives better results than CGS.

---

[2]Because the anomaly detection task should be performed for every frame, we compose t'-th trajectory collections from the trajectories on the current frame.

To achieve an online learning of the proposed topic model, a large set of the trajectory collections for the offline learning needs to be separated by time. Because each separated set of the collections can be an input to the distributed processing, VI method can be a better option for the online learning of our model than CGS. VI assumes that each variational distribution used to approximate the posterior and to treat each document (in our case, collection of trajectories) is independent. For this reason, it is difficult to apply VI directly to our model because the model has the states for each collection which is dependent on the previous state. Moreover, inferring all latent variables all together is not efficient to real-time computation in terms of a search space.

In our greedy inference approach, in order to directly apply VI to the proposed model in Fig. 5, we utilize the fact that the state s_t is hardly changed in a short time for the online inference; thus, θ_t can be inferred without knowing the current state s_t. Also, to reduce the search space for the solution, we assume that each velocity pattern μ, Σ in a cell c of each typical pattern z is inseparable. On the assumption, we can find the typical patterns z based on the cells c at first, and then velocity patterns are mined on the regions of each typical pattern. This assumption is reasonable from the fact that activity regions c are more susceptible to the typical pattern z than precise velocity v. As a result, three simple sub-models are obtained as shown in Fig. 6. The first sub-model in Fig. 6a is the same graphical model of Latent Dirichlet Allocation (LDA) [4], so it is straightforward to adopt online VI [9] to the sub-model. If latent variables z and θ are given from the learning of the first

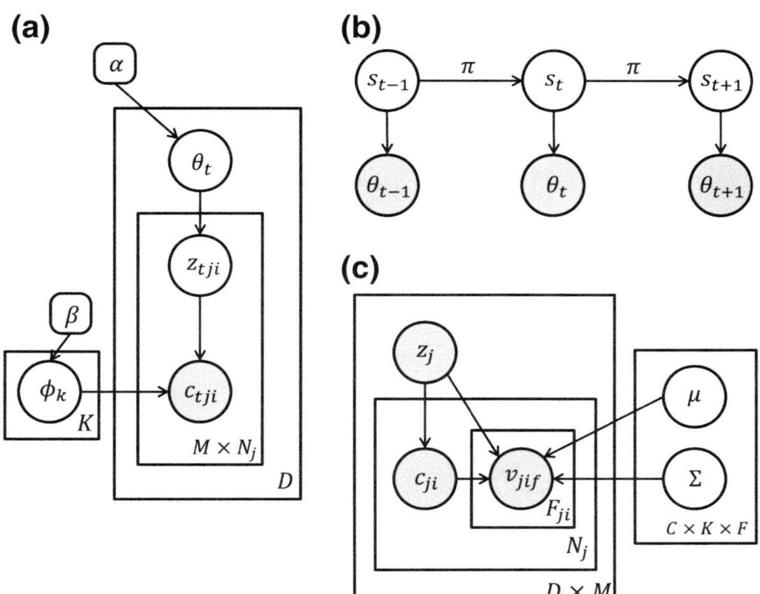

Fig. 6 Three sub-models for two-stage learning

sub-model in Fig. 6a, remaining latent variables $\{s_t\}$ and $\{\mu_{ckf}, \Sigma_{ckf}\}$ in Fig. 5 are conditionally independent by d-separation property [3]. In other words, $\{s_t\}$ does not influence $\{\mu_{ckf}, \Sigma_{ckf}\}$ and vice versa for the given z and θ. Therefore, we can reasonably optimize the sub-model of the first stage and then use these results to optimize the remaining two sub-models in Fig. 6b, c in a greedy manner.

First, we optimize ϕ, θ, and z of the first sub-model in Fig. 6a using LDA. The LDA can be used to cluster trajectories effectively, since it is robust to broken trajectories using the co-occurrence property. To be specific, because the collection is composed of concurrent trajectories in short time duration, the LDA can cluster co-occurring cells (words) in trajectory collections (documents) into the same trajectory patterns (topics). Using the inference result in the first stage, we use $\{\theta_t\}$ as observations to infer hidden variables $\{s_t\}$ and state transition matrix π in Fig. 6b. In addition, the pattern assignments of each trajectory z inferred in the first stage is also used as observations to infer Gaussian parameters per cell c, typical pattern k, and frame f in Fig. 6c. By this procedure, the search space to solve the complex model can be reduced effectively. Detailed description for each sub-model is presented in the following.

A. **Online Trajectory Clustering**

Leaning of the first sub-model takes a role of online trajectory clustering. For the online processing, the entire T collections of trajectories for the proposed model in Fig. 5 should be separated into a small set of collections by time. The small set that consists of the D collections is used as an input for the mini-batch learning whose results allow the model to be updated online. In other words, D is the number of collections for the mini-batch, so $\frac{T}{D}$ times of mini-batches should be performed for the whole video. Because the proposed model in Fig. 5 is assumed to be divided by ignoring the dependence between s and θ and between z and v, the full joint probability of the proposed model in the Eq. (9) can ignore $p(s_t|s_{t-1})$, $p(v_{tjif}|z_{tj}, c_{tji}, \mu, \Sigma)$ and can approximate $p(\theta_t|s_t, \alpha) \approx p(\theta_t|\alpha)$. Thus, the objective function of each mini-batch and joint probability of the first sub-model for the D collections is given by:

$$\phi^*, \theta^*, z^* = \underset{\phi,\theta,z}{\arg\max}\, p(\phi, \theta, z|c, \alpha, \beta), \tag{12}$$

where,

$$\begin{aligned} p(\phi, \theta, z, c|\alpha, \beta) &= \left(\prod_{k=1}^{K} p(\phi_k|\beta)\right) \\ \prod_{t=1}^{D} p(\theta_t|\alpha) \prod_{j=1}^{M} p(z_{tj}|\theta_t) &\prod_{i=1}^{N_{tj}} p(c_{tji}|z_{tj}, \phi). \end{aligned} \tag{13}$$

Also, in order to make Eq. (13) to be the same as the joint probability of LDA, the topic assignment z_{tj} for each trajectory is changed to be assigned for each cell (i.e. z_{tji}), and then z_{tj} is obtained by post-inference using z_{tji}.
To find the optimal points of Eq. (12) in online manner, we use online variational inference (VI) using mini-batch LDA using the small set of D collections coming in as time goes on. For details, refer to [4, 9]. After finding the optimal points, we get z^*_{tji} indicating the topic assignment of each cell as shown in Fig. 6a. This result cannot be directly used in the next stage because the inference result of the full model (of Fig. 5) is the latent variable z^*_{tj} indicating the most typical pattern of the j-th trajectory among the K clustered patterns. To resolve the incompatibility, we consider the mode of the inference results of the first sub-model as the results of the original model. For example, if we have $\{z^*_{tj1}, z^*_{tj2}, z^*_{tj3}, \ldots, z^*_{tjN_j}\}$ and $\{c_{tj1}, c_{tj2}, c_{tj3}, \ldots, c_{tjN_j}\}$ for a j-th trajectory in t-th collection, then we assign z^*_{tj} as

$$z^*_{tj} = \text{Mode}\{z^*_{tji}\}_{i=1}^{N_{tj}}. \tag{14}$$

This is a reasonable assignment since choosing the mode would give least error with respect to maximum likelihood estimation [6].

B. Spatiotemporal Dependency of Activities

The spatiotemporal relationship among the typical patterns is represented in Fig. 6b. From the set $\left\{z^*_{tj}\right\}_{j=1}^{M}$ obtained in the first stage inference, θ^*_t per trajectory collection is also obtained. Given a set of histogram $\{\theta^*_t\}_{t=1}^{D}$, where D is the number of collections, we partition the D observations into S sets $\{\Theta_1, \Theta_2, \ldots, \Theta_S\}$. The objective function to minimize is the within-cluster sum of squares:

$$\underset{\{\Theta_n\}_{n=1}^{S}}{\arg\min} \sum_{n=1}^{S} \sum_{\overline{\theta}^*_t \in \Theta_n} \left\| \overline{\theta}^*_t - m_n \right\|^2, \tag{15}$$

where $m_n \in \mathbb{R}^K$ is the mean of vectors in a set Θ_n and $\{\overline{\theta}^*_t\}$ is the dimension-wise normalized version of $\{\theta^*_t\}$. In the normalization, different observation frequencies in topics are set to the same scale. To minimize the objective function, we perform K-means clustering with $K = S$. Then with the clustering results, we obtain the cluster indices $\{s^*_t\}_{t=1}^{D}$ for all $\{\theta^*_t\}_{t=1}^{D}$, where $s^*_t \in \{1, 2, \ldots, S\}$ corresponds to cluster index of θ^*_t. The state transition matrix π also can be obtained by counting the frequency of transition in the cluster indices. The parameter m_n implies general patterns about spatial co-occurrences of trajectory patterns, such as cars are moving horizontally (m_1) or cars are moving vertically (m_2). The m_n is also used to estimate a current state at the anomaly test phase.

In the online process, only $\{\theta_t^*\}_{t=1}^D$ residing inside a sliding time window is kept so that the model adapts to the changes in time. A size of the sliding window is designed to be bigger than the size of mini-batch for online-LDA in order to increase the clustering performance. As *K*-means performance depends much on initialization, we perform this multiple times with random initial conditions and use the best result. As the *K*-means algorithm is very fast, it scarcely affects entire computational time of the proposed method.

C. Velocity Learning

As in Fig. 6c, given clustered trajectory information $\{z_{tj}^*\}$ and the observations $\{c_{tji}\}$ and $\{v_{tjif}\}$, Gaussian models learn velocities of the trajectory. The velocities can be modeled for each pixel in the scene, but it is a waste of memory and needs extremely large amount of data. Assuming adjacent pixels in the scene have similar motions, we learn these motions based on each cell. In our modeling scheme, Gaussian models exist not only for each cell but also for each typical pattern. Therefore, since multiple typical patterns may exist for the same cell, multiple Gaussian models may exist to describe the complex motions of a single cell. An example of this case would be a cell in the center of an intersection. The Gaussian model learns the statistical information about the position of a trajectory at *f* frame before. Figure 4 is an illustration of obtaining the relative vector $v_{tjif} \in \mathbb{R}^2$ for a trajectory. Then for each Gaussian model, we update the Gaussian parameters $\mu \in \mathbb{R}^2$ and $\Sigma \in \mathbb{R}^{2\times 2}$ with each trajectory using the typical moving average concept. To avoid the model from being overly stiff, we keep lower bound for the learning rate.

2.2.2 Anomaly Detection

The optimization problem of Eq. (10) for anomaly detection is related to find the most appropriate $s_{t'}$, $z_{t'j}$ from the observations $\{c_{t'ji}\}$, $\{v_{t'jif}\}$ and the distribution parameters obtained through learning procedure in Sect. 2.2.1. The distribution parameters are assumed to be fixed in the anomaly detection phase. Since the computational complexity for exact inference for Eq. (10) is heavy with complexity of $O(SK^M)$, we present approximate inference method. For the approximation, we make two assumptions: 1) the typical pattern (topic) of each trajectory is independent from others in a state; 2) activity regions c are more dominant to determine the typical pattern than precise velocity v. Using the first assumption, we can estimate the topic assignment $z_{t'j}$ of *j*-th trajectory without knowing the current state $s_{t'}$; thus, $z_{t'j}$ is not dependent on $s_{t'}$, $\theta_{t'}$. The second assumption makes the dependence between z and v to be ignored; thus μ and Σ can be also ignored. Using the assumptions, a posterior of topic assignment $z_{t'j}$ can be approximately computed by only given regional observations c and the learned multinomial parameters ϕ as follows:

$$p(z_{t'j}|\{c_{t'ji}\}_{i=1}^{N_j}, \{v_{t'jif}\}_{i=1,j=1}^{N_{t'j},F_{t'ji}}, \mu, \Sigma, \phi, \theta, \alpha, \beta) \approx p(z_{t'j}|\{c_{t'ji}\}_{i=1}^{N_{t'j}}, \phi). \tag{16}$$

Also, the approximate posterior can be factorized into likelihood and a prior by Bayes' rule,

$$p(z_{t'j}|\{c_{t'ji}\}_{i=1}^{N_{t'j}}, \phi) \propto p(\{c_{t'ji}\}_{i=1}^{N_{t'j}}|z_{t'j}, \phi)p(z_{t'j}|\phi). \tag{17}$$

Because the likelihood $p(\{c_{t'ji}\}_{i=1}^{N_{t'j}}|z_{t'j}, \phi)$ follows multinomial distribution defined as in Eq. (4) and the prior is uniform, we can find the proper topic assignment $z^*_{t'j}$ given by:

$$z^*_{t'j} = \underset{k\in\{1,\ldots,K\}}{\arg\max}\left[p(\{c_{t'ji}\}_{i=1}^{N_{t'j}}|z_{t'j}, \phi_k)\right]. \tag{18}$$

Likewise, the state $s^*_{t'}$ is estimated by utilizing $\{m_n\}_{n=1}^{S}$ obtained in Eq. (15) and the K-dimensional histogram $\theta^*_{t'}$ calculated from the frequency of $\{z^*_{t'j}\}_{j=1}^{M}$ as follows:

$$s^*_{t'} = \underset{s\in\{1,\ldots,S\}}{\arg\min}\left\|\theta^*_{t'} - m_s\right\|. \tag{19}$$

As a result, the computational complexity of the posterior optimization in Eq. (10) can be reduced from $O(SK^M)$ into $O(KM) + O(S)$ via the proposed approximation.

By using the estimated $s^*_{t'}$ and $\{z^*_{t'1}, z^*_{t'2}, \ldots, z^*_{t'M}\}$, we can assume all latent variables are given, so the observations $\{c_{t'ji}\}$ and $\{v_{t'jif}\}$ are tested based on the trained model in reverse:

$$p(\{c_{t'ji}\}, \{v_{t'jif}\}|s^*_{t'}, \{z^*_{t'j}\}, \mu, \Sigma, \phi, \theta, \alpha, \beta) \propto p(\{c_{t'ji}\}, \{v_{t'jif}\}, s^*_{t'}, \{z^*_{t'j}\}, \mu, \Sigma, \phi, \theta|\alpha, \beta). \tag{20}$$

The right-hand side of Eq. (20) can be factorized into the six pre-defined distributions Eq. (1–6) by conditional independence as in case of Eq. (9). In fact, the probability of learning parameters $p(\phi_k|\beta)$, $p(\theta^*_{t'}|s^*_{t'}, \alpha)$ does not have influence on the Eq. (20). Thus, we check the remaining four conditions in Eqs. (1, 3, 4, 6) to decide whether the current state or each trajectory is normal or not:

(a) For the current state, $p(s^*_{t'}|s^*_{t'-1})$ defined in Eq. (1) is tested using the state transition matrix π and the given the previous state $s^*_{t'-1}$. It is to examine the temporal relation among the typical patterns of trajectories.

(b) For the topic assignment $z^*_{t'j}$ of j-th trajectory in the current scene, $p(z^*_{t'j}|m_{s^*_{t'}})$ defined in Eq. (3) is tested. Even though each trajectory is assumed to be independent of others when the inference of Eq. (10) is approximated, after estimating the dominent current state $s^*_{t'}$, an abnormal trajectory violating the current state can be detected. It can consider the spatial relation among the typical patterns of trajectories.

(c) For a set of cells $\{c_{t'ji}\}_{i=1}^{N_{t'j}}$ passed by j-th trajectory, $p\left(\{c_{t'ji}\}_{i=1}^{N_{t'j}}|z^*_{t'j}, \phi\right)$ defined in Eq. (4) is tested given the topic assignment $z^*_{t'j}$. It is to examine the overall path of the trajectory.

(d) For a set of velocities $\{v_{t'jif}\}_{i=1,f=1}^{N_{t'j},F_{t'ji}}$ obtained as in Fig. 4, the conditional probability $p\left(\{v_{t'jif}\}_{i=1,f=1}^{N_{t'j},F_{t'ji}}\middle|z^*_{t'j}, \{c_{t'ji}\}_{i=1}^{N_{t'j}}, \mu, \Sigma\right)$ in Eq. (6) is tested. It is to detect an trajectory with abnormal speed although its overall path is similar to one of the typical patterns.

If the current state has low probability on the condition (a), the state of the current frame is decided to be abnormal. Also, a trajectory that has low probability under at least one of the conditions (b)–(d) is determined to be abnormal; thus, a cell containing current position of the abnormal trajectory is regarded as an abnormal region.

2.3 Summary of the Proposed Method

Given the observations defined in the Sect. 2.1, the proposed inference method can be summarized as **Algorithms 1** and **2**.

Algorithm 1 Two-stage Greedy Inference (Model Learning)

Input: $\{c_{tji}\}_{i=1,j=1,t=1}^{N_{tj},M,T}$, $\{v_{tjif}\}_{f=1,i=1,j=1,t=1}^{F_{tji},N_{tj},M,T}$ ▷ T is the total number of trajectory collections in the video.

Output: $\{s_t\}, \{\phi_k\}, \{\theta_t\}, \{z_{tji}\}, \{\mu_{ckf}\}, \{\Sigma_{ckf}\}, \{m_n\}$ for all indices.

1: **for** $i \leftarrow 1, \ldots, \frac{T}{D}$ **do** ▷ D is the number of collections for the mini-batch. (In our case, D = 10)
2: For each set of collection for the mini-batch, optimize
3: $\phi^*, \theta^*, z^* = \underset{\phi,\theta,z}{\arg\max}\, p(\phi, \theta, z|c, \alpha, \beta)$ in Eq.(12)
4: Find a topic assignment, $z^*_{tj} = Mode\{z^*_{tji}\}_{i=1}^{N_{tj}}$ by Eq.(14).
5: Using the given θ^*, optimize
6: $\underset{\{\Theta_n\}_{n=1}^S}{\arg\min} \sum_{n=1}^{S} \sum_{\overline{\theta}^*_t \in \Theta_n} \left\| \overline{\theta}^*_t - m_n \right\|^2$ using K-means.
7: Then we obtain $\{s^*_t\}$ and $\{m_n\}$.
8: Using the given z^* from Eq.(14) and observations c and v,
9: update Gaussian parameters of μ and Σ.
10: **end for**

Algorithm 2 Anomaly test

Input: Observations $\{c_{t'ji}\}_{i=1,j=1}^{N_{tj},M}$, $\{v_{t'jif}\}_{f=1,i=1,j=1}^{F_{tji},N_{tj},M}$ and distribution parameters $\{\phi_k\},\{\mu_{ckf}\},\{\Sigma_{ckf}\},\{m_n\}$.
Output: Indices of abnormal trajectory $j \in \{1,...,M\}$.
for every current frame t' **do**
  for $j \leftarrow 1,\ldots,M$ **do**
    $z^*_{t'j} = \arg\max_{k\in\{1,\ldots,K\}} \left[p(\{c_{t'ji}\}_{i=1}^{N_{t'j}}|z_{t'j},\phi_k)\right]$
  end for
  Calculate, $\theta^*_{t'} = histogram\left(\{z^*_{t'j}\}_{j=1}^{M}\right)$
  $s^*_{t'} = \arg\min_{s\in\{1,\ldots,S\}} \left\|\theta^*_{t'} - m_s\right\|$
  Using the estimated $s^*_{t'}$ and $\{z^*_{t'j}\}_{j=1}^{M}$,
  Test $p(s^*_{t'} \mid s^*_{t'-1})$ defined in Eq.(1)
  for $j \leftarrow 1,\ldots,M$ **do**
    Following three probabilities are calculated and compare with the threshold:
    $p(z^*_{t'j}|m_{s^*_{t'}})$ defined in Eq.(3)
    $p\left(\{c_{t'ji}\}_{i=1}^{N_{t'j}}|z^*_{t'j},\phi\right)$ defined in Eq.(4)
    $p\left(\{v_{t'jif}\}_{i=1,f=1}^{N_{t'j},F_{t'ji}}\middle|z^*_{t'j},\{c_{t'ji}\}_{i=1}^{N_{t'j}},\mu,\Sigma\right)$ defined in Eq. (6)
  end for
end for

3 Experiments

We have done experiments on six different videos to analyze motion patterns and to detect abnormal activities. The MIT dataset is from [32], the QMUL Junction dataset is from [10], Wide Intersection (WI) video is our own dataset of an eight-lane road with heavy traffic, the UCSD dataset is from [26], the UMN dataset is from [27], and the level crossing is from [15]. The first three datasets are of intersections used to evaluate the validities of the unsupervised modeling results of our method. In these videos, traffic flows are governed by a trafic signal which has been modeled with state transition in our model. The other three datasets were used to detect abnormal activities in scenes. These videos contain abnormal activities which are hard to detect in case of using quantized directions and conventional topic modeling methods [7, 10, 14, 28, 31, 32].

The cell size of each video was identically fixed to 10 × 10 and the mini-batch size D is fixed to 10 in the all experiments. We equally set the number of topic K to 12 for three intersection videos and K to 3 for other videos. This is because, unlike intersection datasets, the latter three datasets are in narrow field-of-view situations where moving objects have only a few typical patterns. Furthermore, we experimented with different K on the state estimation and the prediction task to be described in Sect. 3.1 and 3.3, but the variation of K did not have a significant impact on the performance as long as K is not significantly far from the actual number of typical patterns. The experiments were conducted on a computer with Intel i5 2500, 3.3 GHz CPU. In spite of non-optimized C++ implementation and single core processing, the proposed method could run on almost real-time (18-20fps), including motion extracting, model learning, and anomaly testing tasks.

3.1 Result of Scene Understanding

WI dataset: Modeling results for the WI dataset are shown in Fig. 7. The number of state S is set to 3, and each state is represented by red, blue, and green. The latent variable set $\{s_t\}_{t=1}^{D}$ inferred by the Eq. (15) is graphically represented with the colored bar on the top of the figure. The horizontal axis of the bar, namely, represents time interval index t of the collection of trajectories. In this bar, we can find that each state changes regularly depending on time. The change of states coincides with the traffic lights which controls movement of vehicles and pedestrians. The state transitions are not well learned at first, but as a result of online learning, the model well describes the state and the transition of states as more data comes in. Our online learning correctly updates the model as more data is observed.

The transition matrix π is shown on the right of the bar. The probability for a transition from state i to state j is π_{ij}. Higher probability is denoted as white, whereas black denotes low probability. The matrix shows that except staying on the same state, the most probable state transition occurs in the order of $1 \rightarrow 2 \rightarrow 3 \rightarrow 1$. Each

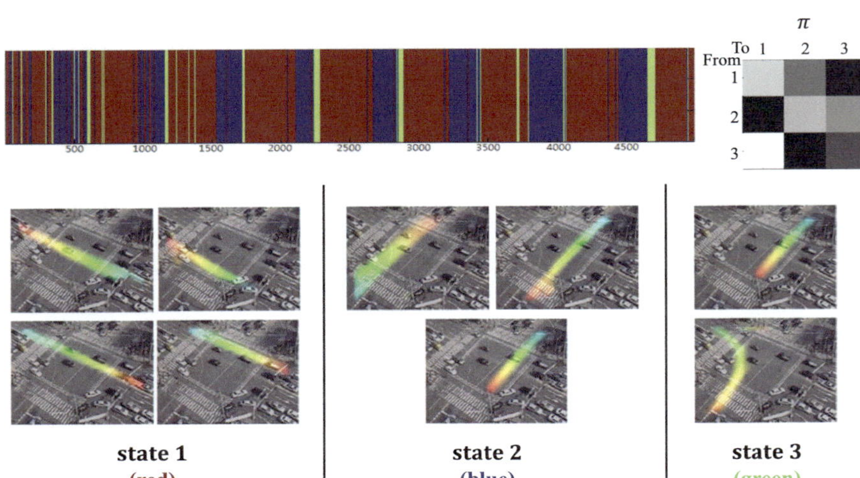

Fig. 7 Typical patterns and their spatiotemporal relationship for the WI video sequence. The *colored bar* on the *top* shows state estimation. The transition matrix is shown on the *top-right*, where higher probability is denoted as white. The typical moving patterns are denoted with *red* and *blue* coloring, where objects move from *red* to *blue* (best viewed in color)

state is represented by a mixture of co-occurring typical activity patterns. Since the width of the road in WI video is wide (eight-lanes), each pattern appears per single or double lane. Typical patterns are shown on the bottom three subfigures in Fig. 7. The patterns are denoted with red and blue coloring, where objects move from red to blue. State 1 is composed of four typical activity patterns: cars coming and going from northwest to southeast. In state 2, cars are coming and going from northeast to southwest, which cannot happen at the same time with state 1. State 3 is a mixture of turning left and going-straight from southwest. During left turn signal, which is state 3, there is no activity going from northeast to southwest. We can also find left turn signal is very short compared to other states as shown in the bar.

In fact, even if $K = 12$ is not strictly the same as the actual number of typical patterns, scene understanding performance of the proposed method is not critically affected. For instance, when K is designed to be smaller than the actual number of typical patterns, co-occurring similar two typical patterns are sometimes merged into one as shown in the first typical pattern in state 2. On the other hand, with a large K, a typical pattern (e.g., go straight) can be split into multiple sub-patterns (e.g., go straight in each lane) as long as K is not significantly far from the actual number. These cases hardly disturb the automatic understanding of traffic patterns. To confirm this, we conducted additional experiments by measuring the state estimation error and by conducting the prediction task with different K, which will be covered later in detail.

QMUL Junction dataset: QMUL Junction Dataset is the footage of objects crossing an intersection which has four-lane and right-turn signals. Three states are used for this experiment. Results are shown in Fig. 8. In the figure, state 1 describes activities with right-turn signal. State 2 includes activities corresponding to vertical movements. Similarly, state 3 captures horizontal movements of cars. As shown in the colored bar and the transition matrix π, states repeatedly change in order of $1 \rightarrow 3 \rightarrow 2 \rightarrow 1$. This transition shows well a change of activity controlled by the signal in the scene. Vertical movements of cars appear when right- turn signal is finished, and the horizontal straight signal starts after the vertical straight patterns.

MIT dataset: We applied two-stage greedy learning to extract two global states from MIT junction dataset. Figure 9 shows the results. Unlike WI and QMUL videos, strict state classification caused by traffic signal is impossible in MIT video because turning and crossing movements is not protected. Hence, we set $S = 2$ for the MIT data so that only rough state assignments (vertical and horizontal moving) could be done. State 1 represents vertical activities and state 2 describes horizontal car movements. These two states are alternately repeated, closely related to the traffic rules in the dataset. In this case, however, KLT tracker performs poorly for objects turning right, which come from bottom and go to right, because they are occluded by the traffic light pole. Although the proposed model can deal with general cases of broken trajectories by co-occurrence property, it still has a limitation in the case that trajectories are always broken at the same position. In this case, a collection cannot often include the trajectories in both sides of the breaking position (e.g., occlusion) at the same time because the collection just covers short duration. Hence, it is difficult to apply co-occurrence property to the consistently

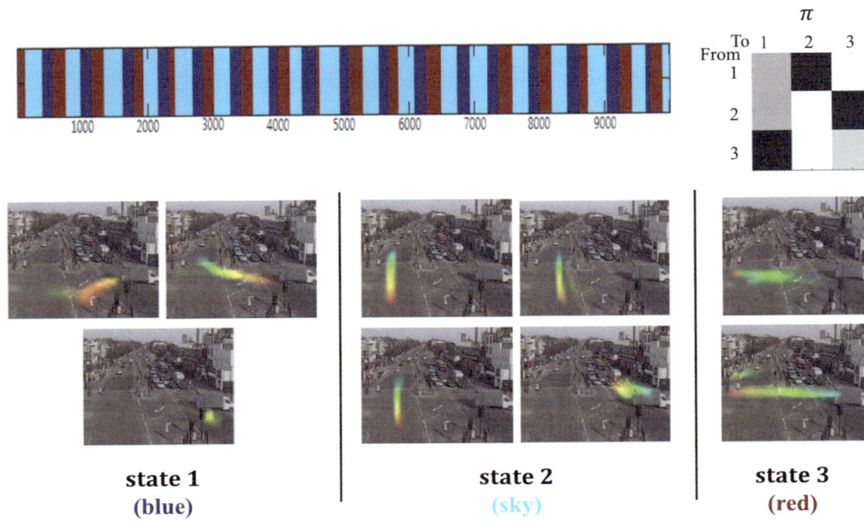

Fig. 8 Typical patterns and their spatiotemporal relationship for the QMUL video sequence. The *colored bar* on the *top* shows state estimation. The transition matrix is shown on the *top-right*, where higher probability is denoted as white. The typical moving patterns are denoted with *red* and *blue* coloring, where objects move from *red* to *blue* (best viewed in color)

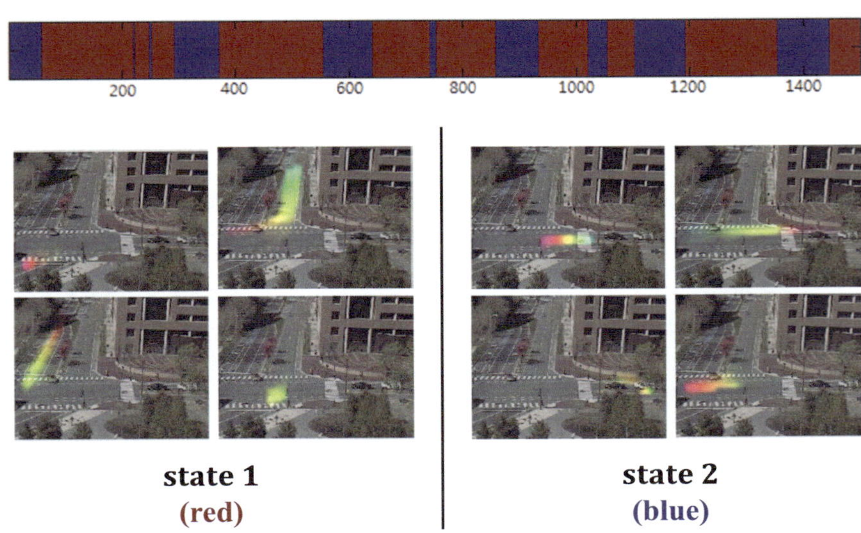

Fig. 9 Typical patterns and their spatiotemporal relationship for the MIT video sequence. The *colored bar* on the *top* shows state estimation. The typical moving patterns are denoted with *red* and *blue* coloring, where objects move from *red* to *blue* (best viewed in color)

broken tracks. Performance improvement is expected if a more robust feature tracker such as [21] is used.

From the above results of three datasets, the proposed method gives an interpretation of activities in the scene (e.g., finding typical activities in unsupervised way, learning spatiotemporal relation among the typical activities), which are essential tasks of topic model based approach [7, 10, 14, 28, 32]. Although the quantitative results of our scene understanding are not so different from the results of the existing methods, there are two main distinctions between the proposed model and the similar methods. First, the proposed method incrementally takes trajectory data with online learning, which is differentiated from the batch learning methods. For example, an existing similar method such as [10] estimates state assignments at once using all data from beginning to end; on the other hand, our method lengthens the state estimation bar as time goes on. In addition, the experiments are conducted with different K, and our method showed consistent results even for the variations of K (= 8,12,16,20). Our online learning method not only enables the adaptation of scene changes but also saves memory because our model does not need to keep old trajectory collections. Second, our novel model utilizes precise velocity as an observation beyond quantized direction. As the merit of adding precise velocity to the model is difficult to display on the scene understanding of results, subsequent sections will show the effect of using velocity observations.

3.2 Applications in Anomaly Detection

This section provides anomaly detection results using the proposed model. Detected abnormal events for MIT, QMUL and WI datasets are as shown in Fig. 10a–d. Figure 10a illustrates a detection of an illegal U-turn action. In Fig. 10b, an ambulance uses improper lanes and goes on the opposite direction. Our method detects these events as abnormal because these activities are not modeled as typical patterns. Figure 10c shows a vehicle ignoring the traffic signal and turning right through opposite traffic. Though this vehicle would be considered normal in state 2 (as Fig. 8), it is detected as abnormal since the activity occurred when state 3 is dominant. Also, bike driving on the opposite direction is detected in Fig. 10d.

In addition to former three videos, we conducted anomaly test for UCSD, UMN and level crossing datasets to confirm the performance of our model. These datasets contain abnormal activities that are difficult to detect when using methods based on conventional topic models with quantized directions (e.g., over-speeding objects, cars stopping on a railroad crossing for a long time, and so on). UCSD dataset captures people, cars, and bicycles showing various velocity patterns. The scene is usually crowded with pedestrians, but bikes and cars drive on pavements rarely. Our method shows good performances by the model with the precise velocity observations. Figures 10e–f illustrate detection of a bike and a car driving on pavement. Since these objects have much faster velocity than others, they are

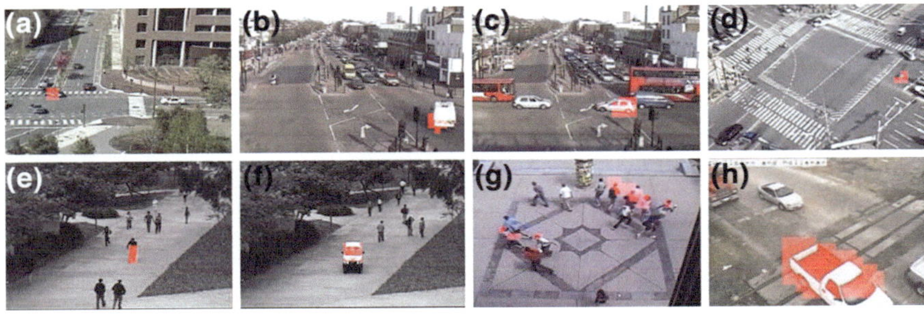

Fig. 10 Example of anomaly detection for various video dataset **a** illegal U-turn; **b**, **d** driving on the opposite direction; **c** disordering the traffic signal; **e**, **f** over-speed on a pavement; **g** unusual crowds speed; **h** a car stops on a railway (best viewed in color)

detected as abnormal. On the other hand, because the quantized directions have no information about speed, the methods based on the quantized direction feature cannot detect an object moving with over-speed. Figure 10g shows a detection result for the UMN dataset. In the video, people are loitering slowly in a square, and then suddenly scatter. The proposed method well detects the event. In Fig. 10h, the result shows detection of a potentially dangerous region, where a car stops on a railway for a long time. Note that other cars stopping before railroad are determined as normal. On the contrary, conventional topic models have difficulty in understanding long-term motion of a single object because they are based on local motions extracted between two frames.

For further analysis of the strength of the velocity observations, we look into the likelihood of trajectories in the scene of Fig. 10f from the UCSD dataset. In this example, we examine six trajectories (two abnormal trajectories and four normal trajectories), and each trajectory is depicted in different color. The first trajectory (blue) and the second trajectory (green) are extracted from a car going from top to bottom, which is faster than usual motion of pedestrians. The third and fourth trajectories (red and sky) are extracted from pedestrians walking from bottom to top, and the fifth and sixth trajectories (purple and yellow) are from a pedestrian walking from top to bottom. In case of the proposed method, which utilizes actual velocities of the trajectories and trains them with Gaussian models, the log-likelihood of trajectory 1 and 2 is lower than that of other trajectories as shown in the first row of Fig. 11. On the other hand, other topic model based methods such as MCTM [10] covert the actual motions between two frames into quantized directions at a grid position. Each quantized direction is depicted as one of the four directions (up, down, right, left) at the grid position as shown in left-bottom of Fig. 11, where the same colored arrows denote that they are extracted from the same object. This motion representation method, however, cannot distinguish over-speed from walking speed. Therefore, all trajectories have similar likelihoods as shown in the lower graph of Fig. 11 because overall paths of the trajectories without velocity information are likely to occur in the scene.

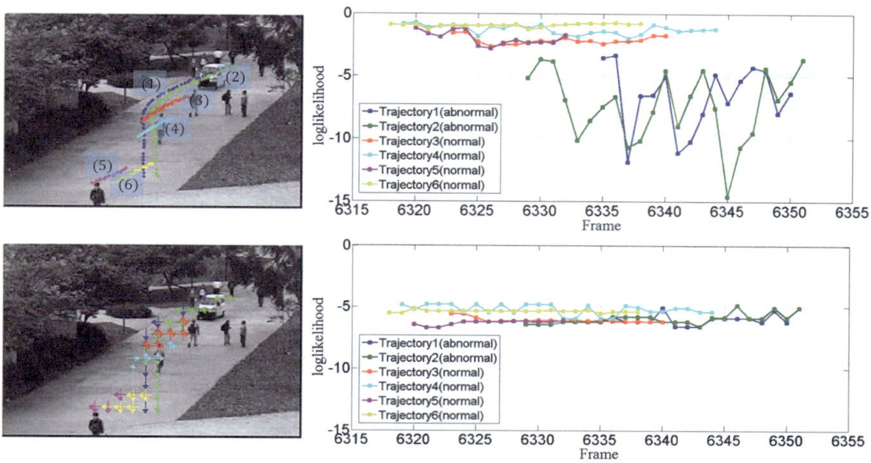

Fig. 11 Comparison of motion likelihood between proposed model (actual velocity of trajectories) and MCTM (quantized direction) [10]. The first row: actual trajectories in the UCSD dataset (*left*) and motion likelihood of each trajectory (*right*). The second row: quantized direction converted from each trajectory denoted with different color (*left*) and their motion likelihood (*right*) (best viewed in color)

3.3 Prediction Task

The number of abnormal activities in the actual traffic video datasets is not enough to give meaningful quantitative results. This is because the model would prefer over-fitting to only a few events, harming the credibility. Therefore, in order to quantitatively compare the performance of our method against other algorithms, we conducted activity prediction tasks presented in [7]. The prediction task can test the whole video sequence although abnormal activities did not happen in the video. For this reason, the prediction tasks can be used for a general evaluation of the model's plausibility. For the task, future observations are estimated using given past observations. For example, if the upward motions are observed in the bottom of the scene and the right-turn pattern is learned at the position, future observations (maybe rightward motions in the right-side of the scene) can be estimated based on the trained model. The estimated future observations are represented as a probability histogram whose summation must be 1, and then the similarities to the actual observations are measured using Bhattacharyya coefficient.

MIT dataset was used for the comparison and the existing methods [7, 10] using 29 past time instances (seconds) to estimate the observations of the 30th time instances. Unlike the existing topic models [7, 10], whose observations are represented by quantized local motions between only two frames, the proposed model utilizes trajectories as observations. This type of observation allows our method to do the prediction task with trajectories from the current frame (not observations obtained from 29 past time instances) and the trained model. Also we validated the

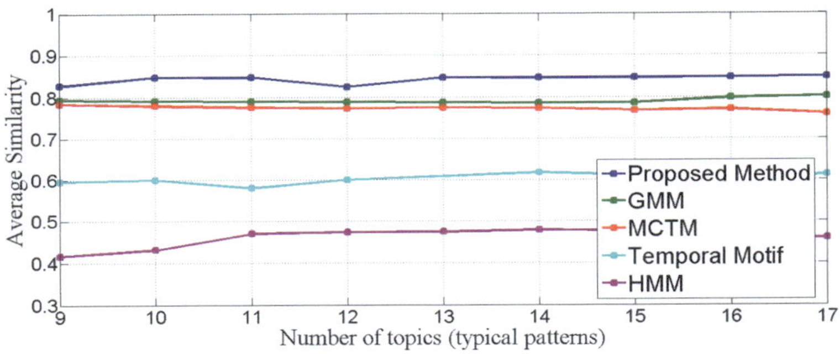

Fig. 12 Comparison of average accuracy on a prediction. *X*-axis indicates number of topics denoted as *K* in our paper. Exceptionally, in case of the GMM based methods, *X*-axis indicates the number of Gaussian components

prediction accuracy on the different design parameter *K*, representing the number of topics. Comparison results are shown in Fig. 12. The figure shows that the proposed method outperforms Temporal Motif [7] and MCTM [10] even though we conduct the prediction task with observations only in the current frame. This result is caused by the fact that Temporal Motif [7] and MCTM [10] utilize quantized local motions, but our model mines actual velocity of trajectories. This provides the validity of the use of accurate velocity observations, allowing more plausible scene model and giving precise predictions.

We also provide the result of comparison with GMM-based trajectory modeling [1], whose trajectory representation method is similar to ours (i.e., it also uses actual velocity observations). The reason why the proposed method is more accurate than [1] is that we have inter-related multi-Gaussian models based on typical patterns (topics). For example, in the center of intersection, the GMM would estimate a future position of the trajectory based on only the previous path. Thus, in some cases, the GMM model may have difficulty in predicting whether an object will go straight or turn right. On the contrary, the prediction of our method (including other topic model based methods) is based on not only previous path but also mutual dependence among typical activities. Therefore, the proposed method can give a confident prediction whether an object will go straight or turn right.

4 Conclusion

This paper introduced a new method for analyzing traffic patterns in a scene and detecting anomalies. By investigation on previous studies we identified the essential requirements for the traffic pattern modeling in actual environments. The proposed method met those requirements by modeling the scene with a graphical inference model which uses the point trajectories of the scene considering the overall path, their spatiotemporal dependency, and their precise velocities. The problem of high dimensionality of the proposed model was relaxed with the proposed two-stage greedy inference, allowing the solutions to be obtained efficiently. This approximate inference strategy is a meaningful attempt to find an alternative outperforming CGS which is conventionally used to learn topic models for scene understanding.

As shown in the experiments, the effects of the proposed approach are summarized as follows. The scene understanding results showed that the proposed method could automatically discover not only typical patterns but also spatiotemporal relations among them. Also, the state estimation results of the proposed online inference maintained comparable performance to the batch learning method. In the experiment on the likelihood evolvement of a trajectory over time, the proposed method was able to distinguish the speed of moving objects, which is impossible with the quantized directions. Using the proposed velocity model with regard to typical patterns, our method also gave outstanding accuracy on the prediction task. On comparison to the online sampling method, the two-stage online inference guaranteed more robust results than the sampling-based learning.

Although we could not find misdetection cases caused by the assumptions for online inference in our experiments, misdetection cases could occur when rigorous validation with more various video is performed. As for the future work, we will validate our sub-model optimization strategy and pursue a relaxation of the assumptions.

Acknowledgments This work was sponsored by Samsung Techwin Co.,Ltd and BK 21 plus program, and also partially supported by the Center for Integrated Smart Sensors funded by the Ministry of Science, ICT & Future Planning as the Global Frontier Project.

References

1. Basharat A, Gritai A, Shah M (2008) Learning object motion patterns for anomaly detection and improved object detection. In: IEEE Conference on CVPR
2. Benezeth Y, Jodoin PM, Saligrama V (2011) Abnormality detection using low-level co-occurring events. Pattern Recognit Lett 32(3):423–431
3. Bishop CM (2006) Pattern recognition and machine learning (Information science and statistics). Springer-Verlag New York Inc, Secaucus
4. Blei DM, Ng AY, Jordan MI (2003) Latent dirichlet allocation. JML. Res. 3:993–1022
5. Canini KR, Shi L, Griffiths TL (2009) Online inference of topics with latent dirichlet allocation. In: AI-STATS
6. Duda RO, Hart PE, Stork DG (2000) Pattern classification, 2nd edn. Wiley-Interscience, New York
7. Emonet R, Varadarajan J, Odobez JM (2011) Extracting and locating temporal motifs in video scenes using a hierarchical non parametric bayesian model. In: IEEE conference on CVPR, pp 3233–3240
8. Griffiths TL, Steyvers M (2004) Finding scientific topics. PNAS 101(Suppl 1):5228–5235
9. Hoffman M, Blei DM, Bach F (2010) Online learning for latent dirichlet allocation. In: NIPS
10. Hospedales TM, Gong S, Xiang T (2009) A markov clustering topic model for mining behaviour in video. In: ICCV, pp 1165–1172. IEEE

11. Hu W, Xiao X, Fu Z, Xie D, Tan T, Maybank S (2006) A system for learning statistical motion patterns. IEEE Trans Pattern Anal Mach Intell 28(9):1450–1464
12. Jeong H, Yoo YJ, Yi KM, Choi JY (2014) Two-stage online inference model for traffic pattern analysis and anomaly detection. Mach Vis Appl 25(6):1501–1517
13. Kratz L, Nishino K (2009) Anomaly detection in extremely crowded scenes using spatio-temporal motion pattern models. In: 2013 IEEE conference on computer vision and pattern recognition **0**, 1446–1453
14. Kuette D, Breitenstein MD, Van Gool L, Ferrari V (2010) What's going on? Discovering spatio-temporal dependencies in dynamic scenes. In: CVPR, pp 1951–1958. doi:10.1109/CVPR.2010.5539869
15. Machy C, Desurmont X, Delaigle JF, Bastide A (2007) Introduction of cctv at level crossings with automatic detection of potentially dangerous situations
16. Mahadevan V, Li W, Bhalodia V, Vasconcelos N (2010) Anomaly detection in crowded scenes. In: IEEE conference on CVPR, pp 1975–1981
17. Morris B, Trivedi M (2008) A survey of vision-based trajectory learning and analysis for surveillance. IEEE Trans Circuits Syst Video Technol 18(8):1114–1127
18. Morris B, Trivedi MM (2009) Learning trajectory patterns by clustering: Experimental studies and comparative evaluation. In: CVPR, pp 312–319
19. Piciarelli C, Foresti GL (2006) Online trajectory clustering for anomalous events detection. Pattern Recognit Lett 1835–1842
20. Qin Z, Shelton CR (2012) Improving multi-target tracking via social grouping. In: IEEE conference on computer vision and pattern recognition
21. Rodriguez M, Ali S, Kanade T (2009) Tracking in unstructured crowded scenes. In: ICCV, pp 1389–1396. IEEE
22. Saleemi I, Hartung L, Shah M (2010) Scene understanding by statistical modeling of motion patterns. In: CVPR, pp 2069–2076. IEEE
23. Saleemi I, Shafique K, Shah M (2009) Probabilistic modeling of scene dynamics for applications in visual surveillance. IEEE Trans PAMI 31(8):1472–1485
24. Stauffer C, Grimson WEL (1999) Adaptive background mixture models for real-time tracking. In: CVPR, pp 2246–2252
25. Tomasi C, Kanade T (1991) Detection and tracking of point features. Techical report, IJCV
26. UCSD (2010) Anomaly dataset. http://www.svcl.ucsd.edu/projects/anoma-ly/dataset.html
27. UMN: Crowd dataset. http://www.cs.ucf.edu/ramin/
28. Varadarajan J, Emonet R, Odobez J (2012) Bridging the past, present and future: Modeling scene activities from event relationships and global rules. In: IEEE conference on CVPR, pp 2096–2103
29. Walk S, Majer N, Schindler K, Schiele B (2010) New features and insights for pedestrian detection. In: Conference on CVPR. IEEE, San Francisco
30. Wang B, Ye M, Li X, Zhao F, Ding J (2012) Abnormal crowd behavior detection using high-frequency and spatio-temporal features. Mach Vis Appl 23(3):501–511
31. Wang X, Ma KT, Ng GW, Grimson WE (2011) Trajectory analysis and semantic region modeling using nonparametric hierarchical bayesian models. Int J Comput Vis 95(3):287–312
32. Wang X, Ma X, Grimson WEL (2009) Unsupervised activity perception in crowded and complicated scenes using hierarchical bayesian models. IEEE Trans PAMI 31(3):539–555
33. Wang X, Tieu K, Grimson E (2006) Learning semantic scene models by trajectory analysis. In: Proceedings of the 9th ECCV, vol Part III, ECCV'06. Springer, Berlin, pp 110–123
34. Zhai K, Boyd-Graber J, Asadi N, Alkhouja M (2012) Mr. LDA: A flexible large scale topic modeling package using variational inference in mapreduce. In: ACM International conference on world wide web
35. Zhao B, Fei-Fei L, Xing EP (2011) Online detection of unusual events in videos via dynamic sparse coding. In: IEEE conference on CVPR. Colorado Springs, CO

Event Detection Module for Low-Power Camera

Byung-geun Lee and Moongu Jeon

Abstract In this chapter, we first propose an effective low-power image sensor system for event detection. The system consisting of a low-resolution auxiliary sensor and a high-resolution main sensor operates in two different modes, sleep and wakeup. In the sleep mode, only the auxiliary sensor works for event detection and the main sensor remains off for power saving. In the wake-up mode, the main sensor turns on based on the data sensed by the auxiliary sensor and into normal operation. Second, a new background subtraction algorithm which can be used for event detection is proposed. The algorithm works much faster than conventional algorithms and requires less computation which is critical for low-power operation. In addition, utilization of depth information in background subtraction for 3-D image applications is also presented. Finally, hardware implementation of a low power low-resolution CMOS image sensor (CIS) is presented. The CIS fabricated in 0.18 μm CIS process is designed to be used as an auxiliary sensor in the proposed system. The CIS generates 4-bit image data and consumes only 1.45 mW out of 3 V supply.

Keywords Low-power image sensor · Low-power event detection · Background subtraction · Depth information · CMOS · FPGA

1 Event Detection Framework

In this section, we introduce our event detection framework in detail. The proposed low-power intelligent image sensor system comprises two modules. The first is the low-power event detection module, as depicted in Fig. 1a, and the other is the main

B. Lee (✉) · M. Jeon
Gwangju Institute of Science and Technology, 123 Cheomdangwagi, Gwangju, Bukgu 500-712, South Korea
e-mail: bglee@gist.ac.kr

M. Jeon
e-mail: mgjeon@gist.ac.kr

C. Kyung (ed.), *Theory and Applications of Smart Cameras*,
KAIST Research Series, DOI 10.1007/978-94-017-9987-4_11

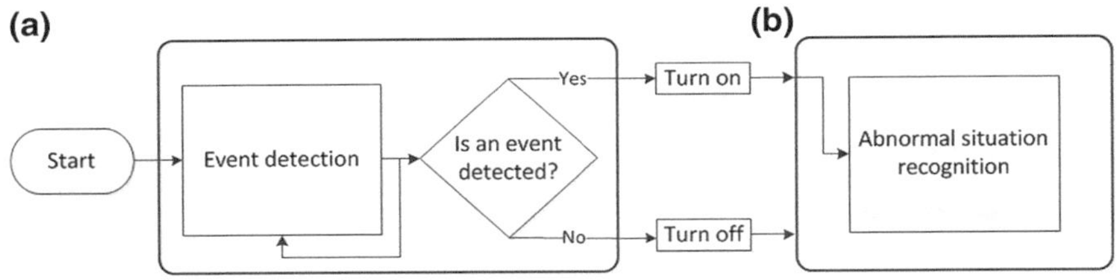

Fig. 1 The flowchart of the proposed low-power event detection system, composed of **a** the low-power event detection module, and **b** the main image sensor module

sensor module, as shown in Fig. 1b. The low-power event detection module is implemented using (1) a field-programmable gate array (FPGA) control chip, (2) a low-power image sensor, and (3) an integration module. The research content for this module is described in detail as follows.

- FPGA control chip: implementation of a developed intelligent image analysis algorithm and an electric power (operating mode) control signal system on the FPGA.
- Low-power image sensor: research and development of a low-power image sensor to receive signals from the electric power control chip as an input and send low-power images as an output.
- Integration module: implementation of the low-power event detection module by integrating the FPGA control chip with the low-power image sensor.

The main sensor module contains an abnormal situation recognition module that is based on color and depth information. The module function comprises (1) image preprocessing and (2) abnormal situation recognition. The process for this module is described in detail as follows.

- Image preprocessing: De-noising and background subtraction from the image (color and depth information) from the main sensor module to make the abnormal situation recognition performance more accurate.
- Abnormal situation recognition: Research and detection of any image sequence likely to contain an abnormal situation and classification of the detected image sequence.

Figure 1 shows the overall procedures used for the proposed low-power image sensor system. When the system is started, the low-power event detection module checks whether an event exists or does not exist in the scene. If an event is detected, then the low-power module turns on the main image sensor module and the abnormal situation recognition process is executed. Otherwise, the main sensor is turned off. We developed a new background subtraction method that uses color and depth information, this method will be described in detail in Sect. 2.1. Also, state-of-the-art recognition techniques can be applied to the main module because we plan to use a commercial embedded system in this module. Finally, by combining this low-power event detection module with the main sensor module, we were able to develop an integrated low-power image sensor system based on color and depth information.

2 Event Detection Algorithm

2.1 *Background Subtraction Using Color and Depth Information*

2.1.1 Introduction

Moving object detection is the essential step in visual surveillance research. Among the various techniques used for moving object detection, numerous background subtraction (BGS) methods have conventionally been used for this purpose. Basic BGS methods, using temporal medians of image frames [4] and statistical approaches using a Gaussian mixture model (GMM) [5] have previously been proposed. More recently, self-organizing maps [6, 7] and multiple features-based methods have been devised [8, 9].

However, these BGS techniques have certain fundamental limitations because they use human perception (visible light)-based color spaces, such as the RGB (red, green, blue) space, the HSV (hue, saturation, value) space, and the YUV space, in which Y and UV represent luminance and chrominance, respectively. Basically, these methods are weak in color camouflage situations and are sensitive to changes in illumination.

To handle these problems, other BGS approaches [10–12] have been proposed using other types of information along with the color information. In particular, depth information from stereo cameras, Microsoft Kinect sensors or time-of-flight (ToF) sensors has been used along with the color information. Harvile et al. [10] proposed the use of foreground segmentation using the YUV color space with additional depth values. However, the method proposed by Harvile et al. [10] has difficulty in determining the correlation between the YUV space and the depth with the covariance matrix. Fundamentally, color and depth are totally different types of information. There are obvious limitations because the method simply adds the depth value to the color vector and uses a single feature vector. In the state-of-the art research, Fernandez-Sanchez et al. [12] and Camplani et al. [11] applied color and depth in each background model (e.g., $P(x_c)$, $P(x_d)$) while not using color and depth in the same model (e.g., $P(x_{cd})$. Camplani and Salgado [11] and Fernandez-Sanchez et al. [12] used background modeling based on a GMM and on a codebook, respectively, and built final classifiers that combine the color model and the depth model. In our paper, we present a fast BGS method based on the GMM using both color and depth. The background/foreground models are estimated in a manner similar to that proposed in [13]. For reasonable combination of the color and the depth, the likelihood of the background/foreground model was determined from the product of the likelihood's of the color and depth models. Also, to achieve fast and real-time implementation, these three-channel color vectors were converted into one-channel gray scale values. Nevertheless, our algorithm demonstrated better performance than conventional BGS techniques for our own dataset, including color, depth, and ground truth images. We used a Microsoft Kinect to obtain color and depth information with OpenNI SDK software, but the Kinect generated noisy data. However, while there is no pre- or post-processing for depth noise reduction, our probabilistic model could deal with this noise.

2.1.2 Background Subtraction Method

Our approach begins from the idea that the correlation between the color and the depth is difficult to determine. This leads us to the question of why we should find the correlation between the color and the depth. Many BGS methods have already been devised based on color information alone, but these color-based BGS techniques have definite limitations. They basically cannot segment objects from similarly-colored backgrounds (i.e., color camouflage) and they are very sensitive to illumination changes. To solve these problems, other approaches [10–12] using depth information have been proposed. For example, Harvile et al. [10] simply extended the dimensions of each pixel vector X = (Y,U,V) to be (Y,U,V,D) by adding the depth (D) value and built a background model based on a GMM using one-dimensional added pixel vectors, in which each pixel corresponds to a GMM, and then, the same number of GMMs as the image resolution (width height) were formed. A GMM can be initialized using clustering methods such as the expectation maximization (EM) algorithm and K-means clustering on the observed T frames. However, the depth information obtained from current devices such as the Kinect, TOF sensors and stereo cameras still contains a lot of noise and thus these are not reliable. Also, clustering requires rather too much processing time to be used in real-time applications, and assumes a situation without any moving object during the initialization time to obtain a good background model. Thus, to reliably combine color with depth, we designed a probabilistic background model based on a Gaussian distribution and used a de-noised depth image with the model. Our algorithm follows two basic steps. The first step is background modeling, and the second is background subtraction using the model obtained.

Background Modeling

Stauffer and Grimson [13] initialized their background models using the recent history of t frames, but we initialized our background models using the first K frames on a pixelwise level. K is the number of Gaussians in a GMM, and Xi indicates a particular pixel located at (x, y) in the ith frame I in Eq. (1).

$$\{X_1, \ldots X_k\} = \{I(x, y, i) : 1 \leq i \leq K\} \tag{1}$$

Our proposed initialization method can skip the clustering process and thus can save a great deal of time. Each pixel has a corresponding mixture of K Gaussians. Heuristically, K varies from 3 to 5. However, we did not add update procedures. The probability function at the given pixel value at time t is then given as follows:

$$P(X_t) = \sum_{i=1}^{K} \omega_i \cdot \eta\left(X_t, \mu_i, \sigma_i^2\right) \tag{2}$$

In Eq. (1), X_t is the observed pixel at time t, and ω_i is the weight related to the ith Gaussian distribution with mean μ_i, t and standard deviation σ_i^2. η is a Gaussian probability density function as in

$$\eta\left(X_t, \mu, \sigma^2\right) = \frac{1}{\left(2\pi\sigma^2\right)^{1/2}} e^{-\frac{(X_t-\mu)^2}{2\sigma^2}} \tag{3}$$

Here, μ_i is the pixel value in the ith image sequence and σ_i is the user parameter. We set the latter parameter in a range from 10 to 20. After all models are initialized, we separate the background and foreground models as follows:

$$B = \mathrm{argmin}_b \left(\sum_{k=1}^{b} \omega_k > T \right) \tag{4}$$

where the first B distributions are selected as the background models when the sum from ω_1 to ω_b exceeds a threshold T, and the remaining distributions are the foreground models. We applied grayscale color (0−255) and depth (0−255) channels to these Gaussian distributions for a single variable. Each channel builds each probability function based on a GMM. The stochastic meaning of each these functions is the likelihood. Let the color model and the depth model be $P(c_t)$ and $P(d_t)$, respectively, as follows:

$$P(x_t) = P(c_t) \cdot P(d_t) \tag{5}$$

We will introduce the method used to subtract the background using this probabilistic model in the next section.

Background Subtraction

In this section, we explain background/foreground segmentation using the probabilistic background models. First, the color model B_c and the depth model B_d classify the observed pixels, c_t and d_t, individually at time t, as shown in Fig. 1. A pixel finds a matched Gaussian model by using the Euclidian distance:

Table 1 Pixel classification according to the matched model

Case	Matched model	Classified result
Case 1	Background (a match is found)	Background
Case 2	Foreground (a match is found)	Foreground
Case 3	No one (no match is found)	Foreground

$$\mathrm{abs}(x_t - \mu_i) \geq k \tag{6}$$

where k is a constant threshold equal to 2.5. A survey of background modeling using a mixture of Gaussians in Bouwmans et al. [14] introduced [13] concepts such as user parameter values in their implementation details [13]. They used the Mahalanobis distance to enable pixel classification. However, we used the Euclidean distance because each feature grayscale color (0−255) and depth value (0 −255) on each model has one dimension in which there is no correlation. Each pixel is classified as one of the three cases as in Table 1.

In the case of the color value c_t, after all the pixels at time t are classified as either background or foreground, the pixel values are then computed probabilistically using the inequality for the matched Gaussian distribution as follows:

$$\theta \cdot \eta(c_t, \mu_i, \sigma_i^2) \geq \text{maximum pixel value} \tag{7}$$

θ is a constant used to scale the Gaussian probability density function values. When the maximum pixel value is 255, θ is equal to 10,000. For example, we assume that the minimum pixel value is 0 and the maximum value is 255. In case 1 and case 2, if the inequality is satisfied, then the pixels have values of 0 and 255, respectively. Otherwise, in case 1 and case 2, the pixels are equal to $\theta \cdot \eta$ and $255-\theta \cdot \eta$, respectively. In case 3, the value becomes 255. Eventually, the BGS result R_c consists of pixel values 0–255, where the probability $P(c_t)$ is higher and the value is closer to 0 (background), as shown in Fig. 2a. However, for the BGS result Rd based on B_d, the probabilistic pixel values allocation is applied in case 2, as shown in Fig. 2b. Thus, the final BGS results R_{final}, as shown in Fig. 2c, are computed as follows:

$$\begin{aligned} &\text{If } R_c(x,y) * R_d(x,y) > \text{minimum pixel value} \\ &\text{then } R_{\text{final}}(x,y) \text{ is foreground} \\ &\text{esle } R_{\text{final}}(x,y) \text{ is background} \end{aligned} \tag{8}$$

Depth Image De-noising

Our method can handle the depth image noise. We discovered the phenomenon where $R_c(x,y)$ correctly classified as background at a location (x,y), where false foreground detection occurs in R_d as a result of depth noise. Thus, we designed the

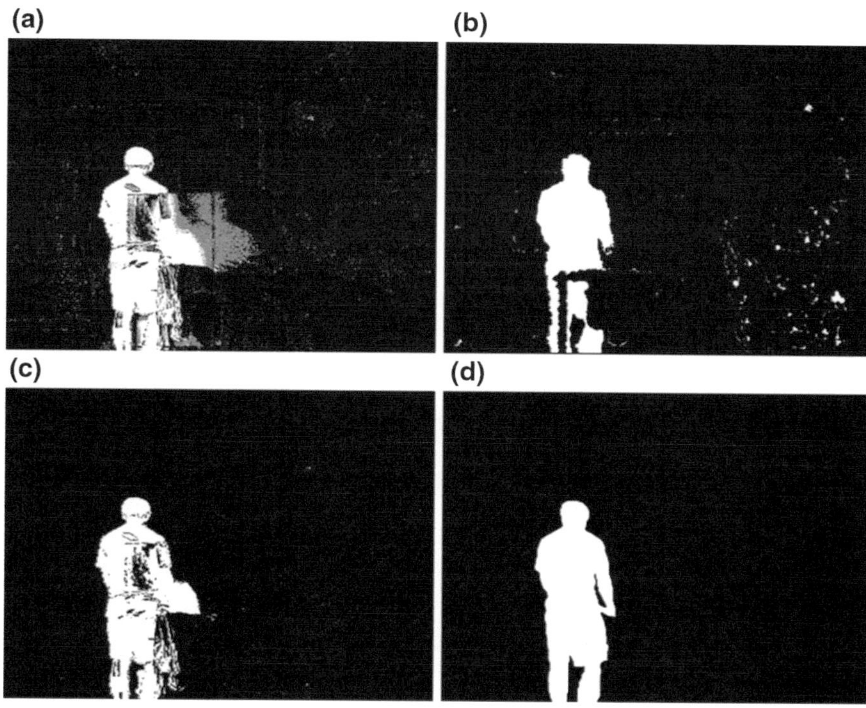

Fig. 2 BGS results based on **a** color model B_c, **b** depth model B_d, **c** the proposed method, **d** the ground truth

pixel classification method shown in Eq. (8). Even if false positive regions exist in R_d because of depth noise, most of these regions belong to the background in R_{final}, as shown in Fig. 2c.

2.1.3 Dataset for Performance Evaluation

We built a new dataset to evaluate the BGS algorithm using color and depth and to compare it with conventional color-based BGS techniques. The dataset contains three categories: (i) normal situation, (ii) color camouflage, and (iii) depth camouflage, which have one scene, two scenes, and one scene, respectively, as seen in Table 2. We focused here on solution of the color camouflage problem, and thus formed one more scene in the color camouflage category. Each scene consists of color, depth and ground truth image sequences, as shown in Fig. 3. The videos were taken in the conditions of resolution 640 by 480 (VGA) and 30 frames per second (fps).

Table 2 Three categories of the dataset: (i) Normal situation, (ii) color camouflage, and (iii) depth camouflage

Category	Color camouflage regions	Depth camouflage regions
(i) Normal situation	X	X
(ii) Color camouflage	O	X
(iii) Depth camouflage	X	O

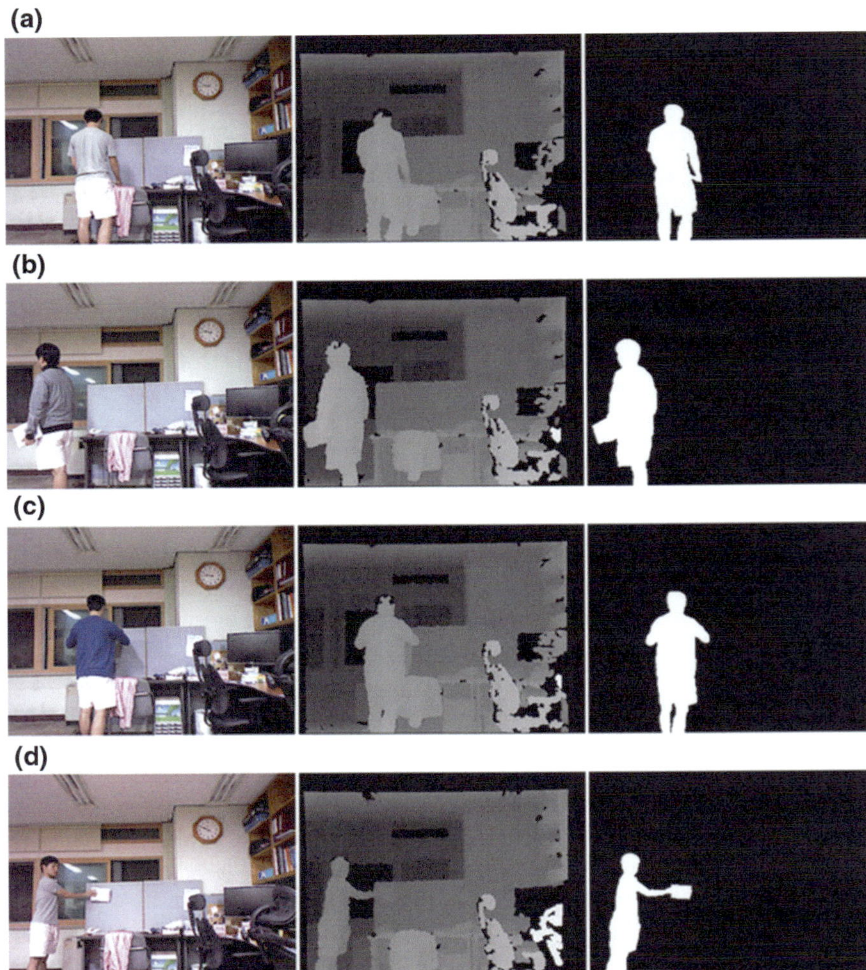

Fig. 3 Dataset samples of **a** color camouflage 1, **b** color camouflage 2, **c** normal situation, and **d** depth camouflage. For each situation (**a**), (**b**), (**c**) and (**d**), the color, depth and ground truth images are on the *left*, *center*, and *right*, respectively

2.1.4 Experimental Results

We introduce the evaluation results for our algorithm found by comparing them with other color-based conventional BGS techniques [4–9] while using our dataset. We used three main measures, precision, recall, and F-measure [16], as follows

$$\text{Precision}(P) = \frac{\text{true positive alarm (TP)}}{\text{TP} + \text{false positive alarm (FP)}} \tag{9}$$

$$\text{Recall}(R) = \frac{\text{TP}}{\text{TP} + \text{false negative alarm (FN)}} \tag{10}$$

$$F\text{−measure} = \text{harmonic mean of P and R} \tag{11}$$

in them TP, FP, and FN denote the true positive rate, the false positive rate and the false negative rate, respectively. F-measure indicates the harmonic mean of precision and recall.

The experimental results show that our approach is the best, as shown in Table 3 in terms of both precision and F-measure. However, from the viewpoint of recall, our algorithm was ranked second. This is because our method detected fewer camouflage regions than the first algorithm. In contrast, the second ranked algorithm [7] in recall was ranked from second to fourth in terms of precision and F-measure because the method segmented the background as the foreground. Although our algorithm was not ranked firsst in terms of all measures, it showed the best overall performance.

2.1.5 Conclusions

We developed a BGS method based on a GMM using both color and depth information to overcome the limitations of color-based BGS, and color camouflage in particular. We built a probabilistic background model to combine color and depth in a reasonable manner. Additionally, we produced a new dataset to evaluate BGS algorithms using color and depth. Using this dataset, we compared our method with several conventional color-based BGS methods [4–9] in terms of precision, recall and F-measure [16]. The proposed method didn't only show better performance than the other methods but also reduced the depth noise. In our future work, our probabilistic background models will be extended to use multi-model data processing, such as that used in thermal imaging cameras and night vision, in addition to depth data. Also, our method can be used for preprocessing to detect regions of interest before high-level applications (e.g., object detection, object tracking and action recognition) in color-based problems such as dynamic color change and low illumination in addition to color camouflage.

Table 3 Quantitative evaluation results table in terms of (a) precision, (b) recall, and (c) F-measure by comparing the proposed method with the methods of [4–9]

(a) The results based on precision				
Algorithms	Normal situation	Color camouflage 1	Color camouflage 2	Depth camouflage
Multi-cues, ACCV2012 [9]	0.754526	0.789207	0.765730	0.768464
Adaptive SOM, TIP2008 [6]	0.678925	0.716150	0.705992	0.748570
Fuzzy adaptive SOM, NCA2010 [7]	0.668636	0.715354	0.711421	0.764075
Adaptive GMM, PRL2006 [5]	0.801492	0.833232	0.764047	0.754855
Temporal median, PAMI2003 [4]	0.652183	0.657870	0.652529	0.702824
Multi-layer, CVPR-VS2007 [8]	0.338086	0.423075	0.480103	0.185146
Proposed method	0.836480	0.877724	0.849011	0.829063
(b) The results based on recall				
Algorithms	Normal situation	Color camouflage 1	Color camouflage 2	Depth camouflage
Multi-cues, ACCV2012 [9]	0.777955	0.770264	0.702065	0.846751
Adaptive SOM, TIP2008 [6]	0.962481	0.939113	0.919591	0.930680
Fuzzy adaptive SOM, NCA2010 [7]	0.943736	0.924956	0.897145	0.919343
Adaptive GMM, PRL2006 [5]	0.772382	0.797471	0.603326	0.845297
Temporal median, PAMI2003 [4]	0.468372	0.408679	0.445172	0.266032
Multi-layer, CVPR-VS2007 [8]	0.092907	0.339863	0.373178	0.101987
Proposed method	0.853854	0.881338	0.842226	0.861781
(c) The results based on *F*-measure				
Algorithms	Normal situation	Color camouflage 1	Color camouflage 2	Depth camouflage
Multi-cues, ACCV2012 [9]	0.752894	0.763279	0.700721	0.805171
Adaptive SOM, TIP2008 [6]	0.795814	0.812047	0.796957	0.829031
Fuzzy adaptive SOM, NCA2010 [7]	0.781971	0.806416	0.789939	0.833832
Adaptive GMM, PRL2006 [5]	0.772826	0.814323	0.627482	0.796503
Temporal median, PAMI2003 [4]	0.479137	0.413890	0.477570	0.376613
Multi-layer, CVPR-VS2007 [8]	0.137997	0.363238	0.406819	0.131379
Proposed method	0.844056	0.879383	0.843082	0.844631

2.2 *BGS Performance Evaluation Software*

In this section, we present new software that was developed to evaluate the performance of various BGS methods at pixel level and at frame level. To evaluate the accuracy of a diverse range of conventional BGS techniques, users should first select a BGS method, its fundamental parameters and a test data set. Then, the software performs the implemented analysis procedures to output qualitatively and quantitatively for evaluation of results in terms of precision, recall and F-measure. The proposed software will be very useful for evaluation of user-provided BGS methods against existing BGS methods for a variety of test data sets.

2.2.1 Introduction

In visual surveillance research, many BGS techniques have been proposed. Basic methods using the temporal medians of the n previous frames [4] and statistical approaches using GMMs [5] have been proposed. More recently, self-organizing map-based [6, 7] and multiple features-based methods are devised [8, 9]. We integrated open source versions [15] of these methods in our software, and designed the user interfaces and functionalities so that the methods can be evaluated under a diverse range of conditions. The following sections describe the developed evaluation software in more detail.

2.2.2 User Interface and Functionalities

There are four types of main functionality in our software, which are shown in Fig. 4.

BGS-Type Selection

Users can select a BGS technique from six types of BGS algorithm. We ported the BGS libraries from open sources [15] into our software (see Fig. 4a).

Main Evaluation Mode Selection

Using this option, users can select whether they will evaluate the chosen BGS algorithm at pixel level or at frame level, as shown in Fig. 4b.

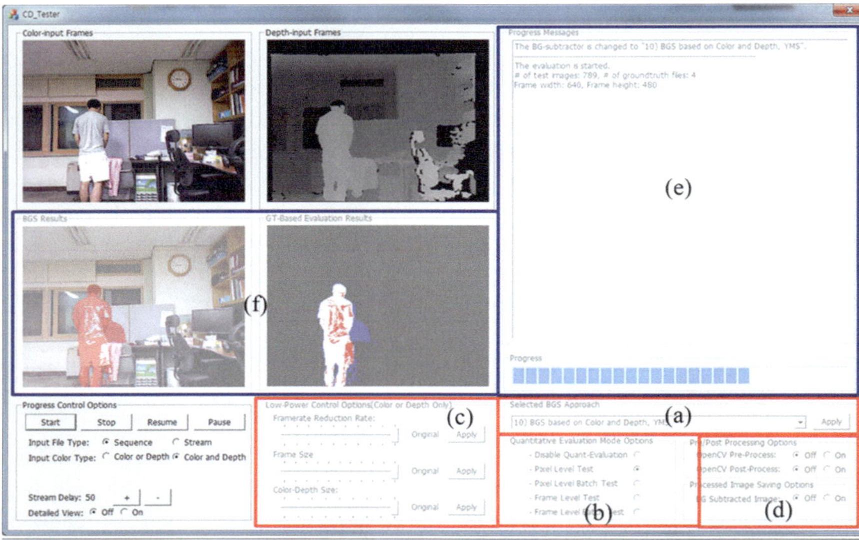

Fig. 4 The user interface: **a** BGS type selection, **b** main evaluation mode selection, **c** input image conversion selection, **d** pre- and post-processing selection, **e** display for quantitative evaluation results, and **f** display for qualitative evaluation results

- Pixel-level mode
  - If this mode is chosen, then the software compares the BGS results with the ground truth in a pixel-wise manner. Both the BGS results and the ground truth images consist of binary pixel values of 0 and 255, where the former and the latter represent the background and the foreground, respectively.
- Frame-level mode
  - In this mode, the performance of the BGS-based change detection algorithm is measured. In this mode, the software considers that changes have occurred in a scene if the number of foreground pixels is more than 0.14*P, where P indicates the total number of pixels in a frame.

Input Image Conversion Selection

The software also offers three user-selective options. The first is to control the temporal resolution, i.e., the frame rate (from the original frame rate to one-tenth of that rate). The second option is to adjust the spatial resolution, i.e., the frame size (from the original image size to 100 × 100). Finally, the color resolution (color-depth) is also variable, from the original resolution to 1-bit grayscale. The details are as shown in Fig. 4c.

Pre and Post-Processing Selection

Before running the selected BGS algorithm, users can activate pre- and post-processing methods (see Fig. 4d). In pre-processing, Gaussian filtering is applied with smoothing-sigma 0.7. If the user turns on post-processing, morphology operations (i.e., erosion and dilation) are performed.

2.2.3 Simulation

We present several simulation results for BGS based on the GMM [5] shown in Figs. 5, 6 and 7.

- Quantitative evaluation
  - Several quantitative results are shown in Fig. 5. We used three main measures, i.e., precision, recall, and *F*-measure [16] to evaluate the quantitative performance.

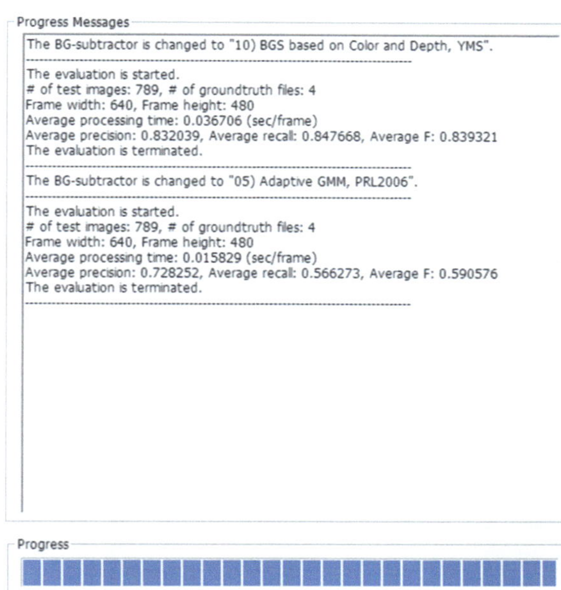

Fig. 5 Detailed example of Fig. 4e

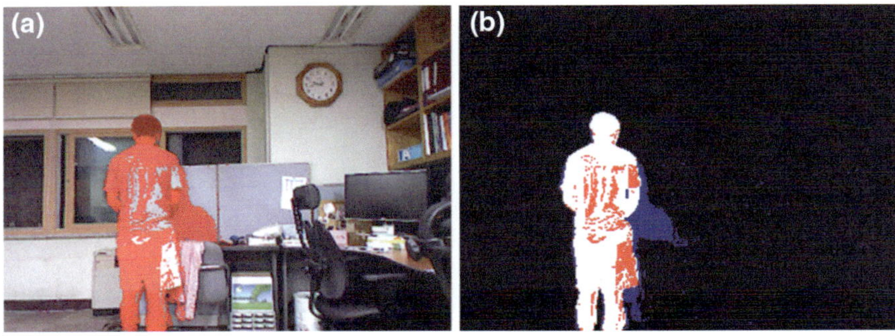

Fig. 6 Detailed example of Fig. 4f in pixel-level mode: **a** BGS results, and **b** evaluation Results

- Qualitative evaluation
  - The BGS results are presented qualitatively in the user interface. When BGS algorithms are evaluated at the pixel level, a display window shows the subtraction results, as shown in Fig. 6a, where the red areas indicate the background regions. Figure 6b shows the ground truth-based evaluation results. Here the white, red, and blue regions indicate true positive, false negative and false positive alarms, respectively. As shown in Fig. 7a, when the evaluation is performed at the frame level, the display window shows the BGS results in a binary format, where the black and white regions represent the background and foreground, respectively. In Fig. 7b, the ground truth based frame-level evaluation results are displayed using text. ON means that change is observed, while OFF means that no changes occurred in the given scenes.

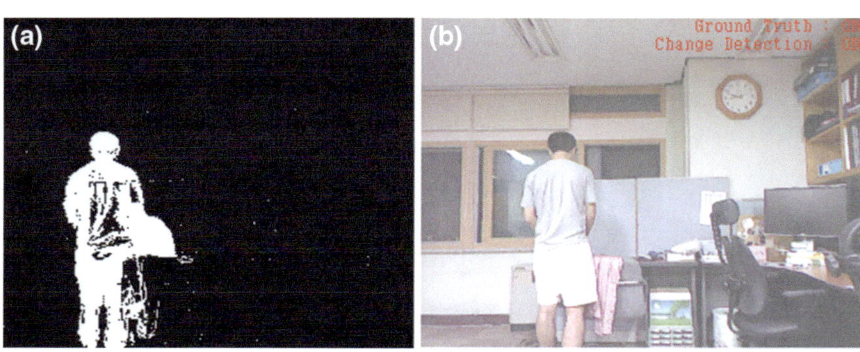

Fig. 7 Detailed example of Fig. 4f in frame-level mode: **a** BGS results, and **b** evaluation results

2.2.4 Conclusions

We developed software to evaluate the performance of various BGS techniques. This software can be useful for evaluation of diverse BGS algorithms [4–9] using a variety of data sets. Also, because this software provides quantitative evaluation results in terms of precision, recall and *F*-measure as the evaluation measures [16], it helps users to easily produce supporting data to analyze a new BGS algorithm and compare the new algorithm with other traditional methods at both pixel level and frame level. Also, because the users can vary the input image conditions, such as the frame rate, image size, and color depth, the evaluation can be carried out more flexibly. In our future work, we will add more algorithms and test data sets to have the quantitative evaluation results automatically drawn from software in the form of statistical charts.

3 Hardware Framework

In this section, a low-power complementary metal-oxide-semiconductor (CMOS) image sensor is designed for event detection. The sensor includes a pixel array, pixel timing controller and analog signal processors. To enhance the sensors noise performance, correlated double sampling (CDS) is adopted in the analog readout circuits. An analog pixel signal is converted to digital values by column shared successive approximation register (SAR) analog-to-digital converters (ADCs). A 64×64 image pixel array is used for motion detection and image acquisition. A low-power, hardware-friendly event detection algorithm is implemented using the FPGA. The proposed design is fabricated in 0.18 μm three-metal one-poly (3M1P) standard CMOS technology, occupying a silicon area of 1.4×1.2 mm^2.

3.1 Introduction

The CMOS image sensor (CIS) has been successfully developed to replace charge coupled device sensors. Because the CIS offers advantages in on-chip functionality, system power reduction, cost, and size aspects. As the CIS can integrate multiple function blocks on a single chip and reduce component and packaging costs, it is preferred for various applications. A desire to reduce cost and optical issues has driven a steady reduction in pixel size. Pixel sizes are becoming smaller, which means that high-resolution image arrays can be integrated. However, when the amounts of data from the pixel array that are processed to produce the final image increases, the image sensors power consumption also increases; research is therefore focused on improving the sensor power performance, particularly for battery operated devices such as cameras for smart phones and for automobile black boxes that record events occurring outside the vehicle. However, limitations exist for long-term recording using conventional architecture of image sensors. Commercial automotive black box products can record videos for 3 days when powered by a car battery. In this article, we propose an image recording system that minimizes the system power consumption by adding an ultralow power image sensor to perform motion detection. So the car black box can record over 2 months with the car battery in extreme case. The proposed system can also be used for battery-operated surveillance cameras to save energy.

3.2 System Overview

A block diagram of the entire system is shown in Fig. 8. The system includes the low-power CIS with an off-chip interface. The CIS consists of a 64 × 64 pixel array, CDS read-out circuits, SAR ADCs, logic and control circuits for timing generation. The image array used is a standard three-transistor (3T) pixel, also called a photodiode-type active pixel sensor (APS), as the photodetector for light intensity sensing. Although a 4T pixel structure which is named a photo gate-type APS has strong points such as an inherent CDS function, low thermal noise and high sensitivity, 4T structure requires additional special processes so a cost of manufacturing is higher. Also the fill factor is lower because of the additional transistor. We decide to use 3T pixel structure because it has enough performance for the event detection and low cost. The CDS read-out is used to reduce fixed pattern noise from the pixel. The data from each 64 × 16 pixel array is processed in a single analog data readout circuit and SAR ADC. Therefore, four readout circuits and four SAR ADCs are required to process all the pixel information. The event detection algorithm is implemented in an off-chip FPGA, and the pixel integration time can be configured from the FPGA board. The parameter influences the exposure of the captured images and thus must be controlled appropriately. The off-chip FPGA also generates a clock signal to operate the CIS. An event detection result is displayed on both, a light-emitting diode (LED) indicator and display devices. The detailed description of the system operation is in next section.

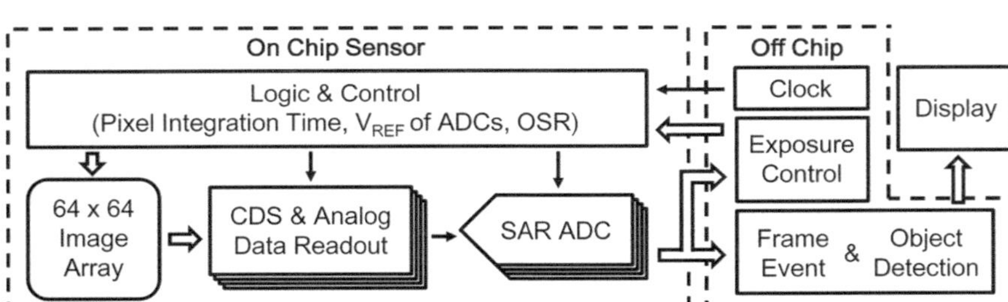

Fig. 8 System block diagram of CIS with an off-chip interface

3.3 *Circuit Implementation*

3.3.1 Image Array and Analog Data Readout

The proposed image sensor has been developed using a 0.18-μm CMOS standard logic process with additional process steps for photodiode fabrication. The pixel size is 10 × 10 μm^2 and the chip area is 1.68 mm^2. The power supply voltages used are 1.8 V for the digital logic gates and 3.3 V for the analog circuits and the pixels. The maximum frame rate is 240 frames per second (fps) with a main clock frequency of 5 MHz .The image sensor enables an analog readout proportional to the light intensity to be produced. The block diagram of the proposed image sensor is shown in Fig. 9. The photodiode-type (PD) APS is composed of one photodiode, one reset transistor (M1), one source follower (M2), and one switch (M3) for the output, which are all integrated in a single pixel [1],[2]. The image sensor consists of pixel circuits, column switching transistors, and a CDS circuit at the output stage for fixed-pattern noise (FPN) reduction [3]. The pixel circuits output both a signal level and a reset level voltage, and the CDS circuits receive and subtract these signals to generate the output signal without the FPN. The pixel operation scheme for this image sensor can be explained as follows. A row of pixel circuits is selected

Fig. 9 Block diagram of the CIS

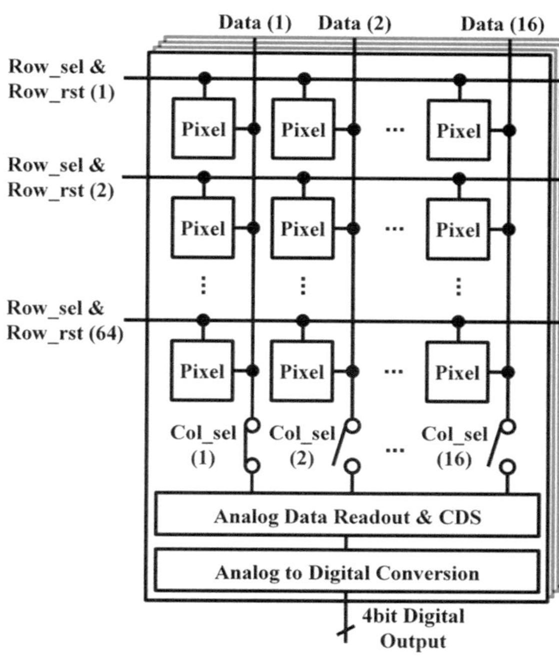

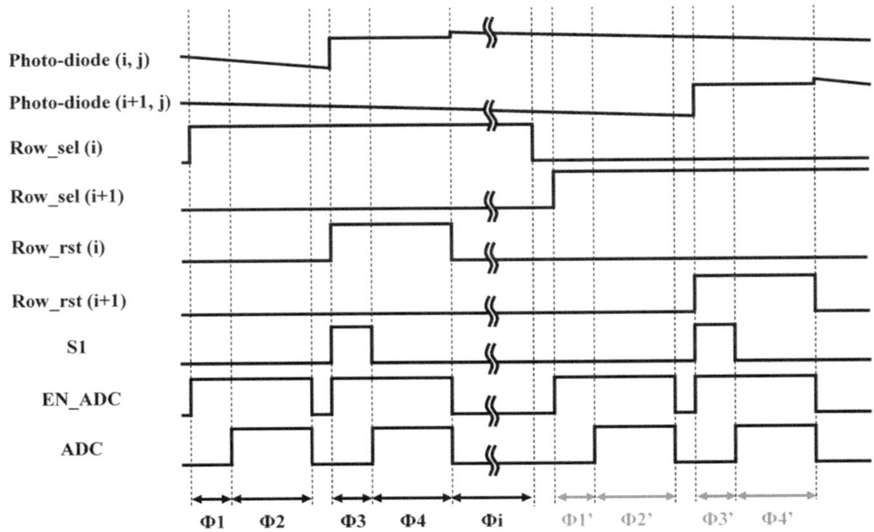

Fig. 10 Timing for CIS operation

in a single period of *Row_sel (i)*. Because a pulse selects a column switching transistor, one pixel in the image area is selected to output its signal to the horizontal signal line in a single pixel period. If a pixel circuit outputs a pixel signal with both a reset level and a signal level in one pixel period, then a CDS circuit at the output stage can subtract the signal level from the reset level of the pixel signal.

As a result, the CDS circuit outputs a signal without pixel signal offset variation, and thus functions as an FPN-reduction circuit. The analog readout is then digitized by the four SAR ADCs. The light intensity acquisition process begins from a reset phase when *Row_ rst (i)* is high, which is indicated as *phase 3* in Fig. 10, and the photodiode voltage has been pulled up to VDD. During *phase 3*, the reset voltage is settled as switch S1 is closed and the SAR ADC is ready for conversion of the pixel reset voltage. 4-bit digital data is then produced during *phase 4*, and this data is used for digital CDS, which can eliminate errors produced by the mismatch of the four SAR ADCs, which originated from process variations. After the reset phase, the photodiode voltage decreases as photon-generated charges accumulate on the photodiode capacitance. The integration phase indicated by *phase i* can be controlled by the integration time control unit. The integration voltage readout process starts when the switch integrated in the pixel *Row_sel (i)* and the column switch *Col_sel (i)* are turned on in *phase 1*. After settling of the integration voltage, the pixel data is converted by the ADCs in *phase 2*.

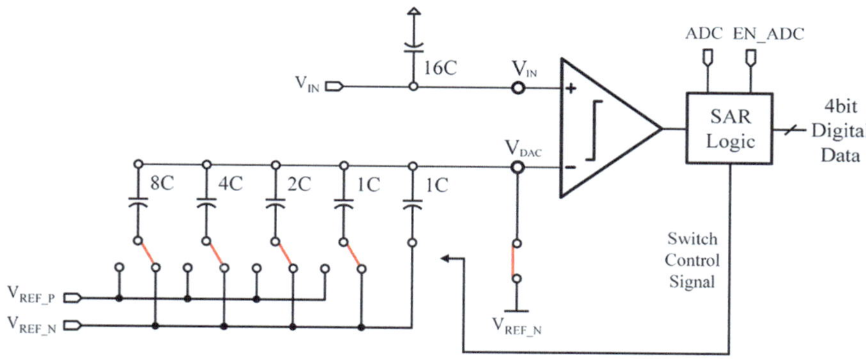

Fig. 11 Schematic of the SAR ADC

3.3.2 ADC Design

The proposed SAR ADC is shown in Fig. 11 and uses a capacitive digital-to-analog (C-DAC)-based structure that is suitable for low power applications. A resistive DAC-based SAR ADC can be implemented in a smaller chip area than a C-DAC-based SAR ADC. However, the resistive DAC-based SAR ADC consumes more power and makes it difficult to adopt the reference voltage control technique. The ADC is set to have 4-bit resolution because the purpose of this sensor is simply to detect events and not to capture high-resolution images. The size of the comparator input transistor and the unit capacitance of the DAC are calculated by noise simulations using the transient noise simulations. Based on the simulation results, the unit capacitor size was chosen to be 50 fF to enable the desired performance to be achieved. The SAR ADC is operated in a synchronous manner. When a digital value is determined, the SAR ADC is designed such that a maximum of only two switches are selected and thus the noise generated by switching is minimized. The input-referred capacitances of the two comparator nodes are matched to achieve accurate comparison results.

3.4 Measurement Results

The prototype chip was implemented in 0.18-μm standard 3M1P CMOS technology, occupying a silicon area of 1.4 × 1.2 mm^2. A microphotograph of the fabricated chip is shown in Fig. 12. Event detection is performed in the FPGA test board, which includes the event detection algorithm. The FPGA receives the digital data from the CIS and calculates whether an event has occurred. All the digital outputs from the chip are also used to communicate with a personal computer (PC) using a universal asynchronous receiver/transmitter (UART) serial port, and the image data can thus be stored by the PC as image files. Figure 13 shows the

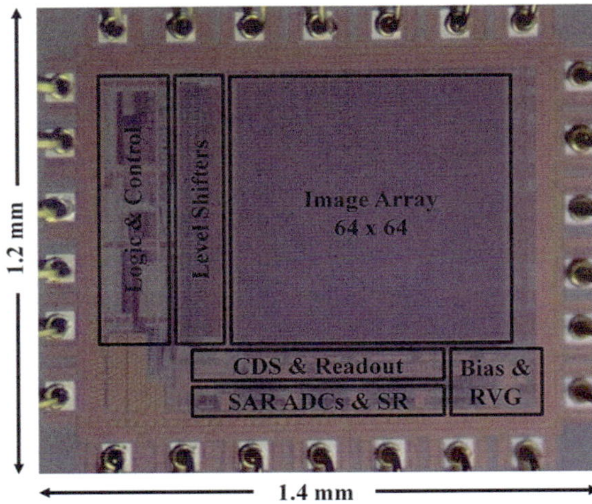

Fig. 12 Microphotograph of the fabricated chip

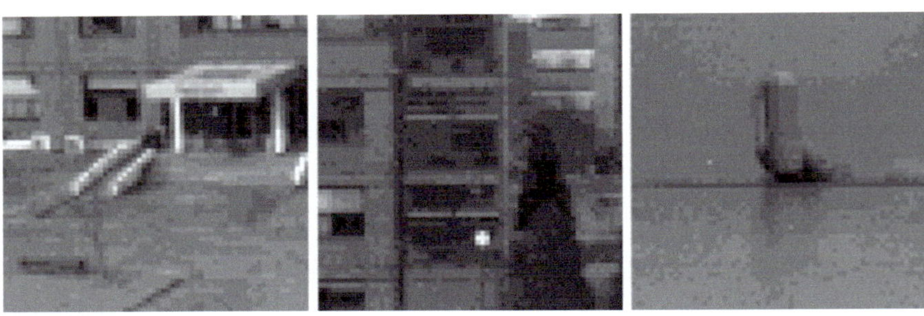

Fig. 13 Images captured by the image sensor

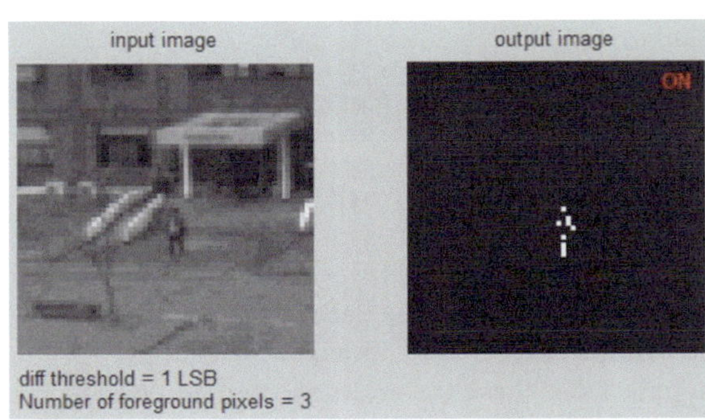

Fig. 14 Measured results using event detection algorithm

images captured by the designed sensor node. The resolution used is 64 × 64, which is high enough for the event detection application. Images captured under various light conditions are shown in Fig. 13. Image captured in a single frame by the event detection algorithm is shown in Fig. 14.

4 Conclusions

A BGS method based on a GMM using both color and depth information to overcome the limitations of color-based BGS such as color camouflage has been presented. A probabilistic background model which utilizes both color and depth information has also been built. To compare our method with conventional color based BGS methods, we first produced a new dataset to evaluate BGS algorithms. Then, using the dataset, we evaluated the proposed algorithms and color-based conventional BGS techniques in terms of precision, recall, and F-measure using the developed software in a pixel-wise manner. The algorithm achieved better performance than the others and reduced the depth noise. The probabilistic background models can be extended to be used in multi-modal data processing, such as thermal imaging cameras and night vision. Also, our method can also be used for preprocessing to define detect regions of interest before high-level applications (e.g., object detection, object tracking, and action recognition) in color-based problems such as dynamic color change and low illumination in addition to color camouflage. Implementation details of CIS as an axillary sensor for event detection have also been explained. The CIS fabricated in 0.18 μm CIS process has a 64 × 64 pixel array and occupies 1.2 mm^2. Column parallel architecture was adapted to convert pixel data into 4-bit digital codes. A SAR ADC architecture has been chosen for a column ADC due to its simplicity and low-power operation and each ADC sequentially processes 16 columns in the array.

Acknowledgments This work is supported by the Center for Integrated Smart Sensors funded by the Ministry of Science, ICT and Future Planning as the Global Frontier Project.

Appendix

See Figs. A.1, A.2, A.3 and A.4

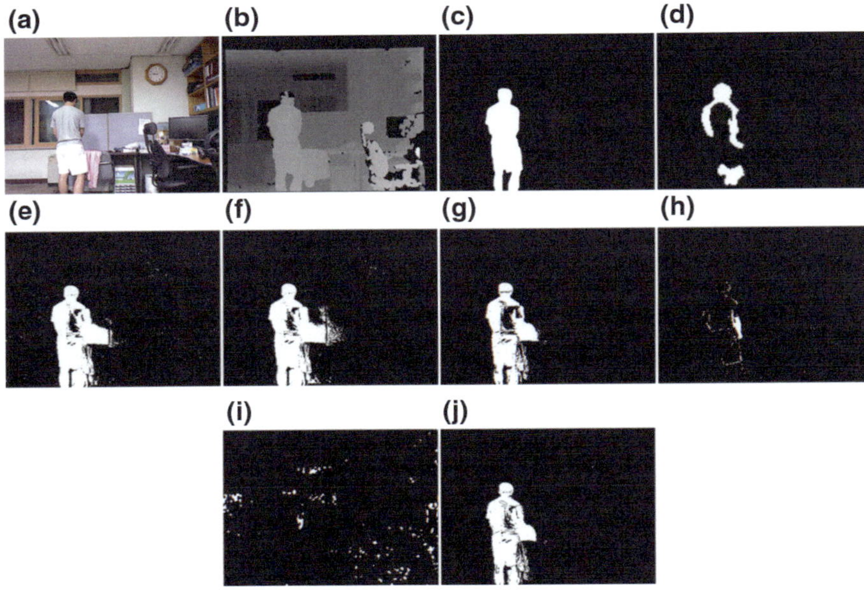

Fig. A.1 Evaluation dataset examples of **a** color, **b** depth image, and **c** ground truth, and qualitative evaluation results of **d** [10], **e** [7], **f** [8], **g** [6], **h** [5], **i** [9], and **j** the proposed method in color camouflage 1, as shown in Fig. 3a

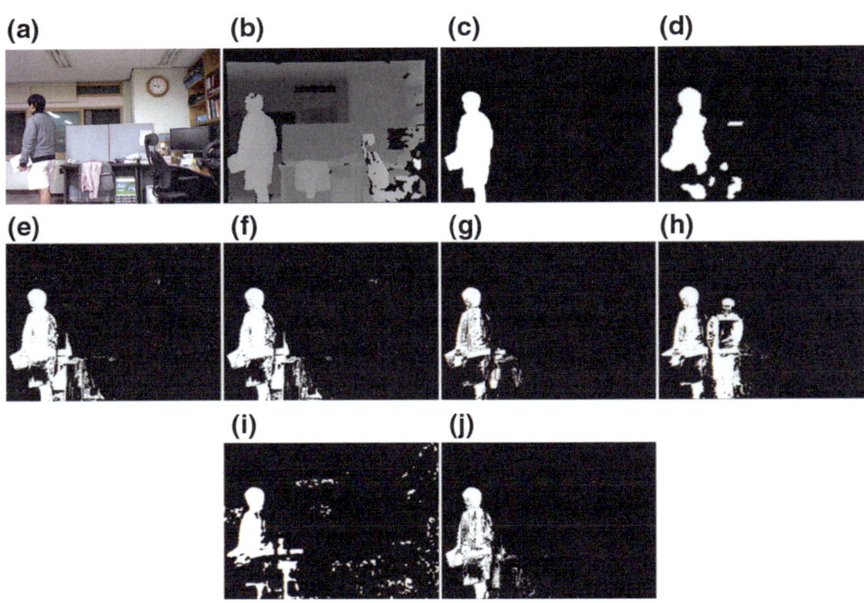

Fig. A.2 Evaluation dataset examples of **a** color, **b** depth image, and **c** ground truth, and qualitative evaluation results of **d** [10], **e** [7], **f** [8], **g** [6], **h** [5], **i** [9], and **j** the proposed method in color camouflage 2, as shown in Fig. 3a

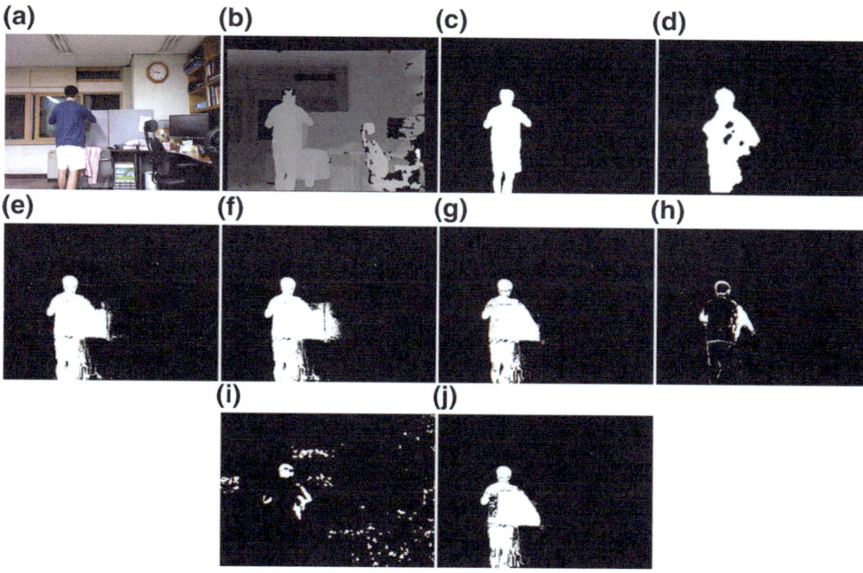

Fig. A.3 Evaluation dataset examples of **a** color, **b** depth image, and **c** ground truth, and qualitative evaluation results of **d** [10], **e** [7], **f** [8], **g** [6], **h** [5], **i** [9], and **j** the proposed method in a normal situation, as shown in Fig. 3c

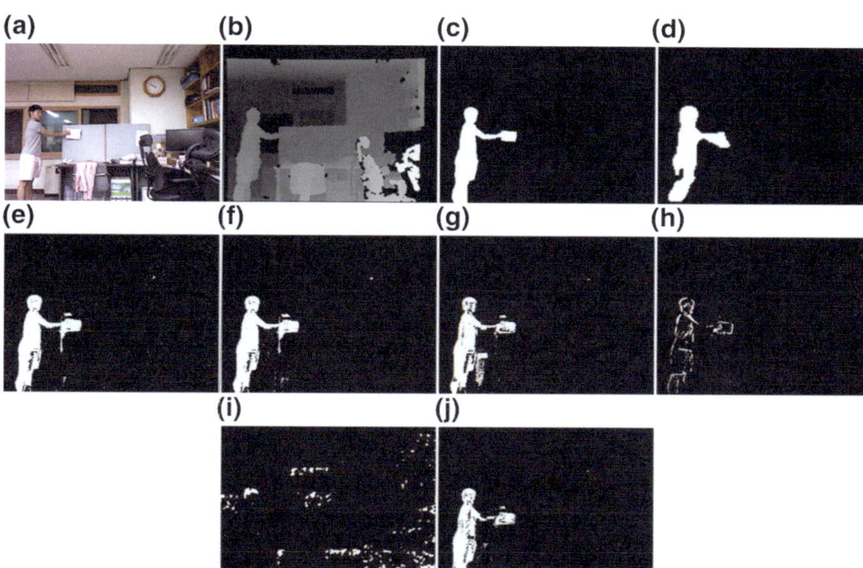

Fig. A.4 Evaluation dataset examples of **a** color, **b** depth image, and **c** ground truth, and qualitative evaluation results of **d** [10], **e** [7], **f** [8], **g** [6], **h** [5], **i** [9], and **j** the proposed method in depth camouflage, as shown in Fig. 3d

References

1. Tang F (2013) Low-power CMOS image sensor based on column-parallel single-slope/SAR quantization scheme. IEEE Trans Electron Devices 60(8):2561–2566
2. Fossum ER (1997) CMOS image sensors: electronic camera-on-a-chip. IEEE Trans Electron Devices 44(10):1689–1698
3. Yonemoto K (2000) A CMOS image sensor with a simple fixed-pattern-noise-reduction technology and a hold accumulation diode. IEEE J Solid State Circ 35(12):2038–2043
4. Cucchiara R, Piccardi M, Prati (2003) A detecting moving objects, ghosts, and shadows in video streams. IEEE Trans Pattern Anal Mach Intell 24:1337–1342
5. Zivkovic Z, van der Heijden F (2006) Efficient adaptive density estimation per image pixel. Patten Recogn Lett 27:773–780
6. Maddalena L, Petrosino A (2008) A self-organizing approach to background subtraction for visual surveillance applications. IEEE Trans Image Process 17:1168–1177
7. Maddalena L, Petorisino A (2010) A fuzzy spatial coherence-based approach to background/foreground separation for moving object detection. Neural Comput Appl 19:179–186
8. Yao J, Odobez J-M (2007) Multi-layer background subtraction based on color and texture. In: IEEE conference on computer vision and pattern recognition, June 2007
9. Noh SJ, Jeon M (2012) A new framework for background subtraction using multiple cues. In: The 10th Asian conference on computer vision
10. Harville M, Gordon G, Woodfill J (2001) Foreground segmentation using adaptive mixture models in color and depth. In: Proceedings of the IEEE workshop on detection and recognition of events in video, IEEE computer society, Los Alamitos, CA, USA, pp 311
11. Camplani M, Salgado L (2014) Background foreground segmentation with RGB-D Kinect data: an efficient combination of classifiers. J Vis Commun Image Represent 25(1):122136
12. Fernandez-Sanchez EJ, Rubio L, Diaz J, Ros E (2013) Background subtraction model based on color and depth cues. Mach Vis Appl 25(5):12111225
13. Stauffer C, Grimson W (1999) Adaptive background mixture models for real-time tracking. In: Proceedings of IEEE conference on computer vision and pattern recognition (CVPR 1999), pp 246–252
14. Bouwmans T, Baf F El, Vachon B (2008) Background modeling using mixture of gaussians for foreground detection-a survey. Recent Pat Comput Sci 1(3):219–237
15. Sobral A, Bender L, Parks D, Yao J, Odobez J-M, Noh SJ "A background subtraction library" in GitHub. https://github.com/andrewssobral/bgslibrary
16. David MW (2011) Evaluation: from precision, recall and f-measure to ROC, informedness, markedness and correlation. Int J Mach Learn Technol 2:37–63

Advanced Human Detection Using Fused Information of Depth and Intensity Images

Gyu-Hong Lee, Dong-Suk Kim and Chong-Min Kyung

Abstract Human detection systems have been applied to many applications such as intelligent vehicles and surveillance cameras with increasing demands on safety and security. The scope of previous works has been confined usually in color (or intensity) images. In this chapter, we present a complete human detection system using the information on both depth and intensity images. First, we apply a segmentation algorithm to a depth image. Then we merge the segmented regions and generate Region-Of-Interests (ROIs) which may contain a human, considering experimentally determined horizontal overlap and aspect ratio, respectively. Second, we use a newly proposed feature descriptor, Fused Histogram of Oriented Gradients (FHOG), to extract feature vectors from the ROIs applied in both depth and intensity images. Finally, we check the presence of humans in the ROIs with linear SVM. Following the basic principles of Histogram of Oriented Gradients (HOG), we develop this FHOG descriptor to utilize both gradient magnitudes of depth and intensity images. With our datasets obtained from Microsoft Kinect sensor, the FHOG descriptor and overall system achieve a miss rate of 1.44 % at 10^{-4} FPPW and of 10.10 % at 1 FPPI, respectively. The computing time of proposed system is also significantly reduced. Experimental results show our system is able to detect humans accurately and fast.

Keywords Human detection · Pedestrian detection · Segment-based ROI generation · RGB-D data

---

G.-H. Lee (✉) · D.-S. Kim · C.-M. Kyung
Smart Sensor Architecture Lab, ITC Building (N1) #314, Korea Advanced Institute of Science and Technology, 291 Daehak-Ro, 305-701 Yuseong-Gu, Daejeon, Republic of Korea
e-mail: ggamid79@naver.com

C. Kyung (ed.), *Theory and Applications of Smart Cameras*,
KAIST Research Series, DOI 10.1007/978-94-017-9987-4_12

1 Introduction

Human detection is one of the most interesting research topics in computer vision. It has traditionally been developed for robotics because the human detection algorithm can extend the perception ability of a system. Further, with growing interest on safety and security, human detection has received considerable attention for surveillance cameras and intelligent vehicles. However, human detection in visible-spectrum images has great difficulty in applying it to those real applications because of insufficient detection performance and high computational cost. According to the pedestrian detection benchmark [1], pedestrian detection performance in visible spectrum images marked over 0.15 of miss rate at 1 false positive per image (FPPI) and took more than 6 s to process a single frame.

It is difficult to address the problems using visible-spectrum images alone, because there are a lot of discouraging factors to build a robust human detector in visible-spectrum images such as illumination changes, complex background, and various human clothing. Further, computing time for human detection is not a negligible issue. Running a heavy algorithm in real time requires additional hardware resources such as GPU (Graphic Processing Unit). This leads to the degradation of power efficiency of overall system. So, the final goal of our research is designing human detection system satisfying both accuracy and computing times. Many researchers have been tried to reach the goal and find various methods for detecting humans. Owing to the development of affordable RGB-D cameras such as Microsoft Kinect and Mesa SR4000, depth information has become a new clue for designing advanced human detection system. (Details in Sect. 2)

In this chapter, we propose complete human detection system using both depth and intensity images. Our contributions are as follows:

- We generate segment-based ROIs on depth images. It reduces the number of candidate windows to classify whether a given image contains human or not. As a result, both false positives and computing times can be reduced.
- We develop a new feature called FHOG which is based on Histogram of Oriented Gradients (HOG). By fusing the depth and the intensity information, contours of human are intensified and thus detection rate can be increased.

The rest of this chapter is organized as follows. Section 2 describes the related work. Section 3 presents our ROI generation method and new feature descriptor, FHOG. Section 4 shows experimental results compared with other approaches.

2 Related Work

A typical human detection system starts detecting regions that are highly likely to contain humans. Then, features that describe a human are derived from the windows and the descriptors are classified by a pre-trained classifier. The advent of affordable RGB-D cameras which provide reliable depth information brought advantages to the human detection systems. The advantages are apparent for two modules: feature extraction and ROI generation.

Typically, features have been mainly extracted from intensity (or color) images. They use texture or gradient information in the image. Haar-like feature is a representative feature of texture-based method [2, 3]. It considers the difference of sum of intensity values in pre-determined rectangles. It works well for face detection but does not work satisfactorily for detecting humans. For human detection, HOG is the well-known descriptor [4]. It focuses on the discontinuities in image intensity. Local gradient orientation histograms are computed and normalized to make the feature more robust to the illumination changes in the image. The shape of head, shoulder, and legs is the most fundamental feature of a human in the HOG descriptor.

After depth information is widely used, new features are introduced which utilize depth information. In [5], Spinello et al. proposed Histogram of Oriented Depths (HOD) descriptor which locally encodes the direction of depth changes. They showed that detection accuracy increases because depth data is not affected by illumination changes unlike visible-spectrum images. Fusing features of intensity and depth images were proposed in [6]. They simply concatenate HOG features with HOD features. As a result, detection performance is improved, but it needs more computing time because feature dimensions are increased.

In visible-spectrum camera-based human detection system, fixed-size window is densely scanned through entire intensity (or color) image to extracting ROIs of including human [4]. To cover various heights of humans, images are scaled up or down, so the number of ROIs is significantly increased. Hence, it decreases overall speed of human detection system. To solve this problem, Q. Zhu et al. capture salient features of humans automatically and discriminate the appropriate regions [7]. In [5], they distinguish compatible scales likely to fit a height of human from a predetermined scale map and test the scaled windows.

Unlike intensity information, depth information is advantageous for extracting ROIs. In [8], they utilize graph-based segmentation algorithm on depth image to generate ROIs by merging segments based on their location.

We take a similar approach with B. Choi et al.'s methods. In our approach, the segments are merged perpendicularly as overlap ratio and ROIs are generated based on the aspect ratio of human body. Then the proposed FHOG descriptor is applied to these ROIs for classification.

3 Proposed Method

The proposed system consists of three stages as shown in Fig. 1. In the first stage, image segmentation is applied to depth images and ROIs are generated by combining the segments. Then features are extracted using fused information of intensity and depth images within ROIs, and finally ROIs are classified using a

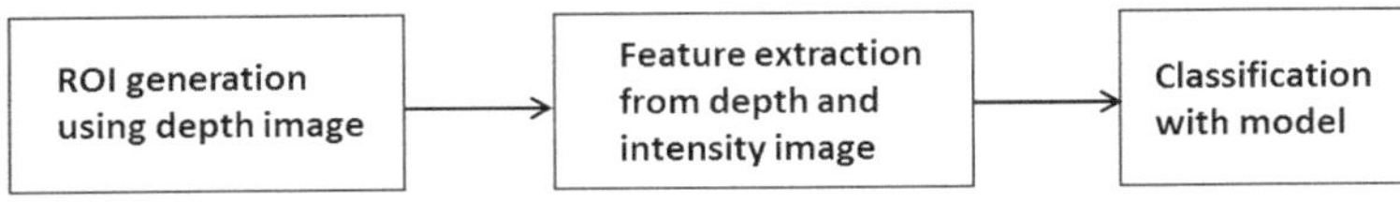

Fig. 1 Flows of the proposed human detection system

linear Support Vector Machine (SVM). Depth images are obtained from Microsoft Kinect sensor. This sensor provides both color and depth images. To use both images in the same image coordinate, image rectification is performed offline.

3.1 ROI Generation Using Depth Images

In depth image, intensity variation within same object is smaller than that of gray (or color) image because depth is not affected by textures on object. So the same object tends to have similar depth values. The characteristic becomes a motivation to utilize image segmentation to depth images for clustering similar depth regions.

(1) Depth image segmentation

We use the mean shift algorithm to segment depth images. The mean shift algorithm is a mode seeking algorithm that was made popular for image segmentation by Comaniciu et al. [9]. The size and the number of segments are decided according to the parameter set. The spatial, range and minimum size parameters for mean shift segmentation are determined experimentally to separate human body from background in depth images. As a result, labeled segments are acquired. Then we find the left-uppermost coordinate and the right-lowermost coordinate of each segment and calculate the width and height. The information is used to merge segments.

(2) Segment merging

In this step, we merge segments to obtain human candidates. Ideal results of the segmentation are that human is represented in one segment. However, human is usually separated into several segments (see Fig. 2). To elicit an intact human candidate from these segments, the horizontal overlap ratio (r_o) is used to combine pairs of segments. The horizontal overlap ratio is defined as

$$r_o = \frac{\text{hlength}(a \cap b)}{\text{hlength}(a \cup b)} \tag{1}$$

where $\text{hlength}(a \cap b)$ and $\text{hlength}(a \cup b)$ are horizontal intersection and union of pairs of segments, respectively. If horizontal overlap ratio between two segments is greater than a threshold, two segments are merged as depicted in Fig. 3. We set the threshold value as 0.35.

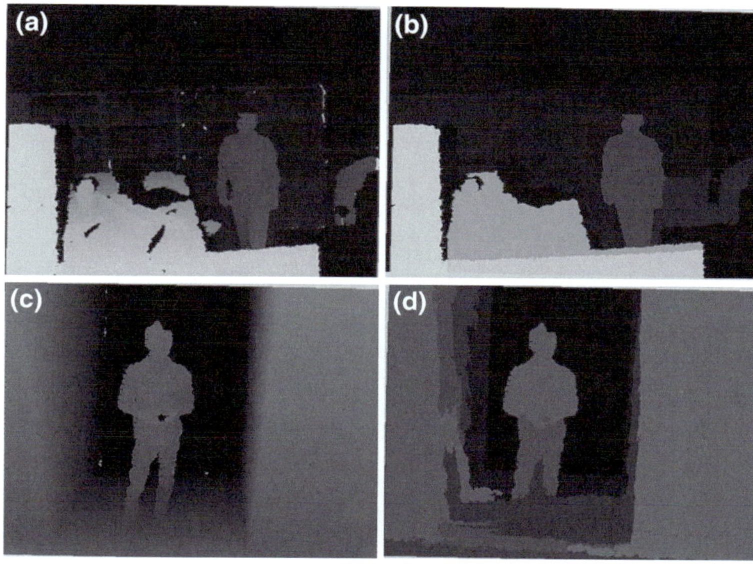

Fig. 2 Image segmentation results. (**a**) and (**c**) are depth images. (**b**) and (**c**) are segmented results of (**a**) and (**c**), respectively. A human in (**b**) is separated as one segment, but (**d**) is not

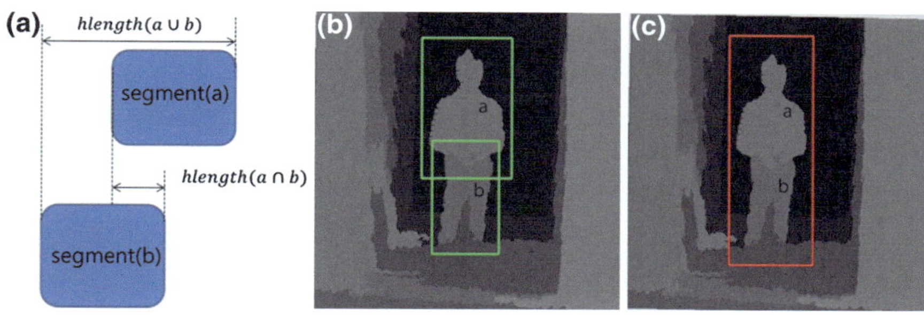

Fig. 3 Concepts for merging segments. **a** Parameters for calculating horizontal overlap ratio, **b** before merging, and **c** after merging

(3) ROI generation

A large set of candidates are generated from previous steps (Fig. 4a). Here, we investigate the aspect ratio (r_a) of candidates to filter out impractical candidates. The aspect ratio of a segment (a) is defined as

$$r_a = \frac{\text{width}(a)}{\text{height}(a)} \tag{2}$$

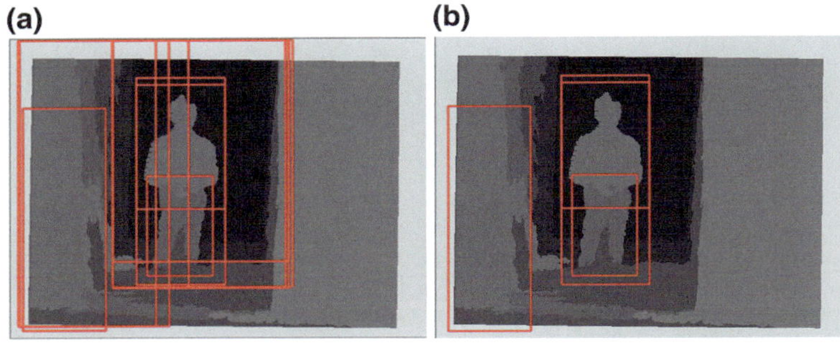

Fig. 4 ROI generation results. **a** Example of candidate generation, **b** final result of ROI generations

where, width (a) and height (a) are the width and height of candidates (a), respectively. All the candidates of satisfying the predetermined aspect ratio are selected as ROIs. The aspect ratio is determined as the ratio of human body (between 0.25 and 1 in our system). The ROIs are confined in a bounding box with their coordinates acquired in step 1. As illustrated in Fig. 4b, ROIs of containing human are successfully generated with a small number of total ROIs.

3.2 Feature Extraction from Depth and Intensity Image

We propose a new feature extraction method which is an extended version of HOG for human classification in depth and intensity images. In this section, we introduce our feature called FHOG after a summary of the HOG descriptor.

(1) Histogram of Oriented Gradients (HOG)

HOG is the most popular feature for human detections. HOG feature expresses a sample image on the basis of its local shape and appearance using histograms of gradient orientation. It computes gradient magnitude and orientation in a fixed-size window called detection window. Then it builds histograms with orientation bins for each cell which is densely subdivided regions in the detection window. Votes of the histograms are accumulated into the orientation bins. The histograms are normalized within a group of cells, which is called a block. The normalization process is necessary to make the feature more robust to the effect of illumination changes. Finally, extract the HOG descriptor by a feature vector concatenation of all the normalized histograms.

(2) Fused Histogram of Oriented Gradients (FHOG)

We developed a new feature extraction method for human classification in depth and intensity images based on HOG descriptor. Our descriptor takes advantage of the information from both depth and intensity image. The reasons for using both images are as follows: Generally, intensity image gives detailed information of an object because it has abundant textures. But it is vulnerable to illumination changes, and also when the background is complex, the rich textures can increase the false positives. In contrast, depth image is robust to illumination changes and can alleviate the effect of the complex background. However, it is sensitive to low return signals and may give insufficient data for detecting humans. Thus, the feature of using both complementary images can be very powerful and promising.

In a depth image, the gradient magnitude of a human contour appears relatively stronger because textures of object and background are ignored. So we can obtain amplified gradient magnitudes around human contours by adding the gradient magnitude of depth image to the intensity image (Fig. 5a). The amplified gradient magnitude (M_s) at pixel(x, y) can be defined as:

$$M_s(x,y) = M_i(x,y) + \omega \cdot M_d(x,y) \tag{3}$$

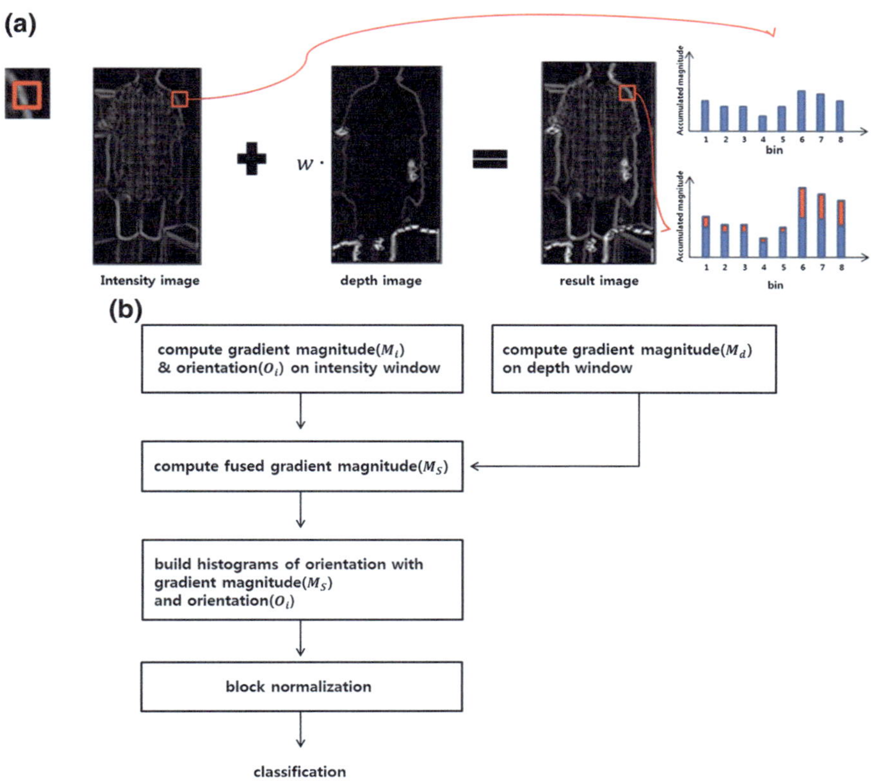

Fig. 5 **a** Procedure used for extracting fused histogram of oriented gradients, **b** an overview of our feature extraction method

Table 1 FHOG parameters

Cell size	12 × 12 pixels
Block size	2 × 2 cells
Overlap of block	1
The number of bins	9 (unsigned)
Vote method for histogram	Gradient magnitude
Normalization factor	L2-norm

where $M_i(x, y)$ and $M_d(x, y)$ are the gradient magnitude at pixel(x, y) of intensity and depth image, respectively, and ω is a weighting factor to control the amount of effect of depth's gradient magnitude. We use the amplified gradient magnitude (M_s) to build histograms for each cell. The overview of our feature extraction procedures is illustrated in Fig. 5b.

In a similar method, S. Wu et al. extract features by combining depth and intensity images [6]; while their method increases the dimensions of feature vectors, our method maintains the dimensions of feature vectors although the descriptors are more discriminative. As a result, we can save memory space and decrease the execution time. The FHOG parameters which are used in this experiment are described in Table 1.

3.3 *Classification with Model*

We use linear SVM for classification. The linear SVM is a binary classifier looking for the most suitable hyperplane as decision function defined as

$$h(x) = \sum w_i x_i + b \tag{4}$$

The optimal $h(x)$ is sum of the inner product of the feature vectors x_i and the weight vectors w_i. Here, w_i and b are obtained from supervised learning with a training set. The sign of $h(x)$ decides whether the features are in-class or out-of-class. We use two sets of training examples which are externally and internally obtained from Kinect sensors.

4 Experimental Results

The proposed human detection system was tested on two different datasets that are externally and internally obtained from Kinect sensors. In this section, we introduce these two datasets and show the evaluation results in terms of the performance of ROI generation, feature extraction, and overall system.

4.1 Dataset

(1) Public dataset

The first dataset is RGB-D dataset provided by Spinello et al. [5]. The dataset has been taken in a university hallway using three vertically mounted Kinect sensors. It includes three sequences of videos and a total of 1648 people in 1088 frames are labeled. As shown in Fig. 6, people in this dataset are upright and completely visible or partially occluded. We use this dataset to compare the performance of our ROI generation method with the method proposed by B. Choi et al. [8]. 1000 positive examples and 4500 negative examples are extracted in depth images for training, and 200 depth images are used for testing (Fig. 6).

(2) Our dataset

We have used Kinect sensor mounted at a height of 1.7 to 2 m from the ground to collect our own RGB-D dataset. The dataset was taken in various indoor places (such as university hallway, laboratory, underground parking-lots, and classroom). Our dataset includes people who are upright and fully visible or partially occluded. We use the dataset to evaluate the performance of feature extraction and overall system. To evaluate feature extraction method, 1025 positive samples and 5000 negative samples are used for training and 1325 positive samples and 11774 negative samples are used for testing. To evaluate the overall system performance, 1175 images containing 1316 people are used. Figure 7 shows some examples of our dataset.

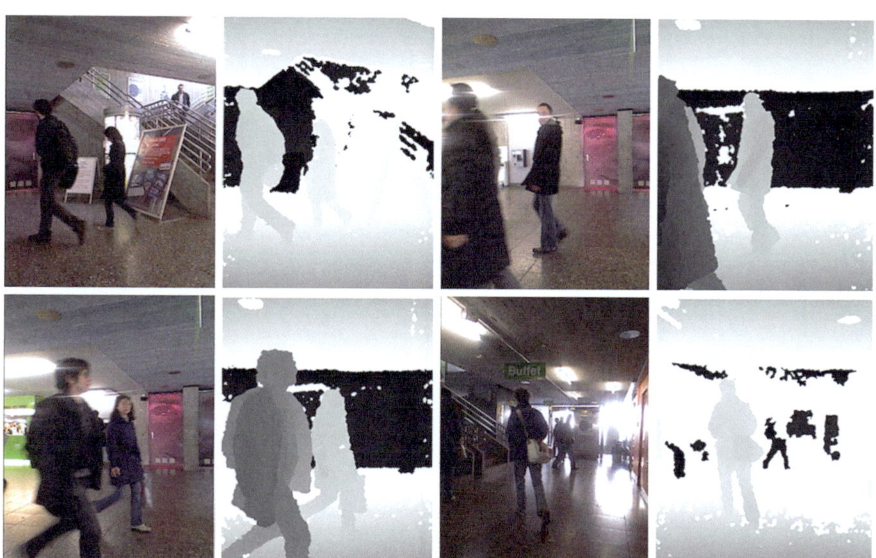

Fig. 6 Examples of public dataset

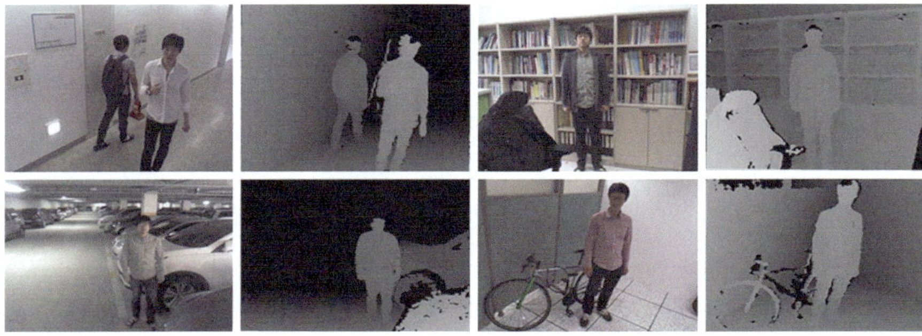

Fig. 7 Examples of our dataset

4.2 Evaluation of ROI Generation Method

In this section, we evaluate our ROI generation performance by comparing to the B. Choi et al.'s method which is based on graph-cut algorithm [8]. They use the HOD descriptor and it is tested on the public dataset which was mentioned in previous section. To compare the performance, we implemented the HOD algorithm and used the same dataset. To quantify performance, we plot Equal Error Rate (EER) curves. The EER is the matching point between recall and precision. In the EER curves, if the matching point is located on the right-uppermost areas, it can be considered that the accuracy of the detection system is relatively high. Our method achieved an EER of 87 %, which performs slightly better than the 84 % of graph-based ROI (Fig. 8).

Fig. 8 Equal Error Rate (EER) curves

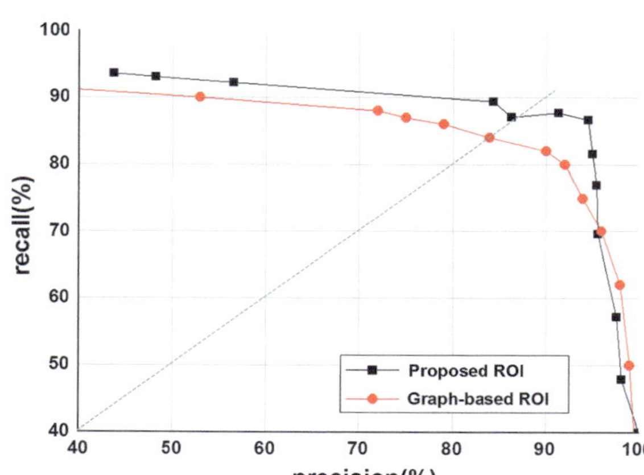

4.3 Evaluation of Feature Extraction

We compare the performance of our FHOG feature with HOG [4], HOD [5] and HOG-HOD [6]. These features are tested on our dataset and we see the per-window performance. Detection Error Tradeoff (DET) curves on a log–log scale are used to evaluate the performance of features, *i.e.,* miss rate (1-recall) versus false positives per window (FPPW). First, we test our FHOG feature to determine the optimal weighting factor (ω) by varying the value from 1 to 100. As shown in Fig. 9a, when the ω is 10, FHOG achieved the lowest miss rate (1.44 %) at 10^{-4} FPPW. For the other features, HOG, HOD, and HOG-HOD achieved a miss rate of 10.94, 8.31, and 6.72 % at 10^{-4} FPPW, respectively (Fig. 9b). It seems that using depth images for detecting human is helpful to improve the detection rate. Further, FHOG reduces the miss rate by 5.28 % as compared to the HOG-HOD. This means that our fusion method strengthens the features better than simple feature concatenation approach.

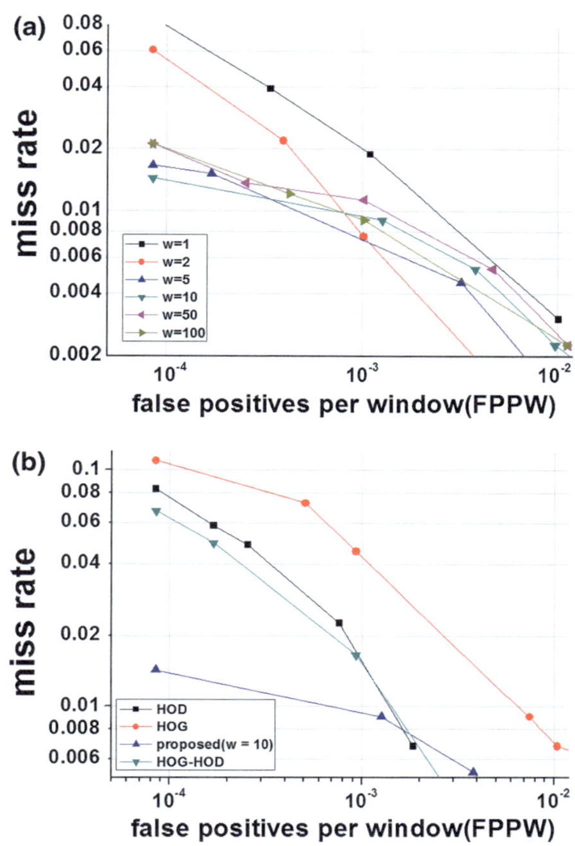

Fig. 9 Detection Error Tradeoff (DET) curves. **a** The effect of ω, **b** different descriptors on our dataset

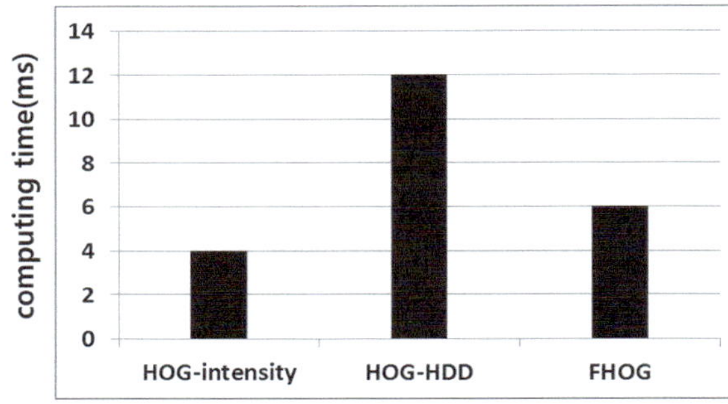

Fig. 10 Computing time of each descriptor

Further, we evaluate the computing time of each features. Figure 10b shows that FHOG is computed faster than HOG-HOD, but slower than HOG (or HOD). Interestingly, our feature can detect better to the partially occluded human (see Fig. 11). This result indicates that fused feature of depth and intensity works robustly on complex background and in the case that the contours of human are lost in an image.

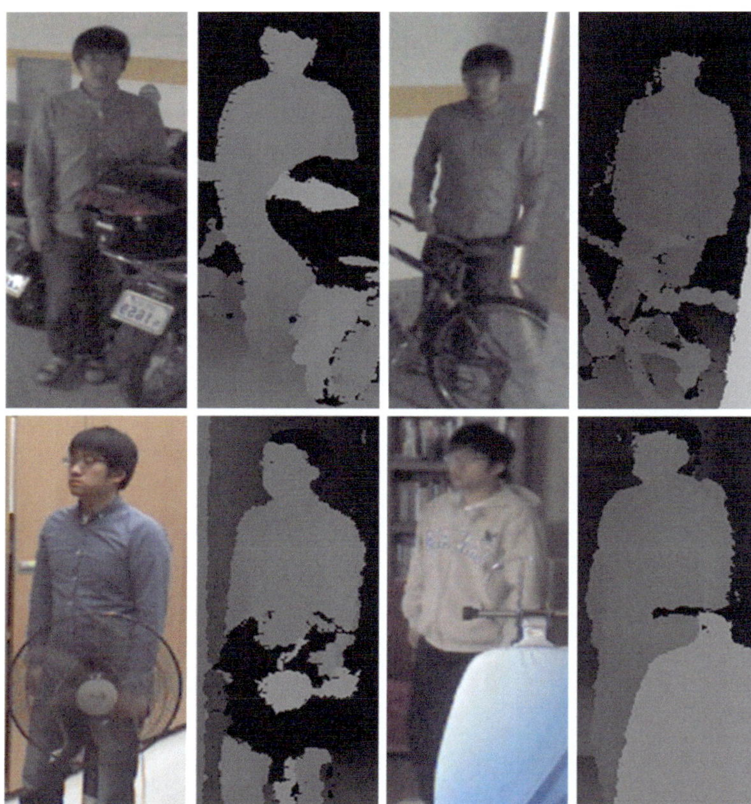

Fig. 11 Examples of detecting a partially occluded human

4.4 Evaluation of Overall System

In this section, our human detection system is compared against a reference system on our data set. Ideally, comparing our system to the system proposed in [8] is more precise since they use a similar strategy (graph-based ROI + HOD) to ours. However, we did not compare our system to [8] since the original implementation of graph-based ROI was not available to us. So we designed a system using FHOG descriptor and sliding window technique for extracting ROIs. To avoid the circumstances that the ROIs do not contain humans, we scanned the image as densely as possible. We plot DET curves by the miss rate versus false positives per image (FPPI) to evaluate per-image performance. As shown in Fig. 12a, our system performs better than the reference system. Further, our system takes 2.65 s to process a frame, while the reference system takes 6.95 s. The experiments were conducted on a computer with an Intel core i5 processor (Fig. 12b). Figure 13 shows the examples of human detection.

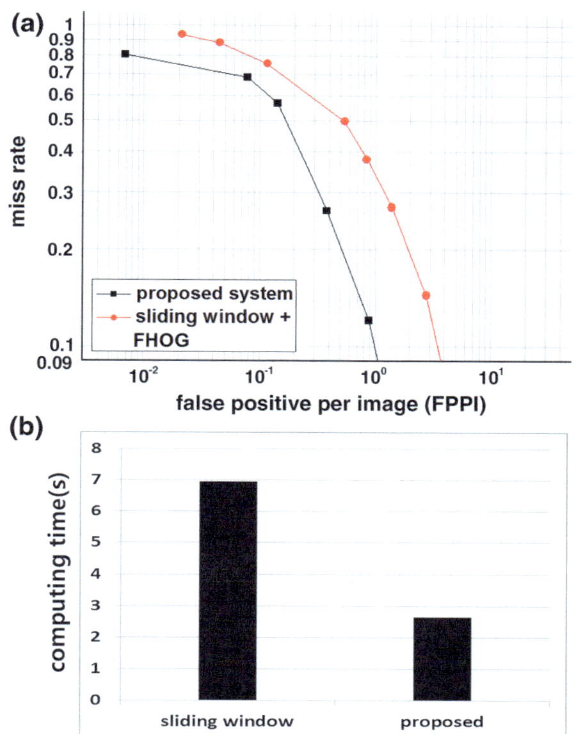

Fig. 12 **a** DET—human detection performance comparison results, **b** computing time of each system

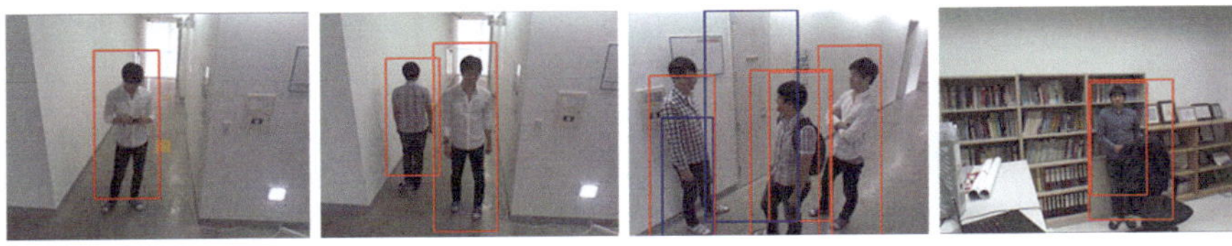

Fig. 13 Examples of human detection. Red boxes—true positives, blue boxes—false positives

5 Conclusions

In this chapter, we introduced an advanced human detection system using depth and intensity images. First, we applied image segmentation to depth images and generated feasible ROIs in consideration of the predetermined aspect ratio. This process significantly reduces the false positives and the computing times. Further, a new descriptor (FHOG) that fusing depth and intensity images is proposed for feature extraction. The FHOG achieved a recall of 98.56 % at 10^{-4} FPPW and it takes about 6 ms for processing a detection window (48 × 96 pixels). Further, the FHOG worked well for detecting partially occluded person. The overall system (mean-shift based ROI + FHOG) achieved a recall of 89.90 % at 1 FPPI and it significantly reduces the computing time by 61.87 % compared to the reference system (Sliding window + FHOG).

Acknowledgments This work is supported by the Center for Integrated Smart Sensors funded by the Ministry of Science, ICT & Future Planning as the Global Frontier Project.

References

1. Dollar P, Wojek C, Schiele B, Perona P (2009) Pedestrian detection: a benchmark. In: IEEE conference on computer vision and pattern recognition, 2009. CVPR 2009, pp 304–311. IEEE
2. Viola P, Jones MJ (2004) Robust real-time face detection. Int J Comput Vision 57(2):137–154
3. Viola P, Jones MJ, Snow D (2003) Detecting pedestrians using patterns of motion and appearance. In Proceedings ninth IEEE international conference on computer vision, 2003, pp 734–741. IEEE
4. Dalal N, Triggs B (2005) Histograms of oriented gradients for human detection. In: IEEE Computer society conference on computer vision and pattern recognition, 2005. CVPR 2005, vol 1, pp 886–893. IEEE
5. Spinello L, Arras KO (2011) People detection in RGB-d data. In: IEEE/RSJ international conference on intelligent robots and systems (IROS), 2011, pp 3838–3843. IEEE
6. Wu S, Yu S, Chen W (2011) An attempt to pedestrian detection in depth images. In 2011 Third Chinese conference on intelligent visual surveillance (IVS), pp 97–100. IEEE

7. Zhu Q, Yeh M-C, Cheng KT, Avidan S (2006) Fast human detection using a cascade of histograms of oriented gradients. In: 2006 IEEE computer society conference on computer vision and pattern recognition, vol 2, pp 1491–1498. IEEE
8. Choi B, Mericli C, Biswas J, Veloso M (2013) Fast human detection for indoor mobile robots using depth images. In: IEEE international conference on robotics and automation (ICRA), 2013, pp 1108–1113. IEEE
9. Comaniciu D, Meer P (2002) Mean shift: a robust approach toward feature space analysis. IEEE Trans Pattern Anal Mach Intell 24(5):603–619

Part III
Wireless Connectivity for Video Sensor Networks

Time Synchronization for Multi-hop Surveillance Camera Systems

Hyuntae Cho

Abstract In recent years, surveillance systems designed for public safety have become more intelligent by providing context awareness. Traditional surveillance camera systems require access to energy and networking infrastructure in order to operate and to transmit the recorded video data. Since such requirements can increase the costs incurred when installing and maintaining surveillance systems, a wireless surveillance camera system is hereby introduced. The system can operate with low power consumption and also provides network connectivity. The battery life of the system is improved by separating the system into master and slave subsystems. The master subsystem provides Wi-Fi connectivity and records video while the slave-subsystem provides low-power event detection with ZigBee connectivity. The system uses Wi-Fi mesh networks to transmit video data and ZigBee networks to define the network topology and to synchronize multiple surveillance camera systems. Time synchronization is a fundamental issue for distributed surveillance camera systems, so this chapter details a method to synchronize time among multiple surveillance camera systems by using ZigBee radio communications.

Keywords Time synchronization · Clock synchronization · Wireless mesh networks · Wireless surveillance systems · Zigbee

1 Introduction

The increase in crime in residential areas and in public spaces has resulted in an increase in demand for surveillance systems, such as those provided by CCTV or by security services [1, 2]. Recently, the market for surveillance camera systems has shifted from CCTVs to IP-based cameras because IP-based cameras offer advantages over CCTVs in terms of resolution, cost, potential applications, etc. [3]. Furthermore,

---

H. Cho (✉)
Center for Integrated Smart Sensors, ITC Building(N1), KAIST, Daehak-ro 291, Yuseong-Gu, Daejeon 305-701, Republic of Korea
e-mail: phd.marine@kaist.ac.kr

C. Kyung (ed.), *Theory and Applications of Smart Cameras*,
KAIST Research Series, DOI 10.1007/978-94-017-9987-4_13

big data, cloud, and IoT services have extended the potential applications of IP-based cameras [4]. However, these camera systems have many technical requirements, so it can be difficult to use them under specific circumstances. In particular, surveillance systems that are based on traditional cameras require access to power and network infrastructure, which results in high installation and maintenance costs for surveillance systems.

This chapter describes a wireless surveillance camera system that provides IP connectivity through use of a wireless networking platform. The proposed system captures images and video and then transmits the recorded data to a remote point through a self-organized wireless network. Since wireless networks consume large quantities of energy to provide network connectivity and to transmit video data, the proposed system is equipped with a dual radio system to conserve energy [5]. The system comprises two subsystems: a master subsystem and a slave subsystem. The master subsystem records and processes video and then transmits the recorded video by using a Wi-Fi mesh network. The slave subsystem turns the master subsystem off to maintain the entire system in a low-power mode as long as possible when no events require video to be captured and transmitted. The slave subsystem also determines the network topology, including synchronizing time, by using a control channel based on energy-efficient ZigBee communications. Time synchronization affects the performance of the entire system and the network for such distributed surveillance camera systems, and so it is a fundamental issue. Time synchronization can be generally achieved by exchanging time information. The exchange of time information via wireless networks can result in uncertainty due to signal delay and jitter, particularly when using a ZigBee network. Therefore, this chapter analyzes uncertainties in the ZigBee network protocol stack and introduces basic techniques to eliminate or minimize them.

This chapter is organized as follows. In Sect. 2, we present conventional approaches used for surveillance camera systems with wireless networking. In Sect. 3, we describe the proposed system and the corresponding network platform, and then we present with basic techniques to synchronize time across distributed wireless surveillance camera systems.

2 Related Work

The recent introduction of low-cost CMOS image sensors has resulted in an increase in the range of applications of wireless video networks [6]. Devices can use these sensors to capture pictures or video from the environment, and one such use case involves wireless sensor networks that provide surveillance and security by using a network of nodes to identify and track objects according to visual information. Wireless video sensor networks can also greatly improve applications in the area of environmental monitoring [7, 8]. Visual information from the environment is important for such applications, including for precision farming or habitat monitoring. Wireless video sensor networks will also enable new forms of entertainment where real-time visual information can be provided at a large scale from a remote location, such as for a digital zoo [9].

Wireless cameras can be used to monitor and track objects in the field, such as in construction sites, harbors, forests, and campuses, and Firetide Inc. [10] and Strix Systems [11] are well-known commercial vendors of traditional wireless mesh products. Current products focus on conventional problems and focus on features that improve performance in terms of coverage, quick mobility, reliability, security, and solid networking. In such products, the camera is simply mounted on a wireless mesh platform.

Several studies [12, 13] have investigated issues relevant to these systems, including video transmission, multi-channel operation, and improvements in network bandwidth. Raniwala and Chiueh [12] presented a multi-channel WMN architecture that effectively improves the bandwidth by exploiting nonoverlapped radio channels available through IEEE 802.11 standards. S. Yang constructed multi-radio, multi-hop wireless mesh networks by developing a Linux-based implementation of a WLAN mesh system [13]. The main design goal for our system is to fully exploit link layer characteristics in order to improve the configuration flexibility as well as the network performance. Although some research has been performed to date for wireless mesh networks for surveillance purposes, such studies have only focused on traditional mesh networking problems.

3 Wireless Surveillance Camera System for Public Safety

Low-power surveillance systems and network platforms require different approaches from those used in conventional systems. This section describes new approaches to reduce power consumption to provide wireless communications for surveillance camera systems. The proposed system consists of two subsystems: a high-performance master subsystem and a low-power slave subsystem, as shown in Fig. 1. The master subsystem is based on an ARM Cortex A9 processor [14] and uses OpenWRT [15] to manage the system. The master subsystem records video and transmits video data by using an FHD camera, a microphone, a high-speed application processor, and Wi-Fi network interfaces. The Ubiquiti Networks SR71 WLAN card [16] and Ath9k are used to provide Wi-Fi communications. For load balancing, the communications radio is separated into four modules: two up-links and two down-links. The master subsystem can also be connected to the Internet via Ethernet, which is more reliable than wireless networks. In addition, the GPU on the main processor helps the system recognize objects and mitigates the load on the main processor core.

The slave subsystem is responsible for topology management and low-power maintenance of the entire system. It includes a VGA camera, a low-power MCU, a microphone, an RTC, memory, and a power management circuit with a solar cell. It also includes gas, temperature, humidity, ozone, UV, and smoke sensors to detect external events and a ZigBee transceiver to provide channel control. The slave subsystem uses an ARM Cortex M4 processor (STM32F407 [17]) as its main processor and FreeRTOS [18] to manage tasks. The detailed architecture is presented in Fig. 1 with (a) the network protocol stack of the entire video sensor system and (b) the hardware block diagram of the system.

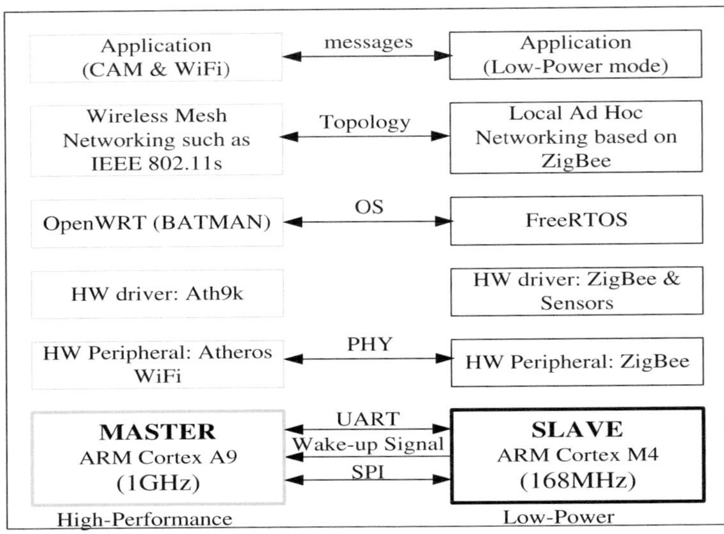

Overview of the system stack

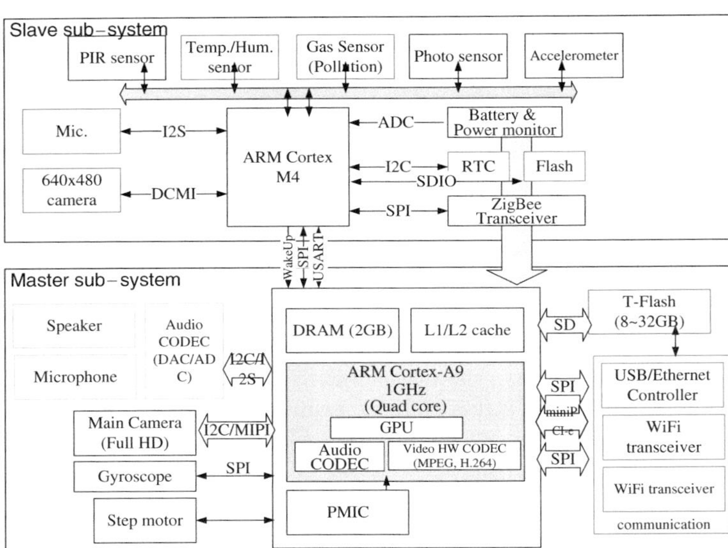

block diagram

Fig. 1 Conceptual overview of the proposed surveillance system

The proposed system transmits recorded video to a remote user through multi-hop communications via the Wi-Fi mesh network. To reduce energy consumption during transmission, we additionally separate the communications channel over dual radios because traditional Wi-Fi mesh networks consume a high amount of energy. Figure 2 shows the conceptual overview of the wireless video sensor network platform. The proposed system uses ZigBee for channel control and a Wi-Fi mesh channel to transmit video data. The system initially operates in its low-power mode by turning the master subsystem off since the master subsystem consumes more energy by several orders of magnitude higher than the slave subsystem does. The network topology and the route to the sink are constructed through the control channel. A number of routing and topology management protocols have been previously developed, including OSLR, AODV, DSR, or

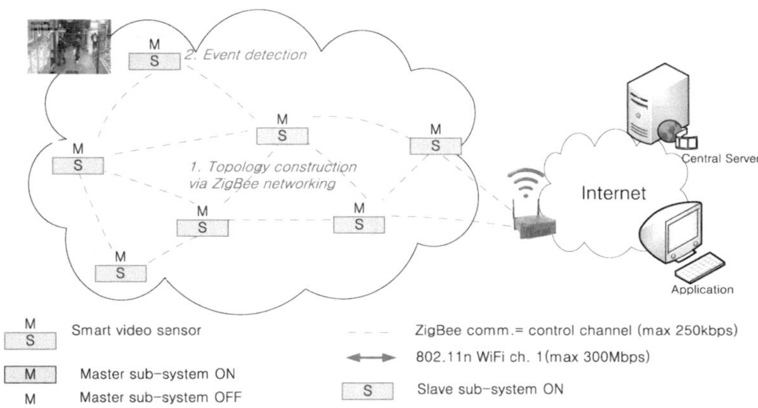

Topology construction and management via the ZigBee control channel

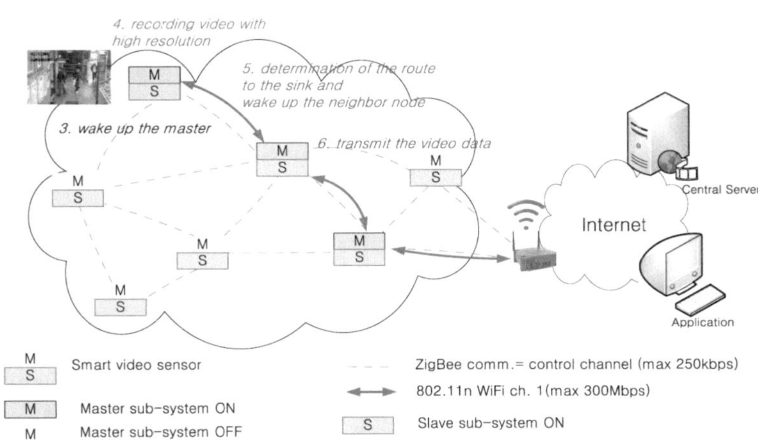

Routing and video data transmission

Fig. 2 Wireless video transmission for low-power communication

BATMAN-advanced [5, 19]. The OSLR and BATMAN protocols are frequently used for Wi-Fi mesh networks.

When the slave subsystem detects an event (whether from the camera, the microphone, the other sensors, or the ZigBee radio), it wakes up the master subsystem to record the ambience and to transmit the recorded data. The master subsystem processes more visual data and extracts much more information than the slave subsystem because the master subsystem has a higher performance processor and a GPU, such as ARM's Mali, with higher data processing capability. The captured video is compressed by system in a manner according to the detected level of importance or the bandwidth available. In addition, the system and the network platform mitigate traffic over the wireless mesh networks by adopting multi-channel, multi-radio, and multi-path approaches.

The system wakes up the neighboring nodes before transmitting the recorded data, and it is important to quickly synchronize time across the nodes to in order to properly organize their schedule in the network. The proposed system would consume a large quantity of energy if Wi-Fi were used to maintain the topology and time synchronization. Therefore, a ZigBee radio is used to perform topology maintenance and to synchronize time, thereby reducing energy consumption. The next section describes the uncertainties introduces by the ZigBee network protocol stack and the methods through which these can be reduced to provide precise time synchronization.

4 Time Synchronization Over ZigBee Networks for Surveillance Camera Systems

Time synchronization is critical for wireless surveillance camera systems as it is for modern computer networks where transmissions must be managed and scheduled while handling contention, among other things. Time synchronization is achieved by sending and receiving time information and frequency over the packet network. A synchronization protocol is used to exchange the time information, such as the *offset* and *propagation delay*, and to synchronize all clocks. This section describes the basic principles for ZigBee-based time synchronization, which is used for the wireless surveillance camera systems.

4.1 Time Synchronization Methods

The global positioning system (GPS) enables precision time-keeping through its satellite clocks because timing is based on a standard atomic clock that uses the oscillations of a particular atom, such as Cesium or Rubidium, as a metronome. Such clocks provide the most stable and accurate time reference. This timing information is obtained by GPS receivers, which require precision timing to compute their distance to each satellite in order to derive their position on Earth.

The Network Time Protocol (NTP) [20] is widely used to synchronize time over computer networks. NTP timestamps are numbered and are exchanged between peers, and messages are then exchanged to calculate the time offset and to synchronize clocks by correcting for the offset. The propagation delay is calculated by using the round trip time.

The IEEE 1588 precision time protocol (PTP) [21] provides a standard method to synchronize devices in a network with submicrosecond precision. The protocol synchronizes slave clocks to a master clock, ensuring that events and timestamps for all nodes use the same timer values. Since a time difference between a master clock and a slave clock is a combination of the clock offset and the message transmission delay, the clock skew is corrected in two phases: offset correction and delay correction. The master node initiates the offset correction by using a sync message and a follow-up message. When the master node sends a sync message, the slave uses its local clock to timestamp the arrival of the sync message. The slave then compares the local timestamp to the actual sync transmission timestamp from the master clock's follow-up message. The difference between the two timestamps represents the offset for the slave, plus the message transmission delay. The second set of messages is necessary to account for variations in the network delay. The slave then timestamps the instant when a delay request message is sent, and the master clock timestamps the arrival of the delay request message. It then sends a delay response message with the delay request arrival of the timestamp. The difference between the timestamps is the slave-to-master delay. The slave averages the two directional delays and then adjusts the clock by the time of the delay to synchronize the two clocks. Since the master and slave clocks drift independently, the offset correction and delay correction are periodically repeated to maintain the clock synchronized (Fig. 3).

In WSNs, sensor nodes synchronize their time according to a reference clock, such as that of the sink node or coordinated universal time (UTC), which is the time standard by which the world regulates clocks and time in time synchronization. For

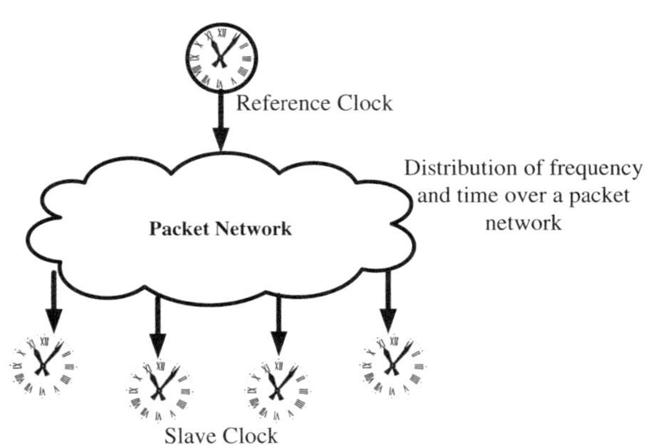

Fig. 3 Time synchronization over a packet network

WSNs, time synchronization requires clocks to be synchronize across a set of sensor nodes connected to one another over single-hop or multi-hop wireless networks. To date, various protocols have been designed to address this problem [22–31]. Time synchronization may be classified into three types: (a) simple unidirectional broadcast, (b) receiver-receiver synchronization, and (c) bidirectional pair-wise synchronization, as shown in Fig. 4.

In the unidirectional reference broadcast method, a reference node simply broadcasts a reference clock signal to other nodes, and these other nodes correct their times to match the reference clock. This method is the oldest and simplest method to synchronize time across a network. The flooding time synchronization protocol (FTSP) [32] is the most well-known approach. FTSP uses a fine-grained clock, media access control (MAC) layer time stamping to reduce jitter and clock drift estimation in order to achieve a relatively high level of precision.

Receiver–receiver synchronization uses an external beacon node that periodically sends beacon messages to the sensor nodes. The sensor nodes that receive the beacon messages exchange the arrival times of the messages among themselves to compare and correct their clocks. Reference broadcast synchronization (RBS) [33] and adaptive clock synchronization (ACS) [34] are receiver–receiver synchronization

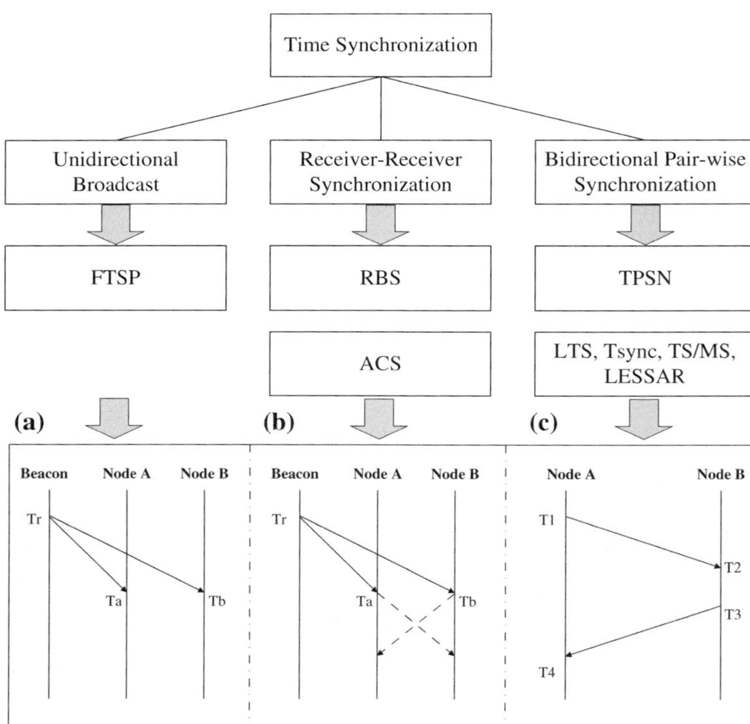

Fig. 4 Classification of time synchronization protocols for WSNs

protocols. RBS does not utilize an explicit timestamp, but rather receivers use the arrival times as points of reference to compare their clocks, as shown in Fig. 4b. This approach directly removes two of the largest sources of non-determinism involved in message transmission: the transmission time and the access time in the network protocol stack. ACS extends RBS and focuses on reducing the number of the messages that are used to exchange the message arrival times. In order to reduce the number of messages, the beacon node is used instead of the sensor node to gather and compare the message arrival times.

Third, bidirectional pairwise synchronization, which can also be referred to as sender–receiver synchronization, uses the round trip time of the message to correct the offset and the propagation delay. This approach uses a handshake protocol between a pair of nodes. That is, sensor nodes achieve clock synchronization with their parent node unlike receiver–receiver synchronization where sensor nodes synchronize their clocks with other sensor nodes on the same level. Figure 4c depicts an example of the basic operation of this method in three sequential phases. First, node A sends its local time at time $T1$, and node B receives the message at time $T2$ and records its local time. Then, time $T2$ is calculated as $T2 = T1 + d + \delta$, where d is the propagation delay between two nodes and δ denotes a clock offset between them. Next, node B responds to node A with an ACK message containing times $T2$ and $T3$. After receiving the ACK message at time $T4$, node A determines time $T4$ as $T4 = T3 + d–\delta$. Finally, node A can calculate the clock offset and the propagation delay between two nodes, as below:

$$\begin{aligned} d &= \frac{[(T2 - T1) + (T4 - T3)]}{2} \\ \delta &= \frac{[(T2 - T1) - (T3)]}{2} \end{aligned} \tag{1}$$

The timing-sync protocol for sensor networks (TPSN) [35], lightweight time synchronization (Tsync) [36], tiny-sync and mini-sync (TS/MS) [37], and level synchronization by sender, adjuster, and receiver (LESSAR) [38] are well-known bidirectional pairwise synchronization protocols for WSNs while NTP is a form of bidirectional pairwise synchronization protocol used over the Internet. TPSN provides synchronization for an entire network. First, a node is elected as a synchronization master, and a spanning tree with the master at the root is constructed by flooding the network. In the second phase, the nodes synchronize to their parent in the tree by means of round-trip synchronization. TSync has a centralized version and a decentralized version. Both protocols use a dedicated radio channel to synchronize messages in order to avoid inaccuracies due to packet collisions. TS/MS uses multiple pairwise round-trip measurements and a line-fitting technique to obtain the offset and drift of two nodes, rather than directly calculating the offset. LESSAR is able to achieve accuracy within a given limitation while also retaining low power consumption, affordable storage, and small computation complexity due to the reduction in packet transmissions.

4.2 Uncertainty in Time Synchronization

Uncertainty is inserted communication within the network protocol stack from the application to the physical layer, including the communications link, as shown in Fig. 5. The time uncertainty in the network protocol stack is dependent on the determination of an instant of time, and such a determination during time synchronization is referred to as time stamping. The time stamping point is critical because it affects the accuracy of the time synchronization procedure. The time stamping point can be determined for any point within the network layers. However, time stamping at an upper layer, such as the application layer, has a disadvantage in that the protocol stack can cause delays that may not be deterministic. The delay between the time stamp and the transmission can vary between a minimal value and a maximal value, depending on the network and protocol states. Transmission can be delayed if it causes a collision, and time stamping by the receiver can be performed at the start of an interrupt, after receiving a frame. The delay in the reception can vary according to the protocol stack and the kernel activity. The delay and jitter can be reduced by performing time stamps as close to the wires as possible [39, 40].

The lowest time stamp point with software is at the MAC layer. However, time stamping at the MAC layer also suffers from delay and jitters. We deal with IEEE 802.15.4 and ZigBee, which is based on carrier sense multiple access (CSMA). Bidirectional pairwise synchronization has an advantage in that uncertainties at the network protocol stack and the propagation delay can be mitigated by using exchange messages. However, this approach requires additional traffic, and the number of messages increases as the scale of the network increases. That is, surveillance cameras contend among themselves to access the channel, as shown in Fig. 6. Thus, a busy channel leads to nondeterministic latency in the MAC layer and finally diminishes the accuracy and precision of the time synchronization. In other words, the MAC verifies whether the channel is clear before it sends a sync or ACK message. If the channel is busy as a result of transmitting other messages, MAC

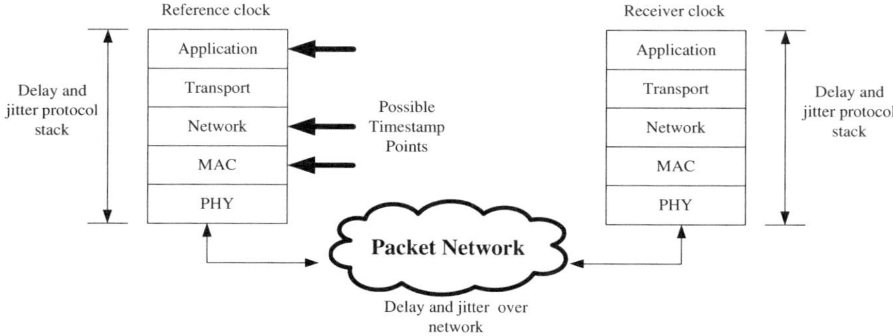

Fig. 5 Uncertainty at the network protocol stack

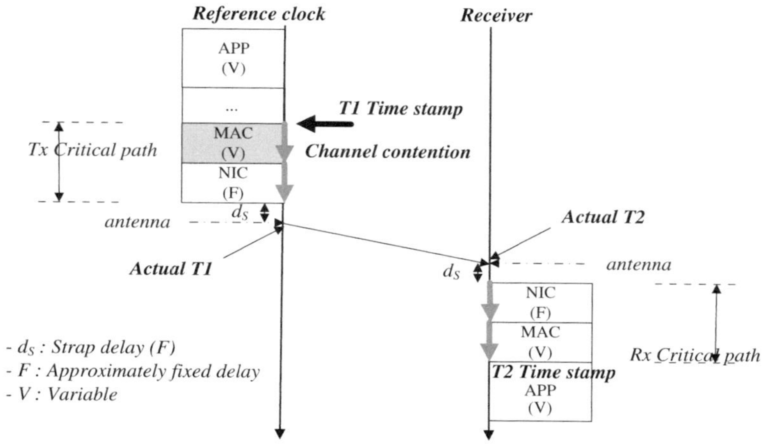

Fig. 6 Uncertainty at the MAC layer

waits for a random back-off period. After waiting for this random back-off period, the node resends the message, including the time stamp value. This delay introduces a serious uncertainty. Thus, the number of messages should be reduced and collisions between the messages must be avoided in order to increase the accuracy and precision of time synchronization.

Tsync [36] and LESSAR [38] are lightweight time synchronization protocols (Fig. 7). These protocols use three message types: *sync*, *delay_req*, and *delay_resp*. A *sync* message is initially sent by the reference clock node, which is defined as level 0 and acts as the root node. The reference node inserts time *T*1 into the *sync* message, and each sensor node receives the packet at time *T*2 and records their local clock. Then, the sensor node determines the clock offset as $\delta = T2{-}T1$. When calculating the delay between the reference node and other nodes, delay calculations from all of the child nodes can produce a high amount of traffic, which results in inaccurate synchronization.

Thus, the uncertainties in the propagation speeds are assumed to be the same in different nodes, and the uncertainty of the propagation delay is less than that of other uncertainties, such as the send, access, receive time, etc. This assumption underlies the proposed method, where only one child node responds in order to calculate the propagation delay from the reference node or the parent node. The reference node determines which node responds to the sink node in order to measure the delay by consulting its neighbor list. This selection is based on a min-ID selection. The information used for the responding node is inserted into the *sync* message, and the node receiving the *sync* message first checks whether it itself is the target for the message. If so, the node sends a *delay_req* message that includes times *T*2 and *T*3. Otherwise, it will be discarded. Then, the reference node receives the *delay_req* message at time *T*4 and records the arrival time for the message. Next, the reference node determines the propagation delay between its

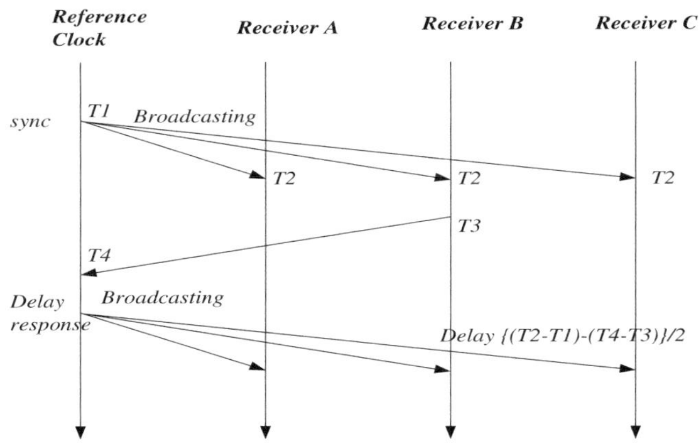

Fig. 7 Lightweight time synchronization

one-hop children nodes and itself, as shown in Eq. (1), and the dealy is broadcast to its one-hop nodes. Finally, the child nodes can correct the propagation delay by receiving the *delay_resp* message from the reference node. As a result, these methods can mitigate random delays at the MAC layer.

However, minimizing the messages is not the optimum solution to remove the random back-off delay. In order to eliminate the delay and the jitter at the MAC layer, it is important to implement hardware-assisted time stamping. Time stamps that use a hardware-assisted stamper can be performed at the media-independent interface between the MAC layer and the Zigbee physical (PHY) layer. When a Zigbee device receives MAC protocol data from the upper layer, it generates a four byte preamble with a one-byte start of frame delimiter (SFD) and a one-byte frame length. Then, the device transfers data to the MAC protocol data unit (MPDU) and performs a cyclic redundancy check (CRC), as shown in Fig. 8. After the last bit of the SFD is transferred at this point, the ZigBee transceiver causes the SFD pin to increase. The time-stamping unit of the sensor node detects the rising edge of the SFD pin, and then the hardware-assisted time stamping unit can detect and store the value of the local clock counter in an internal register.

Figure 9 depicts the time stamping points of the SFD from the time processing unit. Figure 9a illustrates the hardware-assisted time stamping unit and (b) shows that the hardware-assisted time stamping unit eliminates uncertainty at the MAC layer and has the same delay. The time stamping unit is independent of the processor of the main module of the system, and this time stamping unit can be implemented by using an independent processor or FPGA. However, it should be connected to the main processor, which executes the time synchronization protocol.

After hardware-assisted time stamping is performed, the stamped time should be inserted into the message to be transmitted to the receivers. The time captured from the SFD signal is inserted into the MAC protocol data, as shown in Fig. 10a. For the wireless bit-stream, most of the wireless controllers (ZigBee transceiver) provide

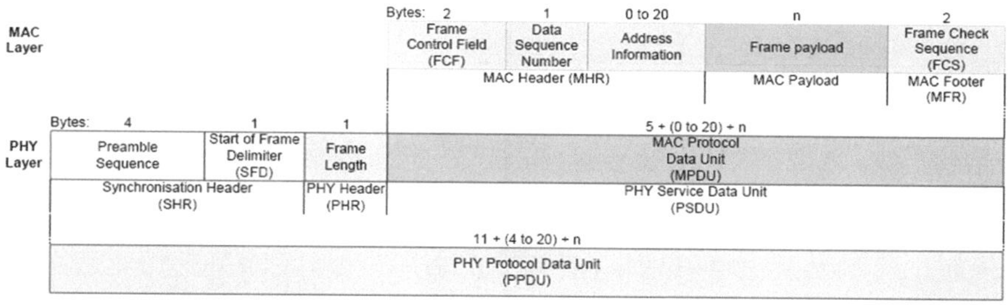

Fig. 8 Frame of IEEE 802.15.4 [41]

FIFOs to transmit and receive data. The processor communicates with the controller (e.g., TI CC2420 and CC2520) by using a synchronous peripheral interface (SPI). Generally, when a message is transmitted, the processor loads up the transmit FIFO with the entire message and then enables transmission.

The timestamp is inserted into the message, and the rest of the message is placed in the FIFO. Assuming that this can all be done quickly enough, the entire message is transmitted properly. If, however, the process is too slow, the FIFO will underrun and the message transmission will abort. This is a real concern since ZigBee specifies a fairly speedy effective bit rate of 250 kbps, where 4 μs is required to transmit 1 bit. In order to insert a time stamp into the payload, the communication speed between the processor and the FIFO of the ZigBee transceiver should be faster than 250 kbps. Generally, the speed of ZigBee transmissions is lower than the actual data rate due to coexistence with Wi-Fi and other radios. Furthermore, since ZigBee-based systems target low-power operation, such systems use low-power processors with a low clock speed. Therefore, these low-speed processors do not achieved the speed required to insert the time stamp into the message.

The message that is used to synchronize time can be separated into two messages, as shown in Fig. 10b. The reference clock sends a sync message, and after passing the *sync* message at *T*1, the time stamping unit reads and stores the local time of the reference clock. The time *T*1 is not inserted into the *sync* message. After receiving the *sync* message, the receiver clock records the value of the local clock counter at *T*2. If the receiver node has information for *T*1 and *T*2, it can calculate the offset between the reference and the receiver clock. However, it does not have information for *T*1, and the reference clock inserts the time stamp for *T*1 into the consequent message. The reference clock may also send the consequent message, which it always associates with a specific *sync* message and contains a more precise estimate for the reference time. The receiver clock uses the information contained in the consequent message to correct its local clock, so as to synchronize time with the reference clock. Such an approach can reduce the uncertainty at the MAC layer during time synchronization.

The propagation delay is also measured, calculated, and corrected according to the round-trip time based obtained using a hardware-assisted time stamping unit, as shown in Eq. (1).

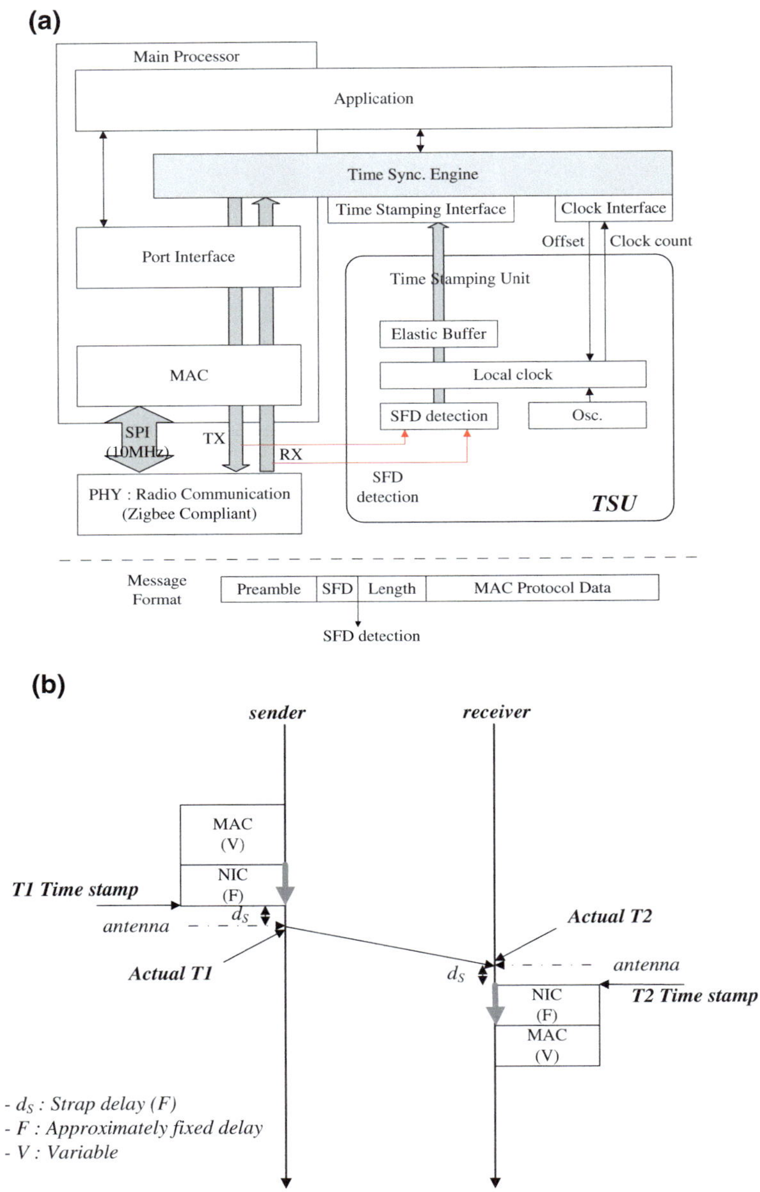

Fig. 9 Hardware-assisted time stamping unit and time stamping points **a** Hardware-assisted time stamping unit, **b** Hardware time stamping point at the sender and the receiver

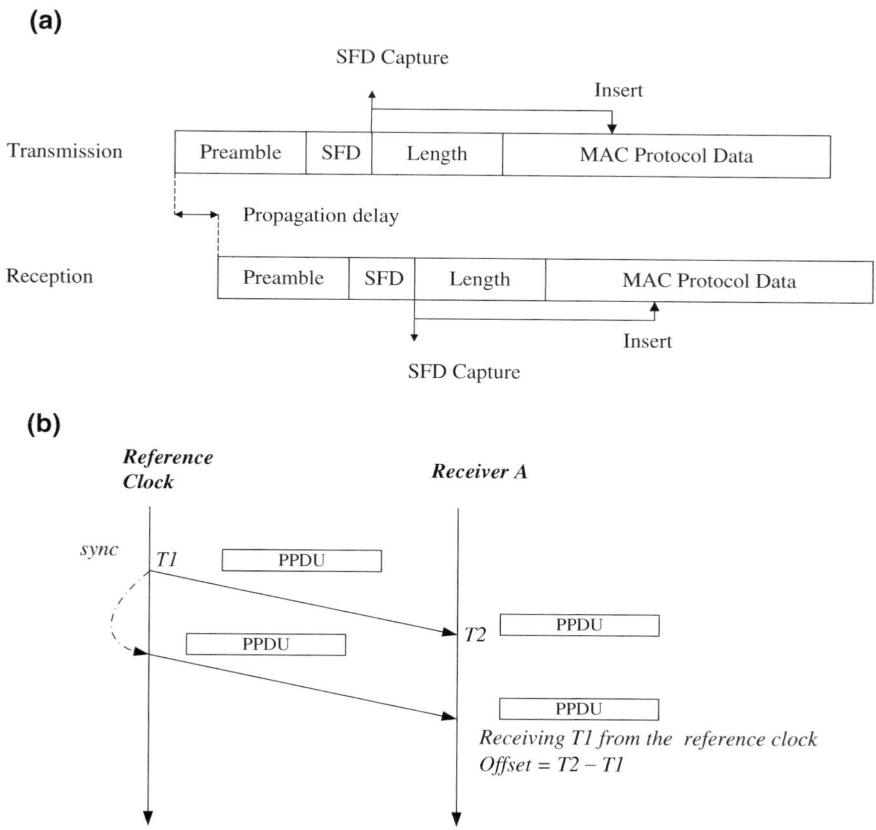

Fig. 10 Precision time stamping and transmission **a** direct insertion of the time stamp and transmission, **b** Consequent transmission of the precision time information

4.3 *Drift and Correction*

A system clock is controlled by a crystal oscillator that operates in a pre-determined manner. Under ideal circumstances, physical clocks oscillate at a constant frequency, but in the real world, manufacturing variations and exposure to out-of-tolerance conditions (e.g., mechanical shock) result in permanent frequency errors in the crystals. In addition, variations in the temperature, age, humidity, etc., result in short-term errors in the crystals. An oscillator with a 1 parts-per-million (PPM) frequency tolerance has a one microsecond drift every second. In general, cheap oscillators have a frequency tolerance from 20 to 50 PPM, where the maximum drift rate is between 20 and 50 microseconds per second. Such a value is inadequate to provide precise time protocol. Thus, for most cases, a temperature-compensated crystal oscillator (TCXO) with a 1.5 PPM frequency tolerance reduces the drift rate. Even if two clocks are initially synchronized by correcting the time offset and delay, a difference can accumulate between them as time progresses [42].

Assume that the local clocks at two nodes, i and k, are $c_i(t)$ and $c_k(t)$. If $c_i(t) = c_k(t)$, the two clocks are synchronized at time t. If the algorithm for time synchronization could know the relative offset between $c_i(t)$ and $c_k(t)$ at time t, $c_k(t)$ can be synchronized to $c_i(t)$ at each epoch by correcting for the relative offset. Figure 11 represents the synchronized clock $c_k^o(t)$. Although $c_k(t)$ is exactly synchronized to $c_i(t)$ through a periodic correction, clock $c_k^o(t)$ pursues a line derived from a variation in clock $c_k(t)$ because this synchronization did not consider clock drift. Thus, LESSAR assumes that clock drift quickly changes, and therefore, the synchronization procedure is frequently conducted. However, this eventually reduces the synchronization accuracy because the channel remains busy with excessive synchronization messages [42, 43].

Frequent sync messages can help calculate the drift from the reference clock, as shown in Fig. 12. Equation (2) presents the drift compensation correction for the receiver nodes.

$$\Delta_m = T_{m+1} - T_m$$
$$\Delta_s = T_{s+1} - T_s$$
$$\Delta_{\text{diff}} = \frac{\Delta_s - \Delta_m}{\Delta_m} \tag{2}$$

where Δ_m is the clock drift of the reference clock node that applies to clocks between T_m (the first time stamp) and T_{m+1} (the consequent time stamp). Δ_s is the clock drift of the sensor node, and it applies to clocks between the arrival time of the *sync* message, T_s, and the arrival time of the consequent sync message, $T2$. Δ_{diff}

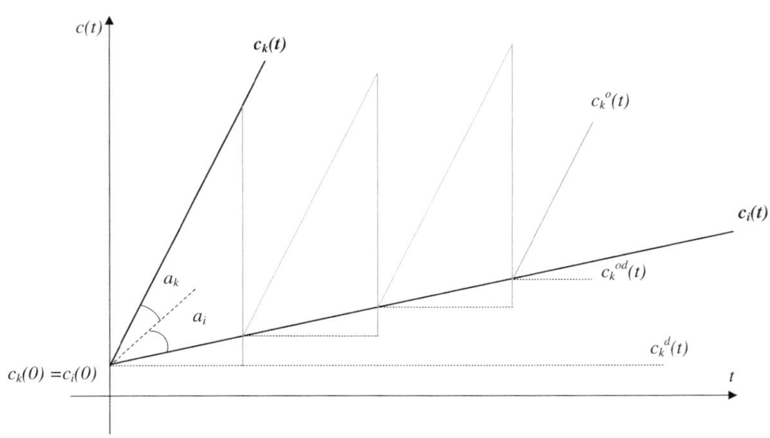

Fig. 11 Clock difference by the local clock drift

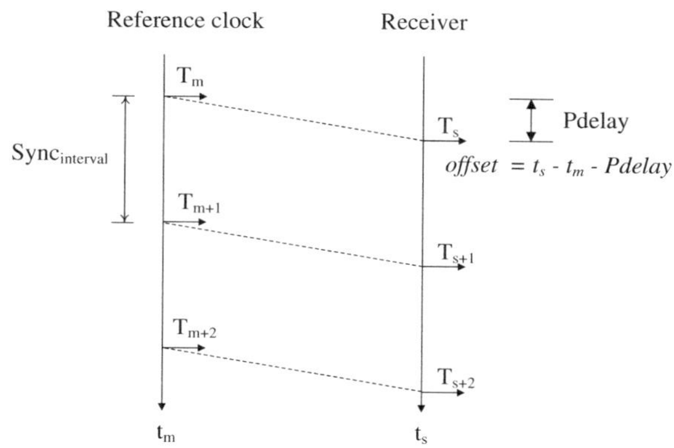

Fig. 12 Frequent time synchronization

indicates the difference between the two nodes that are to be corrected. The approach makes it possible to calculate the drift rate by using only one synchronization procedure, which dramatically reduces the number of messages that are needed for synchronization.

The period of time it takes to correct the clock drift between the two nodes is defined according to Eq. (3).

$$\mathrm{Sync}_{\mathrm{interval}} = \frac{\mathrm{err}_{\mathrm{tolerance}} \times 10^6}{f_{\mathrm{drift}}} \tag{3}$$

where f_{drift} is the drift rate of the crystal oscillator including its stability, and $\mathrm{err}_{\mathrm{tolerance}}$ is the tolerance of the time error between the two nodes.

4.4 *Time Representation Error*

Most of the uncertainty introduced in the network protocol stack can be reduced by using a precise time stamping unit. This means that the accuracy during time synchronization is determined by the time stamping point and the time stamping unit. However, this time stamping unit also contains a time representation error comprised of the delay and jitter in the signal. The time representation error is the difference between the actual time of an event and the nearest time value that can be represented. Figure 13 shows an example of the time representation error. Assume that the time processing unit operates at 1 MHz. The interval between the clocks is of one microsecond, and the time stamping unit detects an event at either the rising edge or at the falling edge of the clock. When the system uses the rising edge, the timer for the stamping unit is also determined at the rising edge. When an event, such as an SFD occurs at T_E, the timer of the time processing unit does not record the time at which the event occurs, but counts it at the consequent rising edge at T_{DE}. The time representation error is a maximum of one microsecond at a 1 MHz clock speed, but it is difficult to remove the time representation error unless a higher

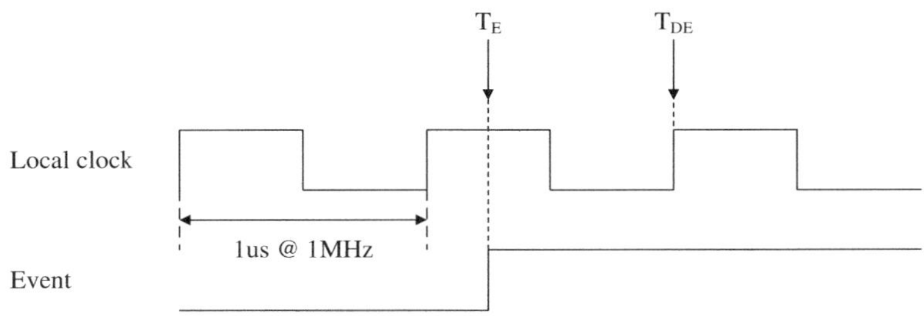

Fig. 13 Time representation error at the time stamping unit

clock speed can be used. For example, a 37.5 MHz oscillator reduces the time representation error to 26.7 ns, which is adequate for submicrosecond accuracy time synchronization. For applications that require a higher time resolution, a higher frequency clock can be used to reduce the time representation error.

The second time representation error occurs at the RF transceiver, as shown in Fig. 14. The RF transceiver requires a certain amount of time to encode and decode the message into electromagnetic waves and vice versa. The encoding time is the time required for the radio chip to encode and transform a part of the message to electromagnetic waves, and this time starts when the radio chip initiates the transfer at an idealized point. The decoding time is the time that is required for the radio chip at the receiver side to transform and decode the message from electromagnetic waves to binary data, and this time ends when the radio chip raises an interrupt indicating reception at the idealized point.

For example, the TI CC2420 radio supports the IEEE 802.15.4/ZigBee standard and has no jitter uncertainty at the transmitter side and a ± 0.125 μs uncertainty at the receiver side because it has an 8 M chip(s). It is impossible to remove the

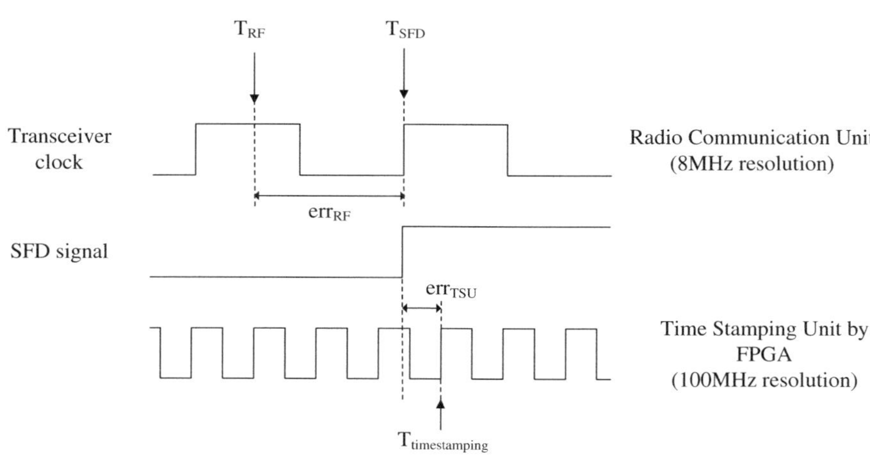

Fig. 14 Time representation error at the ZigBee transceiver

uncertainty at the receiver side. However, a many-to-many message handshake can achieve a reasonable value. Jitter can occur during encoding and decoding, and it can be reduced by using filtering methods, such as a mean, a median, and a learned function of multiple measurements.

The time representation error and the clock skew between the transmitter and the receiver cannot be eliminated. The Kalman filter is an algorithm that operates recursively on streams of noisy input data to produce a statistically optimal estimate of the underlying system state [44]. Although the Kalman filter can produce a better time estimate, we do not describe how to use a Kalman filter for time synchronization in detail. Figure 15 illustrates a general example where a Kalman filter is

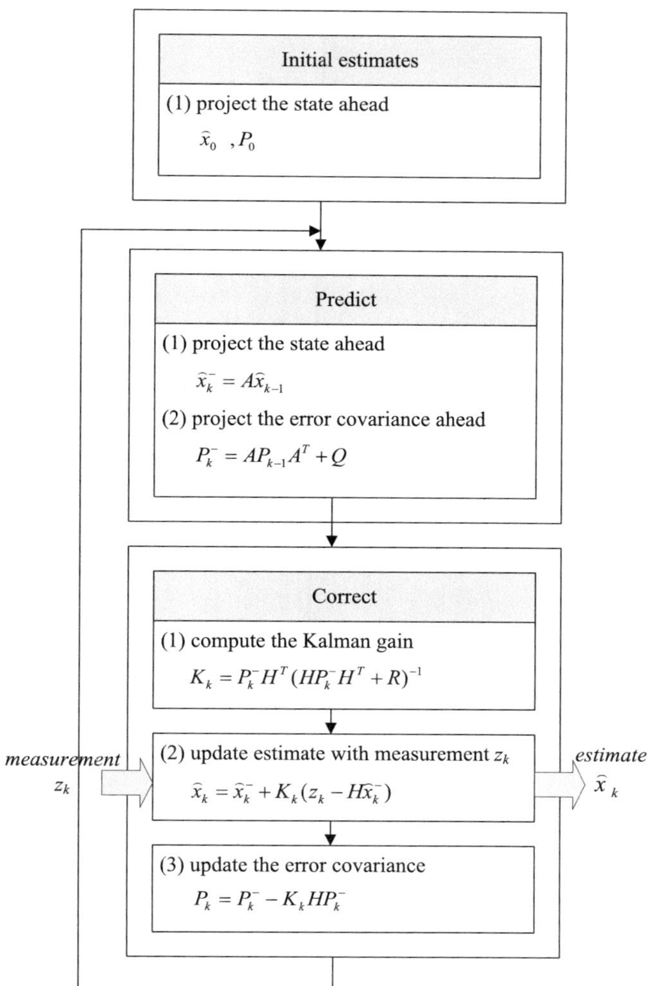

Fig. 15 The Kalman filter architecture

used for precise time synchronization. The Kalman filter is also used in NTP for accurate compensation.

We evaluate the performance of using a Kalman filter during time synchronization (Fig. 16). We minimize most of the uncertainties in the ZigBee network protocol stack and obtain precision time synchronization with sub-microsecond accuracy, as shown in Fig. 16a, where the standard deviation is of approximately 53 ns. However,

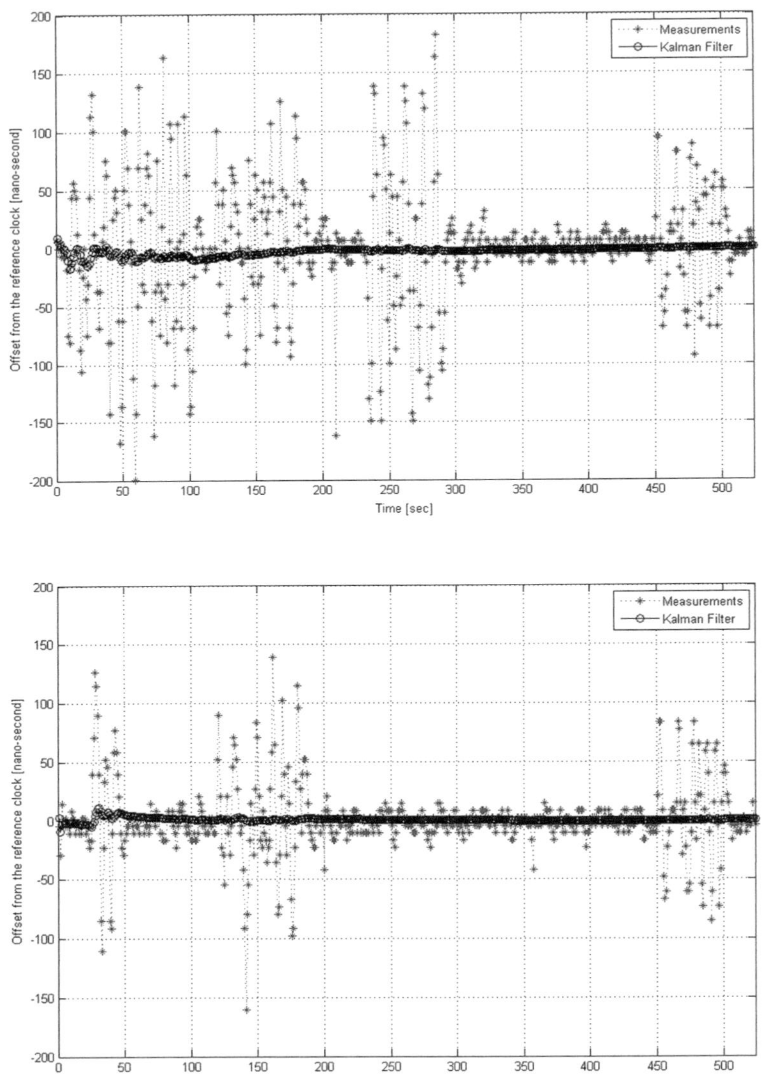

Fig. 16 Performance evaluation: precision time synchronization with and without a Kalman filter

when using a Kalman filter, we obtain a standard deviation of 2.9 ns in single-hop communications. Figure 16b exhibits the case where the time representation error is minimized down to half. The result indicates a value of approximately 29.65 ns when a Kalman filter is not used, and 1.76 ns when the Kalman filter is used. As a result, Kalman filtering is an important procedure that is necessary to reduce uncertainty during time synchronization.

5 Conclusion

This chapter discussed a wireless surveillance camera system and its corresponding time synchronization. The proposed system is decomposed into a multi-sensor environment, video and audio surveillance, and wireless sensor networks. We also described the essential constraints for wireless surveillance cameras in terms of the time synchronization. In particular, we provided an analysis of the uncertainties introduced in the ZigBee network protocol stack, and we also described how to minimize uncertainties during precision time synchronization.

Acknowledgments This work is supported by the Center for Integrated Smart Sensors funded by the Ministry of Science, ICT and Future Planning as the Global Frontier Project.

References

1. Raty TD (2010) Survey on contemporary remote surveillance systems for public safety. IEEE Trans Syst Man Cybern Part C Appl Rev 40(5):493–515
2. Huang G, He J, Ding Z (2008) Wireless video-based sensor networks for surveillance of residential districts. Lect Note Comput Sci 4976:154–165
3. Aruba networks (2011) White paper: using wireless mesh networks for video surveillance version: 1. http://www.arubanetworks.com
4. Atzori L, Iera A, Morabito G (2010) The internet of things: a survey. Comput Netw 54:2787–2805
5. Cho H, Baek Y, Kyung C-M (2014) Wireless video sensor network platform and its application for public safety. In: IEEE international conference on embedded software and systems, Aug 2014
6. Akyildiz IF, Su W, Sankarasubramaniam Y, Cayirci E (2002) Wireless sensor networks: a survey. Comput Netw 38:393–422
7. Akyildiz IF, Melodia T, Chowdhury KR (2006) A survey on wireless multimedia sensor networks. Int J Comput Telecommun Netw 51(4):921–960
8. Akyildiz IF, Melodia T, Chowdhury KR (2007) Wireless multimedia sensor networks: survey. IEEE Wirel Commun 14(6):32–39
9. Karlsson J (2010) Wireless video sensor network and its applications in digital zoo. Doctoral thesis of UMEA University
10. Firetide Inc. (2014) Hot port mesh. http://www.firetide.com
11. Strix Systems (2009) White paper: wireless mesh networks for distributed video surveillance. http://www.strixsystems.com

12. Raniwala A, Chiueh T (2005) Architecture and algorithms for an IEEE 802.11-based multi-channel wireless mesh network. In: INFOCOM 2005, pp. 2223–2234
13. Yang S-C, Yoon M-K, Kim D-H, Kim J-D (2010) Implementation of a multi-radio, multi-hop wireless mesh network using dynamic WDS based link layer routing. In: International conference on information technology: new generations (ITNG)
14. Freescale (2014) i.MX 6 datasheet. http://www.freescale.com
15. OpenWRT (2014) OpenWRT documentation. http://openwrt.org
16. Ubiquiti Networks (2014) SR71 datasheet. http://www.ubnt.com
17. STMicroelectronics (2014) STM32F407 datasheet. http://www.st.com
18. FreeRTOS (2014) http://freertos.org
19. Open Mesh (2014) B.A.T.M.A.N. advanced documentation overview. http://www.open-mesh.org
20. Mills DL (1992) Network time protocol (version 3) specification, implementation and analysis, RFC 1305
21. IEEE 1588-2008 (2008) IEEE standard for a precision clock synchronization protocol for networked measurement and control systems. IEEE Instrumentation and Measurement Society
22. Yicka J, Mukherjeea B, Ghosal D (2008) Wireless sensor network survey. Comput Netw 52:2292–2330
23. Dong J, Gu L, Zheng C (2011) Research on fault-tolerant strategy of time synchronization for industrial wireless sensor network. In: Proceedings of the 3rd international conference on measuring technology and mechatronics automation, Shanghai, China, pp 1146–1149, 6–7 Jan 2011
24. Rhee IK, Lee J, Kim J, Serpedin E, Wu YC (2009) Clock synchronization in wireless sensor networks: an overview. Sensors 9:56–85
25. Elson J, Romer K (2003) Wireless sensor networks: a new regime for time synchronization. ACM Comput Commun Rev 33:149–154
26. Sundararaman B, Buy U, Kshemkalyani AD (2005) Clock synchronization for wireless sensor networks: a survey. Ad Hoc Netw 3:281–323
27. Sichitiu ML, Veerarittiphan C (2003) Simple, accurate time synchronization for wireless sensor networks. In: Proceedings of the 2003 IEEE wireless communications and networking, New Orleans, LA, USA, 20 March 2003
28. Cox D, Jovanov E, Milenkovic A (2005) Time synchronization for Zigbee networks. In: Proceedings of the 37th annual southeastern symposium on system theory, Tuskegee, AL, USA, pp 135–138, March 2005
29. Noh K, Serpedin E, Qaraqe K (2008) A new approach for time synchronization in wireless sensor networks: pairwise broadcast synchronization. IEEE Trans Wirel Commun 7:3318–3322
30. Akyildiz IF, Su W, Sankarasubramaniam Y, Cayirci E (2002) Wireless sensor networks: a survey. Comput Netw 38:393–422
31. Lim H, Kim C (2001) Flooding in wireless ad hoc networks. Comput Commun 24:353–363
32. Maroti M, Kusy B, Simon G, Ledeczi A (2004) The flooding time synchronization protocol. In: Proceedings of the 2nd international conference on embedded networked sensor systems, SenSys 2004, Baltimore, MD, USA, pp 39–49, 3–5 Nov 2004
33. Elson J, Girod L,. Estrin L (2002) Fine-grained network time synchronization using reference broadcasts. In: Proceedings of the fifth symposium on operating systems design and implementation (OSDI), Boston, MA, USA, pp 147–163, 9–11 Dec 2002
34. Palchaudhuri S, Saha AK, Johnsin DB (2004) Adaptive clock synchronization in sensor networks. In: Proceedings of the international symposium on information processing in sensor networks, Berkeley, CA, USA, 26–27 April 2004
35. Ganeriwal S, Kumar R, Srivastava MB (2003) Timing-sync protocol for sensor networks. In: Proceedings of the 1st international conference on embedded networked sensor systems, SenSys 2003, Los Angeles, CA, USA, pp 138–149, 5–7 Nov 2003
36. Dai H, Han R (2004) Tsync: A lightweight bidirectional time synchronization service for wireless sensor networks. ACM SIGMOBILE Mob Comput Commun Rev 2004(8):125–139

37. Greunen J, Rabaey J (2003) Lightweight time synchronization for sensor networks. In: Proceedings of the second ACM international conference on wireless sensor networks and applications, WSNA 2003, San Diego, CA, USA, pp 11–19, 19 Sept 2003
38. Ye Q, Zhang Y, Cheng L (2005) A study on the optimal time synchronization accuracy in wireless sensor networks. J Comput Netw 48:549–566
39. Weibel H, Bechaz D (2004) Implementation and performance of time stamping techniques. In: Proceedings of the 2004 conference on IEEE 1588, Gaithersburg, MD, USA, 27–29 Sep 2004
40. Cho H, Jung J, Cho B, Jin Y, Lee SW, Baek Y (2009) Precision time synchronization using IEEE 1588 for wireless sensor networks. In: Proceedings of the IEEE international conference on computational science and engineering, Vancouver, BC, Canada, pp 579–586, 29–31 Aug 2009
41. Texas Instrument (2013) CC2420 datasheet. http://www.ti.com
42. Ren F, Lin C, Liu F (2008) Self-correcting time synchronization using reference broadcast in wireless sensor network. IEEE Wirel Commun 15:79–85
43. Song P, Shan X, Li X, Qi G (2009) Highly precise time synchronization protocol for ZigBee networks. In: Proceedings of the IEEE/ASME international conference on advanced intelligent mechatronics 2009 (AIM2009), Singapore, 14–17 July 2009
44. Welch G, Bishop G (2006) An introduction to the Kalman filter, TR 95-041, University of North Carolina

Distributed Medium Access for Camera Sensor Networks: Theory and Practice

Hojin Lee, Donggyu Yun and Yung Yi

Abstract Camera sensor networks (CSN) have recently emerged as an important class of sensor networks, where each node is equipped with a camera and has a capability of visually detecting events in its neighborhood. The applications of CSN are highly diverse, including surveillance, environmental monitoring, smart homes, and telepresence systems. In this article, we focus on one of the key unique characteristics of CSN: An event detected by a sensor node can trigger a large amount of sensing data generation, which should be delivered in a distributed manner, whereas in "traditional" sensor networks the volume of sensing data is typically small. Networking protocols to convey the captured image from sensors to decision making modules consist of from distributed and energy-efficient layers accessed via a high-throughput and low-delay MAC to fancy routing and transport protocols. In this article, we focus on the MAC layer and survey the theory and the practical implementation efforts of CSMA-based MAC mechanisms, referred to as optimal CSMA, that are fully distributed with the goal of guaranteeing throughput and delay.

Keywords Camera sensor network · Optimal CSMA · Throughput optimization · Utility optimization · Multi-channel

H. Lee (✉) · D. Yun · Y. Yi
Electrical Engineering, KAIST, Daejeon, South Korea
e-mail: hojin.lee.79@gmail.com

D. Yun
e-mail: dgyun@lanada.kaist.ac.kr

Y. Yi
e-mail: yiyung@kaist.edu

C. Kyung (ed.), *Theory and Applications of Smart Cameras*,
KAIST Research Series, DOI 10.1007/978-94-017-9987-4_14

1 Introduction

1.1 Camera Sensor Network

1.1.1 Definition and Applications

CSN are also called visual sensor networks, whose definition is presented in Wikipedia [36] as follows (Fig. 1):

> A visual sensor network is a network of spatially distributed smart camera devices capable of processing and fusing images of a scene from a variety of viewpoints into some form more useful than the individual images.

CSN can be applied to many types of useful applications, including:

- *Surveillance*: Surveillance has been the primary application of camera-based networks, where the monitoring of large public areas (such as airports, subways, etc.) is performed by a large number of security cameras. Cameras themselves usually produce just raw video streams. Thus, obtaining important and meaningful information from collected images necessitates a huge amount of local processing in the sensors as well as post-processing of them by delivering the images to the processing servers. This implies that both high-throughput

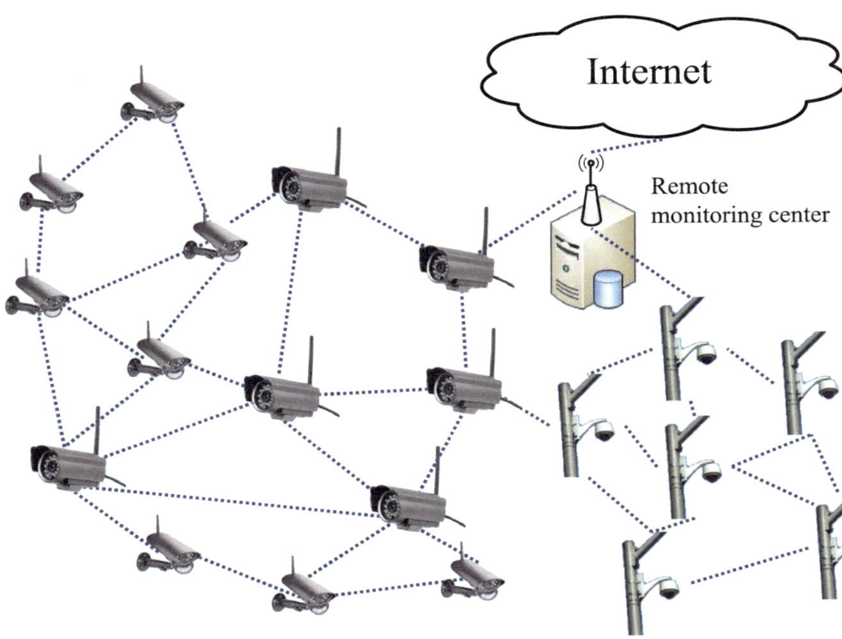

Fig. 1 Camera sensor network

wireless networks and smart processing engines are necessary to run CSNs efficiently.

- *Environmental monitoring*: CSN can be used to monitor the areas that are remote and inaccessible, in which case energy-efficient operations, e.g., by duty cycling sensor nodes as in the conventional wireless sensor networks, to lengthen the lifetime of the networks. Traffics are generated on either event or time basis, depending on which the mechanism of operating the network should be different.
- *Telepresence*: Telepresence systems are the ones that enable remote users to virtually visit some location sensed by cameras. Examples include virtual museum or exhibition rooms equipped with live video cameras that are connected to the Internet and controlled by remote users. This case differs from the earlier two applications in that traffic patterns are "bi-directional" between sensors and users, although the traffic volume may be asymmetric (Fig. 2).

Example: *CitySense* [23]

As a nice example of CSN, we take *CitySense* project [23] that is an open, urban-scale wireless networking testbed with the goal of supporting the development and evaluation of novel wireless systems that span an entire city. CitySense consists of about 100 Linux-based embedded PCs outfitted with dual 802.11a/b/g

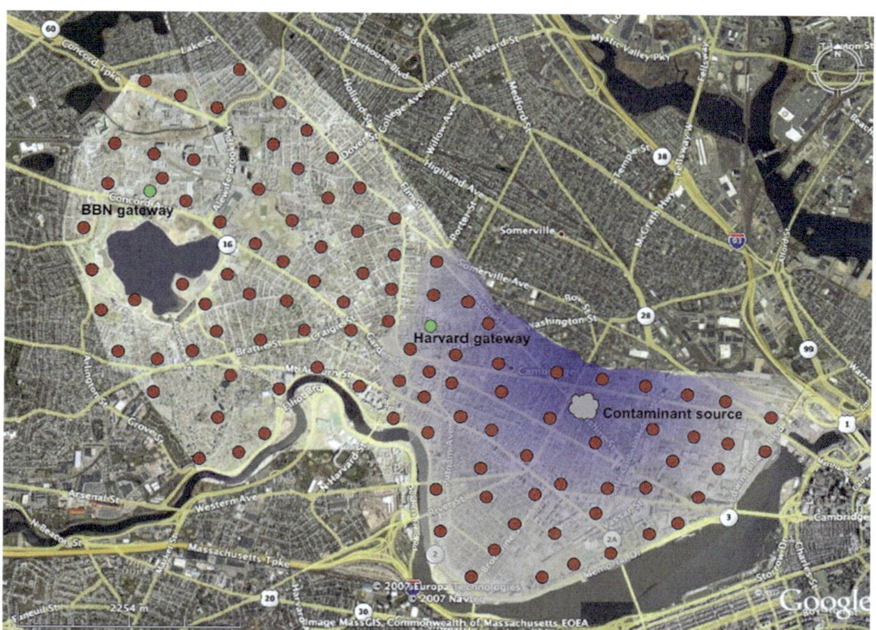

Fig. 2 Node deployment in CitySense. *Source* [23]. Copyright©2008 IEEE

radios and various sensors, mounted on buildings and streetlights across the city of Cambridge. The goal of CitySense is explicitly not to provide public Internet access, but rather to serve as a new kind of experimental apparatus for urban-scale distributed monitoring systems and networking research efforts.

1.1.2 Networking and Data Delivery

The key difference of CSN from other conventional sensor networks is the nature and the amount of information generated by each sensor. The captured visual data can be generated either periodically or on an event basis. In particular, sensor nodes capture a large amount of visual information which may be partially processed with the visual data from other cameras in the network, and thus changing the volume and the information from individual sensors. However, despite such in-network data processing, the volume of sensed data often still remains high, requiring high-performance wireless sensor networks. Also, it is often the case that the end-to-end data transmissions should satisfy low latency, thus requiring stable routing paths.

Figure 3 shows a reference architecture of CSN, proposed by [1], where a variety of connection types can be designed. Sensors can form a single-tier, flat, or clustered network. A multi-tier architecture is also possible, where a group of sensors

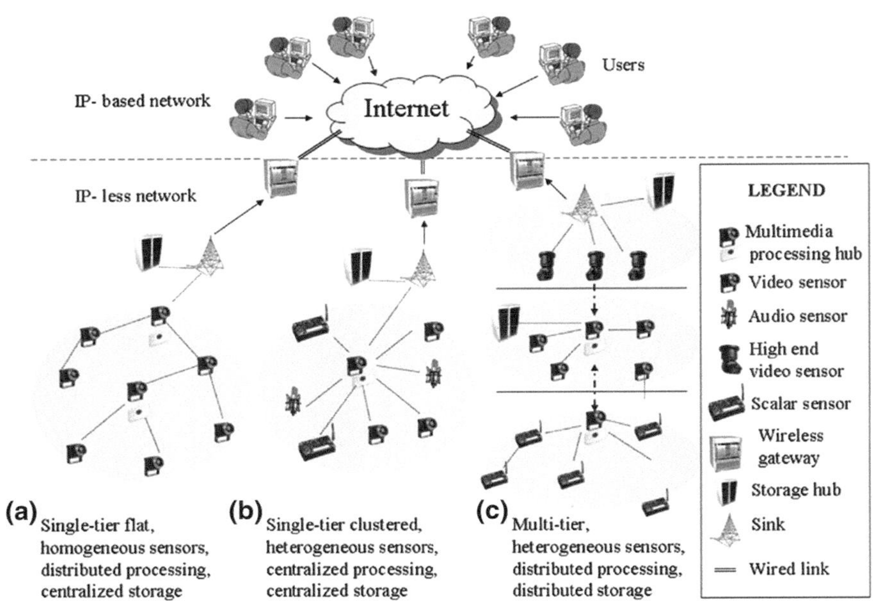

Fig. 3 Reference architecture of camera sensor network. *Source* [1]. Copyright©2007 Elsevier

form one tier, connected to another tier through a gateway node. The network architecture can be selected differently, depending on different target applications, resource budget, and the scale size of the network.

1.2 Focus of This Chapter

Motivated by the fact that in CSNs a high volume of data is injected to the network by asynchronous events or periodic visual monitoring, and sensors should work in a fully distributed manner, we focus on how to deliver such large amount of traffic using a CSMA-based MAC, which is one of the famous, fully distributed MAC in the current practice. The popular 802.11 DCF, which can be a nice candidate MAC for CSNs, is a good example based on CSMA. However, this chapter's focus is on providing the fundamental theories of running CSMA parameters, which guarantees a sense of optimal performance in terms of throughput and delay. These approaches have been extensively attempted in the name of *optimal CSMA*, as will be elaborated shortly. We note that in this chapter we do not explicitly consider energy-efficiency, but it can have high potential to be easily merged with optimal CSMA due to its fully distributed operation.

1.3 Optimal CSMA

1.3.1 Motivation

CSMA (Carrier-Sense Multiple Access)

Carrier Sense Multiple Access (CSMA) is one of the most popular random access protocols in practice, which we see in most wireless textbooks. The key feature of CSMA is that each link with a pair of transmitter and receiver senses the medium and transmits a packet only if the medium is sensed idle. Due to its simple and distributed nature, it has been regarded as one of the most practical MAC protocols in wireless networks, e.g., CSMA is a basic medium access algorithm in IEEE 802.11. Thus, there exists a vast array of research results on CSMA in terms of its analysis under various settings and its applications to practical systems.

Wireless Scheduling: A Rough History

CSMA is referred to as the class of algorithms to schedule links over time in wireless networks. There are also numerous other types of algorithms in the area of wireless link scheduling, where their performances are often measured by various metrics, e.g. throughput, delay, fairness, etc. In the year 1992 a seminal paper by

Tassiulas and Ephremides [34] was published, in which so-called throughput optimality was defined, and a scheduling algorithm achieving throughput optimality, referred to as *Max-Weight*, was presented. Despite its provable optimality, Max-Weight requires to solve a computationally intractable problem, called Maximum Weight Independent Set problem, over each time, which has been a major obstacle to implement it in practice.

Since the work on Max-Weight, a surge of papers on MAC scheduling, which essentially follows the philosophy of Max-Weight, have been published. They partially or fully guarantee the performance, typically in terms of throughput and utility, where the efforts have been classified into (i) ones which trade-off between complexity and efficiency, (ii) ones which achieve optimality at the cost of increasing delay, and (iii) random access style algorithms with suboptimality but worst-case performance (e.g., lower bound of the performance) guarantee, see e.g., [37] for a survey. A single sentence summary of the key ideas of all the above-mentioned research would be: Balancing the supply–demand differential by prioritizing links with larger differentials in scheduling algorithms, where differentials are quantified by link queue lengths.

However, many aforementioned algorithms still require heavy message passing or computations, thus remain just theoretical rather than being made practical. Therefore, it has been a long-standing open problem to find simple random access (hopefully, without message passing) achieving *full* optimality in the research community. About 15 years after Max-Weight, in 2008 a simple CSMA with no message passing was shown to be provably optimal in terms of throughput and utility. Since then more and more research interests in this so-called *optimal CSMA* area have been taken in the community, whose survey is the major content of this paper. For convenience, we survey the research results on optimal CSMA based on the following criteria reflecting different models, proof techniques, and research methodologies (e.g., theory or implementation).

1.3.2 Taxonomy

Saturate Versus Unsaturated

In unsaturated cases, there is arrival of traffic with finite workload to each link, and stability is a key metric, whereas in saturated cases, there is infinite backlog behind each link, and the utility value of equilibrium rate is often the objective to be maximized. In terms of potential applications in CSN, unsaturated cases correspond to when sensing traffic is periodically generated, where periods can be deterministic or random, whereas event-driven visual sensors are well modeled by saturated cases, where when event occurs, a large volume of data traffic is generated so as to saturate the network temporarily.

Synchronous Versus Asynchronous

Synchronous systems have a notion of *frames*, each of which typically consists of a control phase and a data phase, where frames are synchronized, whereas in asynchronous systems, each link independently accesses the medium after sensing other links' transmissions.

Continuous Versus Discrete

This criterion can also be called with versus without collisions. For mathematical tractability, continuous models are often used, where backoff and holding times can be arbitrary real numbers. In practice, the systems are actually discrete, where the systems evolve over discretized time slots (e.g., 20 μs in IEEE 802.11b) and collisions will inevitably occur, when two links contend at a same time slot.

Time-Varying Channels Versus Static Channels

Static channels are often assumed mainly for analytical simplicity, where every link capacity is set fixed. Wireless channels, however, are time varying in practice, where the results on optimal CSMA may significantly change, depending on the timescale difference between the speed of channel variations and CSMA parameter controls.

Time-scale Separation Versus not

As will be discussed later in more detail, the behavior of optimal CSMA is modeled by a Markov chain, and this timescale separation assumption corresponds to whether the Markov chain reaches a stationary distribution immediately or not. Results based on this "fake" assumption have been accepted in the community without much criticism, especially when analyzing the CSMA Markov chain becomes mathematically intractable.

Theory Versus Implementation

Most of the work in the literature has produced theoretical results with emphasis on discovering CSMA's ability toward optimality. There are also some recent researches which implement and evaluate optimal CSMA, in conjunction with several redesign proposals to bridge the gap between theory and practice.

Following these six criteria, we summarize the key features of the research papers on optimal CSMA in Table 1. The rest of the paper is devoted to explaining their key concepts and brief summaries.

Table 1 Taxonomy of optimal CSMA

	Work	Sat/unsat	Cont/disc	Sync/async	TSS	Summary and comments
Theoretical work	[7]	Unsat	Cont	Async	O	The first optimal CSMA with partial proofs
	[6]	Unsat	Cont	Async	×	More complete proof of [7]
	[8]	Unsat	Disc	Async	×	Throughput optimal with collision
	[28]	Unsat	Cont	Async	×	Queue based approach with full optimality proof without TSS
	[31]	Unsat	Disc	Async	×	Connecting Max-weight and CSMA with maximum queue size estimation
	[30]	Unsat	Cont	Async	×	Continuous time version of [31]
	[21]	Sat	Cont	Async	×	Utility optimal CSMA based on stochastic approximation with Markovian noise
	[26]	Sat	Cont	Async	×	Utility optimal CSMA under multiple channels
	[25]	Unsat	Disc	Sync	O	Queue based approach under synchronous system
	[5]	Unsat	Disc	Sync	×	Bounding delay based on parallel update of transmission aggressiveness
	[10]	Unsat	Disc	Async	O	Throughput optimal for imperfect carrier sensing
	[12]	Unsat	Cont	Async	×	Delay of optimal CSMA algorithms based on asymptotic variance
	[27]	Unsat	Cont	Async	O	MIMO and SINR-based interference model
	[18, 38]	Unsat	Cont	Async	×	CSMA over time-varying channel
	[11, 13]	Sat	Disc	Sync	×	Delay optimality of a throughput optimal CSMA
	[4]	Sat	Cont	Aync	×	Game-theoretic understanding of optimal CSMA
	[9, 39]	Sat	Cont	Aync	×	Approaching optimal CSMA with belief propagation in the theory of stochastic mechanics
	[19]	Unsat	Disc	Sync	×	

(continued)

Table 1 (continued)

	Work	Sat/unsat	Cont/disc	Sync/async	TSS	Summary and comments
						Throughput optimal CSMA with worst-case delay guarantee
Impl.	[17, 24]		Disc	Async		Evaluation of optimal CSMA
	[2, 16]		Disc	Async		Study of interaction between CSMA and TCP
	[15]		Disc	Async		A new MAC and experimental validation on 802.11 hardware

TSS: timescale separation. This table is a extended version of that in [40]

2 CSMA: A Theoretical Perspective

2.1 *Model*

In wireless networks, each link shares the wireless medium with other neighbor links that interfere with the link. To model this, a wireless network topology is represented as an interference graph, where links are vertices and undirected edges are generated between two interfering links. Let $G = (\mathscr{L}, E)$ denote the interference graph, where $\mathscr{L}$ and E are the set of links and the set of edges between interfering links, respectively. We define by $\vec{\sigma} \triangleq [\sigma_i : i \in \mathscr{L}]$[1] a scheduling vector for links in G. Since interfering links cannot successfully transmit a packet simultaneously, $\vec{\sigma}$ is called feasible (i.e., there is no collision) if $\vec{\sigma}_i + \vec{\sigma}_j \leq 1, \forall (i,j) \in E$, where (i,j) denotes the edge between link i and j. Thus, the set of all feasible schedules is defined as

$$\mathscr{I}(G) \triangleq \{\sigma \in \{0,1\}^n : \sigma_i + \sigma_j \leq 1, \forall (i,j) \in E\}, \tag{1}$$

where n is the number of links. The feasible rate region (or capacity) $C = C(G)$ is convex hull of $\mathscr{I}(G)$, namely,

$$C(G) \triangleq \left\{ \sum_{\sigma \in \mathscr{I}(G)} \alpha_\sigma \sigma : \sum_{\sigma \in \mathscr{I}(G)} \alpha_\sigma = 1, \alpha_\sigma \geq 0, \forall \sigma \in \mathscr{I}(G) \right\}.$$

[1]Let $[x_i : i \in \mathscr{L}]$ denote the vector whose ith element is x_i. For notational convenience, instead of $[x_i : i \in \mathscr{L}]$, we use $[x_i]$ in the remainder of this paper.

Under CSMA, prior to trying to transmit a packet, links check whether the medium is busy or idle, and transmit the packet only when the medium is sensed idle. To control the aggressiveness of medium access, a notion of backoff timer is used, which is reset to a random value when it expires. The timer ticks only when the medium is idle. With the backoff timer, links try to avoid collisions by the following procedure: each link does not start transmission immediately when the medium is sensed idle, but keeps silent until its backoff timer expires. After a link grabs the channel, the link holds the channel for some duration, called holding time. Intuitively, the probability that link i is scheduled is decided by the average backoff time and the average holding time. Let the backoff and holding times be denoted by $1/b_i$ and h_i, respectively.

For tractability, if we assume that backoff and holding times follow memoryless (i.e., exponential) distributions, the scheduling process $\{\sigma(t)\}$ of CSMA protocols becomes a time reversible Markov process. Then, the stationary distribution of a schedule σ is defined by $b = [b_i]$ and $h = [h_i]$:

$$\pi_{\vec{\sigma}}^{b,h} = \frac{\prod_{i \in \mathscr{L}} (b_i h_i)^{\vec{\sigma}_i}}{\sum_{\vec{\sigma}' \in \mathscr{I}(G)} \prod_{i \in \mathscr{L}} (b_i h_i)^{\vec{\sigma}'_i}}, \tag{2}$$

which is a function of the product $b_i \times h_i$, for all $i \in \mathscr{L}$. Let $r_i = \log(b_i h_i)$ and $r = [r_i]$, where r implicitly denotes transmission aggressiveness of links. From (2), the probability $s_i(r)$ that link i is scheduled for r, which is the link i's throughput, is computed as follows:

$$s_i(r) = \sum_{\sigma \in \mathscr{I}(G):\sigma_i=1} \pi_{\sigma}^{b,h} = \frac{\sum_{\sigma \in \mathscr{I}(G):\sigma_i=1} \exp(\sum_{i \in \mathscr{L}} \sigma_i r_i)}{\sum_{\sigma' \in \mathscr{I}(G)} \exp(\sum_{i \in \mathscr{L}} \sigma'_i r_i)}.$$

In the discrete time model, where geometric distributions are used for backoff and holding time instead of exponential, due to collisions, the stationary distribution is slightly different from (2). However, the stationary distribution becomes close to (2) when the holding time h is large enough so that the collision time become ignorable, since the time fraction of collision period declines as the holding time increases for the same transmission aggressiveness r.

2.2 *Objectives*

2.2.1 Unsaturated System

When a CSMA-based algorithm can stabilize any feasible arrival rate $\lambda \in C(G)$, the algorithm is called *throughput optimal*. Intuitively, when $s_i(r^*) > \lambda_i$ for all link i, the arrival λ can be stabilized with transmission aggressiveness r^*. A question to address is:

(**Q1**) For any $\lambda \in C(G)$, is there any transmission aggressiveness r such that $s_i(r) \geq \lambda_i$ for all link i? If there exists such r, what are the CSMA algorithms that provide the transmission aggressiveness r over long-term without any message passing and explicit knowledge of the given arrival rate λ?

2.2.2 Saturated System

In this case, each link is assumed to be infinitely backlogged. Thus, CSMA algorithms are exploited to control the service rate of each link so as to make the long-term service rate close to some point of the boundary of $C(G)$, formally, a solution of the following optimization problem:

$$\max_{\gamma} \sum_{i \in \mathscr{L}} U(\gamma_i) \qquad \text{subject to} \quad \gamma \in C(G) \tag{3}$$

where $U(\cdot)$ denotes a utility function with the nice properties such as concavity and differentiability. The question to address in this case is:

(**Q2**) Let the solution of (3) be γ^*. How can we make each link have transmission aggressiveness to r_i^* so that $s_i(r^*) = \gamma_i^*$?

3 Optimal CSMA: Survey

The research papers on optimal CSMA to date directly or indirectly address the questions (**Q1**) and (**Q2**). In this section, we summarize them, starting the first two subsections by summarizing the results which can be arguably representative in terms of models and algorithms, followed by more extensions according to the criteria mentioned in Sect. 1. Note that our presentation in terms of positioning and sequencing the papers cited here may be biased by the authors to some degree, and there may also be some missing references.

3.1 *Basic Results: Unsaturated*

In [7], it is shown that, for any feasible arrival rate λ, there exists a finite transmission aggressiveness r^* such that $s_i(r^*) \geq \lambda_i, \forall i \in \mathcal{N}$. From this, the authors conjectured that *throughput optimality* can be achieved by CSMA. We summarize the results on throughput-optimal CSMAs by classifying them into rate-based and queue-based approaches.

3.1.1 Rate-Based Approach

The authors in [7] propose a simple rate-based approach which allows transmission aggressiveness r to converge to the r^* with a timescale separation assumption that the schedules from CSMA immediately follow a stationary distribution at each time slot. Later, Jiang et al. [6] show that without the timescale separation assumption, the proposed rate-based approach converges to r^* for any strictly feasible arrival. The algorithm operates as follows:

Step (1): Each link i investigates packet arrival and schedule duration for a sufficient long time interval. Let link i adjust its transmission aggressiveness $r_i(j)$ at time $T(j)$ for $j \in \mathbb{Z}^+$.[2] Let $\{A_i(t)\}$ and $\{S_i(t)\}$ be arrival and scheduling process of link i, respectively. Then, the empirical arrival and service rates at $T(j+1)$, denoted by $\hat{\lambda}_i(j)$ and $\hat{s}_i(j)$, respectively, are calculated by

$$\hat{\lambda}_i(j) = \frac{1}{T(j+1) - T(j)} \int_{T(j)}^{T(j+1)} A_i(t)\mathrm{d}t$$

$$\hat{s}_i(j) = \frac{1}{T(j+1) - T(j)} \int_{T(j)}^{T(j+1)} S_i(t)\mathrm{d}t.$$

Step (2): Link i adjusts its transmission aggressiveness r_i according to the empirical packet arrival and service rates as follows:

$$r_i(j+1) = r_i(j) + \beta(j)(\hat{\lambda}_i(j) - \hat{s}_i(j)), \tag{4}$$

where $\beta(j)$ is a decreasing step size.

---

[2]We use j to index the state updates, and $T(j)$ is the time of j-th update.

3.1.2 Queue-Based Approach

The rate-based approach is summarized as the scheme which directly estimates the demand and then provides the service rates to balance the demand and supply. A different approach can be developed by implicitly quantifying the supply–demand differential using a queue-length information, which we call queue-based approach. This queue-based CSMA can be regarded as an algorithm which emulates Max-Weight in a sluggish manner. By sluggish, we mean that the Markov chain induced by CSMA requires a time to reach a stationary distribution (close to what Max-Weight achieves).

In [25], the authors propose a scheme called Q-CSMA where $r_i = f(Q_i)$, where Q_i is the queue length of link i and f is a weight function. They prove that Q-CSMA is (throughput) optimal for any increasing function f under the timescale separation assumption. Although they use a discrete time model, no collision exists due to synchronous operations (see Sect. 3.4). Thus, the probability that a schedule is selected at each time slot follows the stationary distribution (2). In other words, due to the choice of $r_i = f(Q_i)$, the probability to schedule $\vec{\sigma}$ is proportional to $\exp(\sum_{i \in \mathcal{N}(G)} \vec{\sigma}_i f(Q_i))$, which becomes negligible if the weight $W(\vec{\sigma}) = \sum_{i \in \mathcal{N}(G)} \vec{\sigma}_i f(Q_i)$ is far from its maximum value (Max-Weight always chooses a schedule maximizing the weight).

The queue-based approach without timescale separation was first proposed and justified in [28] for special choices of weight function f, e.g., $f(x) = \log \log(x)$. The key challenge in the work is to analyze a nontrivial correlation between queueing and scheduling dynamics (operating in the same timescale) induced by a queue-based algorithm such as Q-CSMA. The authors in [28] resolve the correlation by (i) sufficiently slowing down the speed of the queueing dynamics using a slowly increasing weight function f, such as $f(x) = \log \log(x)$ and (ii) showing that scheduling dynamics run in a much faster timescale than queueing dynamics in a certain sense. Due to some technical issues, we note that the CSMA in [28] requires a slight message passing to broadcast certain global information (e.g. the number of queues, the maximum queue-size) over the network. In the following work [30], the authors refine their approach toward removing the message passing. However, the maximum queue-size information still remains to be broadcasted, which was conjectured to be not necessary. The conjecture has been recently resolved in [31] using a certain distributed 'learning' mechanism: each node runs it to infer the maximum queue-size information without explicit message passing (and only using sensing information).

3.1.3 Comparison

The common goal of rate- and queue-based approaches is to control the CSMA parameters for the desired high performance, where they use the arrival rate or queue-size information for the control, respectively. The performance guarantees of rate-based algorithms are inherently sensitive to the assumption that the arrival rate is fixed (or very slowly changing), while queue-based ones are more robust against this issue, i.e., the queue-based results [28, 30, 31] hold even under time-varying arrival rates. However, analyzing queue-based algorithms are technically much harder, and hence the timescale separation assumption or the information of the maximum queue length has been often used for technical convenience.

3.2 Basic Results: Saturated

If each link has infinite backlog, the object of CSMA algorithms is to maximize network utility rather than stabilize the queues of links. In [8], utility optimality is considered for flows under the timescale separation assumption. The algorithm in [8] considers a joint scheduling (via CSMA) and congestion control problem as follows:

$$\max_{\mu\in\Omega,\lambda\in[0,1]^n} \sum_{i\in\mathscr{L}} U_i(\lambda_i) - \frac{1}{V}\left(\sum_{\sigma\in\mathscr{I}(G)} \mu_\sigma \log \mu_\sigma\right) \text{s.t.} \mathbb{E}\{\sigma_i\} \geq \lambda_i, \forall i \in \mathscr{L}, \quad (5)$$

where V is some constant and Ω is set of all probability measure on $\mathscr{I}(G)$. Then, the optimal solution turns out to be close to the utility optimal within $\frac{\log|\mathscr{I}(G)|}{V}$ bound.

The formal proofs for saturated case without timescale separation assumption are proposed in [6, 21]. In [21], the authors provide an algorithm motivated by stochastic approximation controlled by Markov noise.

Time is divided into *frames* of fixed durations, $j = 1, 2, \ldots$. At the starting time instance of each frame, similarly with (4), transmission aggressiveness is updated as follows: Each link i maintains its own virtual queue q_i, updated by

$$q_i(j+1) = q_i(j) + \alpha(j)\left(U'^{-1}\left(\frac{q_i(j)}{V}\right) - \hat{s}_i(j)\right), \quad (6)$$

where V is some constant and $\alpha(j)$ is a decreasing step size. Then, based on $q_i(j)$, CSMA runs with the backoff and holding times satisfying $b_i(j+1)h_i(j+1) = \exp(q_i(j+1))$.

Similar to (5), V controls the distance from optimality. The virtual queue length is a Lagrange multiplier that appears from the dual decomposition of the original objective (3), quantifying the demand-supply differential.

In [6], they also show that without timescale separation, the optimal solution of the problem (5) can be achieved by primal–dual relationship as follows:

$$
\begin{aligned}
r_i(j+1) &= \max\{0, r_i(j) + \alpha(j)(\lambda_i(j) - \hat{s}_i(j))\} \\
\lambda_i(j+1) &= \arg \max_{y \in [0,1]} V \cdot U(y) - r_i(j+1)y.
\end{aligned} \tag{7}
$$

Note that the algorithms in [6, 21] are essentially the same, from the definition of $r_i = \log(b_i \times h_i)$, but there exists minor difference in their proof details.

The key rationale for the saturated case lies in the fact that the transmission aggressiveness is updated by quantifying the supply–demand differential, and the new aggressiveness is applied to the system with more modest updates with the belief that the system approaches to what is desired. The extension to multi-channel networks is provided in [26] without timescale separation based on a much more simpler optimality proof. For faster convergence, a steepest coordinate ascent algorithm is proposed in [3]. Under this algorithm, at each time slot j, the transmission aggressiveness of link i is set to be proportional to the first derivative of utility function at empirical service rate, such that $r_i = k \cdot U'(\gamma_i(j))$ where $\gamma_i(j) = \frac{1}{j+1}\sum_{t=0}^{j} \hat{s}_i(t)$.

3.3 Timescale Separation Assumption

In a Markov chain, it takes some time for a state to be close to a stationary regime. This time is called mixing time. In optimal CSMA algorithms, the transmission aggressiveness $r(t)$, which determines the transition rates (in continuous cases) and probabilities (in discrete cases), is time varying, Thus, the main challenge in performance analysis of the optimal CSMA algorithms lies in the fact that the mixing time can be much longer than the change of transmission aggressiveness. In some papers, e.g., [7, 10, 25, 27], timescale separation assumption, i.e., the assumption that a Markov chain can immediately reach a stationary distribution, has been made, which removes all the dirt in the proof.

As briefly mentioned in Sects. 3.1 and 3.2, two optimality proof techniques exist when no timescale separation is assumed. First, the change of transmission aggressiveness is slowed down by taking a function of the parameter that affects the aggressiveness. For example, in [28, 30, 31], the queue length is such a parameter, where to represent the link weight, $\log\log(Q_i)$ is used to make the regime that the speed of weight changes (thus, the speed of aggressiveness changes) becomes much slower than that of the mixing time. Another approach is to have an explicit device such as a step-size, which decreases with time. Examples include the work by [6, 21] for the saturated case, where the step-size $\alpha(j)$ plays such a role.

3.4 Continuous/Discrete and Synchronous/Asynchronous

The assumption of continuous distributions of backoff and holding times, where most of work based on the continuous setting assumes exponential distributions, conveniently removes the need to consider collisions, leading to simple analysis. However, a real system is not continuous. For example, 802.11 operates based on the notion of a slot whose duration is 20 μs. In this discrete system, collisions naturally occur when two links contend at a same slot. Then, a link i's throughput becomes characterized in more complex way by considering the transmission loss due to collisions. Note that in the discrete case, geometrically distributed backoff and holding times are used in the modeling because of its memoryless property.

Two directions are taken for discrete time systems in the papers. First, since the stationary distribution for the given backoff and holding times is decided by their product, not their individual values, the holding time can be arbitrarily large as long as the product is chosen as planned. This implies that the throughput loss by collisions can be sufficiently reduced by enlarging the holding times, so that their performance is almost close to what has been obtained in the continuous case. However, this may not be practical, because long holding times are very bad for short-term fairness. In [20, 21], the tradeoff between throughput and short-term fairness is asymptotically analyzed, where it is indeed required that a high cost of short-term fairness should be paid to increase throughput; where short-term fairness is defined as the inverse of the average delay between two successive successful transmissions. In [8, 31], for a desired transmission aggressiveness r_i for each link i, the authors propose throughput optimal algorithms with collisions, where the holding time of link i is proportional to $\exp(r_i)$ with a fixed backoff time, so that the holding time consequently increases if a larger aggressiveness is needed. This approach shares the idea, mentioned earlier, that the enlarged holding time can reduce the throughput loss due to collisions. Second, as in [25], a synchronous system with frames, consisting of separate control and data phases, is designed so that, through slight message passing in the control phase, collisions is resolved.

When links operate under a common clock, the control actions can be time-synchronized, and thus, more efficient design is possible. Continuous systems, where continuity is assumed for theoretical purpose, is by nature asynchronous. More serious issues on synchronization are raised in discrete systems, for example, slots can be skewed, where guard time needs to be allocated, and loss of efficiency due to guard time overhead etc. requires more study. However, so far all discrete time-based papers assume perfect synchronization.

3.5 Channel: Time-Varying Versus Fixed

In modeling channels, most of the work assume that channel capacity is fixed. However, the channels are often time varying in practice. Optimal CSMA over time-varying channels have been recently investigated [18, 38]. In [18], CSMA under time-varying channels has been studied only for complete interference graphs, when the arbitrary backoff rate is allowed. The proof is based on the timescale separation assumption, which does not often hold in practice and extremely simplifies the analysis (no mixing time-related details are needed). In [38], the authors consider a channel model that the link capacity is randomly varied between 0 and 1 and the channel varying process is independent across links. Under this model, two canonical CSMA algorithms are considered: (i) A-CSMA which transmits a packet only if the capacity is 1 and (ii) U-CSMA which operates independently of the channel variation. Despite the intuition that A-CSMA may outperform U-CSMA due to its channel tracking ability, it is proved that U-CSMA can guarantee more throughput than A-CSMA, depending on the speed of channel variations, in particular, when the speed of channel variation is fast. However, for slowly varying channel, A-CSMA achieves throughput optimality, whereas U-CSMA is suboptimal. Such performance difference comes from the mixing time of Markov chain, i.e., when the channels change faster than mixing time, A-CSMA may behave in an undesirable manner.

3.6 Imperfect Sensing and MIMO

More practical situations start to be considered for optimal CSMA. First, in [10], the authors consider the case when sensing is imperfect. An example of imperfect sensing is the famous hidden terminal nodes. Other examples include false alarm (resp. miss detection), where a link can sense the idle (busy) medium as busy (idle) with a positive probability. False alarm is not highly critical to throughput optimality, but miss detection could reduce throughput since it generates collisions. In [10], the protocol, which overcomes miss detection, is proposed, which is provably throughput optimal, by letting each link operate with small backoff rate and long holding time.

In most of the aforementioned research, the physical layer is abstracted. For example, for interference model, the protocol model is used, assuming that packet transmission of a link depends on neighbor links only. In practice, success of a transmission is decided by whether its SINR is above a threshold or not, called SINR model. In [27], SINR model is considered with MIMO transmission. Under this model, each link can select a data rate and the transmission is successful when total interference is less than the marginal interference for the transmission rate. Even for the MIMO and SINR model, the authors propose an algorithm that achieve throughput optimality with an assumption where each link has to have topological information.

4 Optimal CSMA: Multi-channel/Multi-radio

So far, we have discussed optimal CSMA for the basic setup, which is the single-channel/single-radio. However, to cope with a high volume of sensing traffic in CSN, the networks with more capacity may be necessary. A natural way of enlarging capacity is to build a network on top of multiple channels over multiple radios. This multi-channel/multi-radio system is not only important for widening the network capacity, but also for significantly reducing the delay. It has been reported that the naive optimal CSMA in general suffers from poor delay performance [22], because to achieve high throughput, once a CSMA schedule is determined, it needs to be frozen for a long time, i.e., high correlation of schedules. However, once channels are various, links can be "interleaved" appropriately so as to reduce correlation. In Sect. 4.1, we provide the model and the optimal algorithm for multi-channel/multi-radio systems, and then in Sect. 4.2, we will present that such multi-channel systems can significantly decrease delay, even achieving the order-wise delay optimality.

4.1 Optimal CSMA for Multi-channel/Multi-radio

4.1.1 Model and Objective

Network Model

The network consists in a set $\mathscr{V}$ of V nodes and a set $\mathscr{L}$ of L links.[3] Denote by $s(l) \in \mathscr{V}$ and by $d(l) \in \mathscr{V}$ the transmitter and the receiver corresponding to link l. We also use the notation $v \in l$ if either $v = s(l)$ or $v = d(l)$. Node v has c_v radio interfaces or *radios* for short. On each link, data transmissions can be handled on any channel of a set $\mathscr{C}$ of C channels. These channels are assumed to be orthogonal in the sense that two transmissions on different links and different channels do not interfere. We model interference by a symmetric boolean matrix $A \in \{0,1\}^{L \times L}$, where $A_{kl} = 1$ if link k interferes link l when transmitting on the same channel, and $A_{kl} = 0$ otherwise.[4] A node uses a radio interface to transmit or receive data on a given channel. Denote by R_{cl} the rate at which $s(l)$ can send data to $d(l)$ on channel c.

---

[3]Note that the notations on the network model in this Sect. 4.1 slightly differ from those in other sections, e.g., in Sects. 2.1 and 4.2. For example, in Sects. 2.1 and 4.2, we use $\mathscr{L}$ to refer to the set of nodes in the interference graph G, and $\mathscr{V}$ was not used there.

[4]The results can be readily extended to the case where the interference matrix may be different on different channels. In such case, interference would be modelled by $A \in \{0,1\}^{L \times L \times C}$ where $A_{klc} = 1$ iff link k and l interfere each other on channel c.

Feasible Schedule Set and Feasible Rate Region

Interference and the limited number of radios at each node impose some constraints on the set of possible simultaneous and successful transmissions on the various links and channels. We capture these constraints with the notion of schedule. A schedule $\sigma \in \{0,1\}^{C \times L}$ represents the activities of the various links on the different channels: by definition, $\sigma_{cl} = 1$ if and only if link l is active on channel c (i.e., $s(l)$ is transmitting on channel c). A schedule m is *feasible* if all involved transmissions are successful, i.e., if for all $k, l \in \mathscr{L}$ and all $v \in \mathscr{V}$,

$$(\sigma_{ck} = 1 = \sigma_{cl}) \Rightarrow (A_{kl} = 0) \qquad \text{(Interference constraint)}$$
$$\sum_{l \in \mathscr{L}: v \in l} \sum_{c \in \mathscr{C}} \sigma_{cl} \leq c_v \qquad \text{(Radio interface constraint)}$$

We define by $\mathscr{I}(G) \subset \{0,1\}^{C \times L}$ the set of the M feasible schedules, which corresponds to the set of all feasible schedules in (1) for the single channel/single radio case.

We are now ready to define the *feasible rate region* $C = C(G)$ as the set of achievable long-term throughputs $s = (s_l, l \in \mathscr{L})$ on the various links:

$$C(G) = \left\{ s : \exists \alpha \in [0,1]^M, \sum_{\sigma \in \mathscr{L}} \pi_\sigma = 1, \ \forall l \in \mathscr{L}, \ s_l \leq \sum_{\sigma \in \mathscr{L}} \pi_\sigma \sum_{c \in \mathscr{C}} \sigma_{cl} R_{cl} \right\}. \qquad (8)$$

In the above expression, π_σ may be interpreted as the fraction of time schedule m is activated.

Objective: Saturated Case

Naturally, we can study the optimal CSMA under multi-channel/multi-radio for both saturated and unsaturated cases, but in this section we focus only on the saturated case. As mentioned earlier, when the transmitters are saturated (i.e., they always have packets to send), the objective is to design a scheduling algorithm maximizing the network-wide utility, as formally given by

$$\max \quad \Sigma_{l \in \mathscr{L}} U(\gamma_l), \qquad \text{subject to} \quad \gamma \in C. \qquad (9)$$

4.1.2 Optimal CSMA for Multi-channel/Multi-radio

Multi-channel/Multi-radio CSMA with $(\lambda_{cl}, b_{cl}, c \in \mathscr{C}, l \in \mathscr{L})$

The following extension of random back-off CSMA protocols can be considered for multi-channel/multi-radio systems. The transmitter of link l has C independent

Poisson clocks, ticking at rates λ_{cl}, $c \in \mathscr{C}$. When a clock c ticks, if the transmitter does have an available radio or if it is already transmitting or receiving on channel c, it does not do anything. Otherwise, it senses channel c, and checks whether the receiver has an available radio. If the channel is idle and if the receiver can receive data, it starts a transmission on channel c, and keeps the channel for an exponentially distributed period of time of average b_{cl}. Define $\lambda_{\cdot l} = (\lambda_{cl}, c \in \mathscr{C})$ and $b_{\cdot l} = (b_{cl}, c \in \mathscr{C})$, and denote by CSMA$(\lambda_{\cdot l}, b_{\cdot l})$ the above access protocol. We also introduce $\lambda = (\lambda_{\cdot l}, l \in \mathscr{L})$ and $b = (b_{\cdot l}, l \in L)$. When each link l runs CSMA $(\lambda_{\cdot l}, b_{\cdot l})$, the network dynamics and performance can be analyzed using the theory of reversible Markov chains.

Let $\sigma^{\lambda,b}(t)$ be the active schedule at time t. Then $(\sigma^{\lambda,b}(t), t \geq 0)$ is a continuous-time reversible Markov chain whose stationary distribution $\pi^{\lambda,b}$ is given by

$$\forall \sigma \in \mathscr{I}, \quad \pi_{\sigma}^{\lambda,b} = \frac{\prod_{l \in \mathscr{L}, c \in \mathscr{C}} (\lambda_{cl} b_{cl})^{\sigma_{cl}}}{\sum_{\eta \in I} \prod_{l \in \mathscr{L}, c \in \mathscr{C}} (\lambda_{cl} b_{cl})^{\eta_{cl}}},$$

where by convention $\prod_{l \in \emptyset}(\cdot) = 1$. Moreover, the link throughputs are given by

$$\forall l \in \mathscr{L}, \quad s_l^{\lambda,b} = \sum_{\sigma \in \mathscr{I}} \pi_{\vec{\sigma}}^{\lambda,b} \sum_{c \in \mathscr{C}} \vec{\sigma}_{cl} R_{cl}.$$

Optimal Algorithm

We now describe a generic algorithm that dynamically adapts these parameters so as to approximately solve the utility-maximization problem (9). Similarly to the optimal CSMA for the saturated case, time is divided into *frames* of fixed durations, $j = 1, 2, \ldots$, and the transmitters of each link update their CSMA parameters (i.e., λ_{cl}, b_{cl}) at the beginning of each frame. To do so, they maintain a virtual queue, denoted by $q_l(j)$ in frame j, for link l. The algorithm operates as follows:

1. During frame j, the transmitter of link l runs CSMA$(\lambda_{\cdot l}(j), b_{\cdot l}(j))$, and records the sum $\hat{s}_l(j)$ of the services received during this frame over all channels;
2. At the end of frame j, it updates its virtual queue according to

$$q_l(j+1) = \left[q_l(j) + \alpha(j) \left(U'^{-1}\left(\frac{q_l(j)}{V}\right) - \hat{s}_l(j) \right) \right],$$

and sets the $\lambda_{cl}(j+1)$'s and $b_{cl}(j+1)$'s such that their products are equal to $\exp\{R_{cl} q_l(j+1)\}$.

The above algorithm is highly similar to that of the single-channel/single-ratio, except that each transmitter of a link runs a multi-channel/multi-radio CSMA algorithm. Virtual queues are maintained per link, but per link/radio CSMA parameters are updated by those per link virtual queue length.

4.2 *Delayed CSMA: Virtual Channel Approach*

4.2.1 Description for Delayed CSMA

The main idea of the *delayed* CSMA is to use multiple schedulers in a round-robin manner in order to effectively reduce the correlations between the link state process, in an attempt to alleviate the so-called *starvation problem*, i.e., once a schedule is chosen, it keeps being scheduled without any change for a large number of slots. Note that the algorithm and the setting in this section is for the case of single-channel/single-radio systems, which, however, shows that virtual multi-channel idea is able to reduce latency significantly. This gives a conjecture that physical multi-channel systems would have highly good delay performance. Different from the model in the time in the optimal CSMA for single- and multi-channel/radio systems, we take a *discrete* time-slotted model, indexed by $t = 1, 2, \ldots$. for convenience. Delayed CSMA [11] is described as follows:

Initialize: for all links $i \in \mathscr{L}$, $\sigma_i(t) = 0, t = 0, \ldots, T-1$.
At each time $t \geq T$: links find a decision schedule,
  $\mathscr{D}(t) \in \mathscr{I}(G)$ through a randomized procedure, and
for all links $i \in \mathscr{D}(t)$ **do**
  if $\sum_{j \in N_i} \sigma_j(t-T) = 0$ **then**
    $\sigma_i(t) = 1$ with probability $\frac{r_i}{1+r_i}$
    $\sigma_i(t) = 0$ with probability $\frac{1}{1+r_i}$
  else
    $\sigma_i(t) = 0$
  end if
end for
for all links $i \notin \mathscr{D}(t)$ **do**
    $\sigma_i(t) = \sigma_i(t-T)$
end for

Here, $N_i = \{j \in \mathscr{L} : (i,j) \in E\}$ as the set of neighbors of link i. In the *delayed* CSMA, at each time slot, a decision schedule is chosen $\mathscr{D}(t) \in \mathscr{I}(G)$, which corresponds to a selection of an independent set of G. The active links in the decision schedule become the candidate links which may change their state. There are various ways to choose a decision schedule $\mathcal{D}(t) \in \mathcal{I}(G)$ at each time slot. For example, each link simply attempts to access the medium with a fixed access probability a_i and then $i \in \mathcal{D}(t)$ with probability $a_i \prod_{j \in N_i}(1 - a_j)$, or a randomized scheme with light control message exchanges can be used, as in [25]. In general, we assume that $\{\mathcal{D}(t)\}$ is a set of independent identical random variables such that $\Pr\{i \in \mathcal{D}(t)\} > 0$ for all i.

As we mentioned in Sect. 2.1, given the transmission aggressiveness $r = [r_i]$, the schedule $\vec{\sigma}(t) : t \equiv k(T)$[5] forms a (discrete-time) irreducible and aperiodic Markov chain for $k = 0, 1, \ldots, T - 1$, e.g., the kth Markov chain is $\{\vec{\sigma}(uT + k) : u = 0, 1, 2, \ldots\}$. The common stationary distribution $\pi = [\pi_{\vec{\sigma}}]$ is given by

$$\pi_{\vec{\sigma}} = \frac{1}{Z} \prod_{i \in \mathcal{L}} r_i^{\vec{\sigma}_i}, \tag{10}$$

where $Z = \sum_{\sigma \in \Omega} \prod_{i \in \mathcal{L}} r_i^{\sigma_i}$ is a normalizing constant. Hence, one can think that the algorithm utilizes multiple T independent Markov chains (or schedulers). From their ergodicity, we know that for all $i \in \mathcal{L}$,

$$\lim_{t \to \infty} \frac{1}{t} \sum_{s=0}^{t-1} \sigma_i(s) = \Pr{}_{\pi}\{\sigma_i = 1\}.$$

There are several ways to find an appropriate transmission aggressiveness $[r_i]$ such that the long-term link throughput $\lim_{t \to \infty} \frac{1}{t} \sum_{s=0}^{t-1} \sigma_i(s)$ is greater than the arrival rate λ_i, as we mentioned in Sect. 3.

Thus, we assume that links initially start with the desired transmission aggressiveness here. Formally speaking, for given ε-admissible arrival rate λ, we assume that

$$\lim_{t \to \infty} \frac{1}{t} \sum_{s=0}^{t-1} \sigma_i(s) = \Pr{}_{\pi}[\sigma_i = 1] \geq \lambda_i + \varepsilon, \text{ for all } i \in \mathcal{L}. \tag{11}$$

4.2.2 Delay-Optimality of Delayed CSMA

For $\lambda \in C(G)$ and given $\varepsilon > 0$, we say that λ is *ε-admissible* if $\lambda_i + \varepsilon < \mu_i$, for all $i \in \mathcal{L}$ and some $\mu = [\mu_i] \in C(G)$. When the arrival rate is ε-admissible, we can define the notion of delay-optimal scheduling algorithm as follows.

---

[5]We say $t \equiv k \pmod{T}$ if $t - k$ is an integer multiple of T. It is called congruent modulo.

Definition 1 (Delay-Optimality) A scheduling algorithm is called *per-link delay-optimal* (or simply delay-optimal),[6] if for any ε-admissible arrival rate λ with $\varepsilon = \omega(1)$,

$$\limsup_{t\to\infty} \mathbb{E}[Q_i(t)] = O(1), \quad \text{for all } i \in \mathscr{L},$$

where $Q_i(t)$ is the queue length of link i at time t. In the above definition, the orders $\omega(1)$ and $O(1)$ are with respect to the network size $|\mathscr{L}|$, i.e., delay-optimality means that the *per-link* queue-size remains 'constant' as the network size grows.

To describe the analysis for the performance of *delayed* CSMA, we first introduce the necessary definitions of the *total variation distance* and the corresponding *mixing time* of the CSMA Markov chain. The total variation distance between two probability distributions $\eta = [\eta_i]$ and $v = [v_i]$ on state space Ω is

$$\| \eta - v \|_{TV} = 12 \sum_{i\in\Omega} |\eta_i - v_i|.$$

Using this distance metric, the *mixing time* of the kth CSMA Markov chain $\{\sigma(uT + k) : u = 0, 1, 2, \ldots\}$ is defined as follows:

$$M^{(k)}(\delta) = \inf\{s : \max_{\mu(k)} \|\mu(uT + k) - \pi\|_{TV} \le \delta, \ \forall u \ge s\},$$

where $\delta > 0$ is some constant and $\mu(t)$ denotes the probability distribution of random variable $\sigma(t)$. The mixing time measures how long it takes for the kth CSMA Markov chain to converge to the stationary distribution for arbitrary initial distribution $\mu(k)$. Since we assume the fixed common transmission aggressiveness across the Markov chains, the mixing time $M^{(k)}(\delta)$ is identical for $k = 0, 1, \ldots, T - 1$. Hence, we use $M(\delta) = M^{(k)}(\delta)$.

The following theorem states the delay-optimality of the delay-optimality of the *delayed* CSMA algorithm.

Theorem 1 *For any ε-admissible arrival rate λ, there exists $T^* = O\left(\frac{1}{\varepsilon^3} \log M(\varepsilon/2)\right)$ such that for all $T > T^*$, the corresponding delayed CSMA algorithm is delay-optimal, more formally,*

$$\lim_{t\to\infty} \mathbb{E}[Q_i(t)] = O\left(\frac{1}{\varepsilon^4}\right), \qquad \text{for all i} \in \mathscr{L}.$$

[6]This per-link optimality is much stronger than the 'network-wide' optimality defined by the averaged delay over all links.

The above theorem states that the per-link average queue-size is bounded by a constant for sufficiently large T, the number of independent CSMA schedulers. The purpose of choosing large T is to effectively reduce the dependency among consecutive link states, which promotes much faster link state changes and hence alleviates the starvation problem. For the proof of the Theorem 1, refer to [13].

4.2.3 Related Work on Delay Reduction

In addition to the "first-order" metric such as throughput or utility, the delay performance of optimal CSMA has been studied recently. Delay in optimal CSMA has been largely under-explored, where only a small set of work has been published with emphasis on the asymptotic results. Shah et al. [32] show that it is unlikely to expect a simple MAC protocol such as CSMA to have high throughput and low delay. Thus, to achieve $O(1)$ delay, in [22, 29], modified CSMA algorithms are proposed. In [29], a modified CSMA requiring *coloring operation* achieves $O(1)$ delay for networks with geometry (or polynomial growth). A *reshuffling* approach, which periodically reshuffles all on-going schedules under time synchronized CSMA, leads to both throughput-optimality and $O(1)$ delay for torus (inference) topologies [22].

Without any modification, the algorithms that split the holding and backoff times for a desired transmission aggressiveness determine the delay. In this approach, mixing time has been a popular toolkit for delay analysis [5, 29]. Jiang et al. [5] proved that a discrete-time parallelized update algorithm achieves $O(\log n)$ delay for a limited set of arrival rates. However, it was shown very recently [33] that mixing time based approach may not be the right way to capture delay dynamics even in the asymptotic sense. In [12], asymptotic variance is used for the other metric that measures delay. In this work, they arrange the CSMA algorithms by asymptotic variance and show that the algorithm reducing asymptotic variance enhances delay performance.

5 Practical Protocol and Implementation

5.1 Research on Optimal CSMA Practice

A limited number of work on the implementation of optimal CSMA exists, mainly with focus on evaluation [17, 24]. They show that multiple adverse factors of practical occurrence not captured by the assumptions behind the theory can hinder the operation of optimal CSMA, introducing severe performance degradation in some cases [24]. In [2, 16], the interaction between TCP and optimal CSMA has been investigated due to the window based congestion control of TCP. Two algorithms each based on multiple sessions [2] or virtual queue mechanism [16], respectively was proposed. Very recently, a protocol, called O-DCF [15], reflecting the rationale of optimal CSMA, has been designed and implemented on the legacy 802.11 hardware, and shows significant performance improvement over the 802.11 DCF. Recently, an enhanced version of O-DCF, called A-DCF [14], was proposed to work better with TCP.

5.2 *O-DCF*

This subsection describes O-DCF [15], which effectively bridges the gap between practice and theory in optimal CSMA. In O-DCF, a product of access probability (determined by contention window (CW) size in 802.11) and transmission length is set to be proportional to the supply-demand differential for long-term throughput fairness. A combination of access probability and transmission length is smartly taken, where an access probability is initially selected as a sigmoid function of queue length and searched by Binary Exponential Backoff (BEB) in a fully distributed manner to adapt to the contention levels in the neighborhood. Then, transmission length is suitably selected for long-term throughput fairness. The explanation of O-DCF is elaborated in the following.

5.2.1 System Architecture of O-DCF

In O-DCF, each node runs a per-neighbor control for accessing the medium by maintaining per-neighbor states, as shown in Fig. 4. Those states are used to determine how aggressively the node should access the medium in transmitting frames in a (link-level) destination-dependent manner. To this end, O-DCF maintains two per-neighbor queues: CQ (Control Queue) and MAQ (MAC Queue). CQ has the role of buffering the packets from upper layers, where each packet from upper layers is first classified according to its destination, and then enqueued into its per-neighbor CQ as frames. MAQ functions as a per-neighbor state that is importantly used to determine frames' medium access aggressiveness. A notion of Rate Controller (RC) resides between a CQ and a MAQ, and controls the dequeuing rate from the CQ to the MAQ. How the dequeuing rate is decided is critical in achieving fair medium access in O-DCF (see Sect. 5.2.2). Then, the service from a MAQ occurs when the HOL (Head-Of-Line) frame of the MAQ is moved into IQ (Interface Queue). 802.11 DCF parameters such as CW_{min} and TXOP are appropriately set for controlling access aggressiveness. For multiple neighbors, the longest MAQ is served first; If the chosen transmission length exceeds a single frame size, multiple frames from the same MAQ are scheduled in succession.

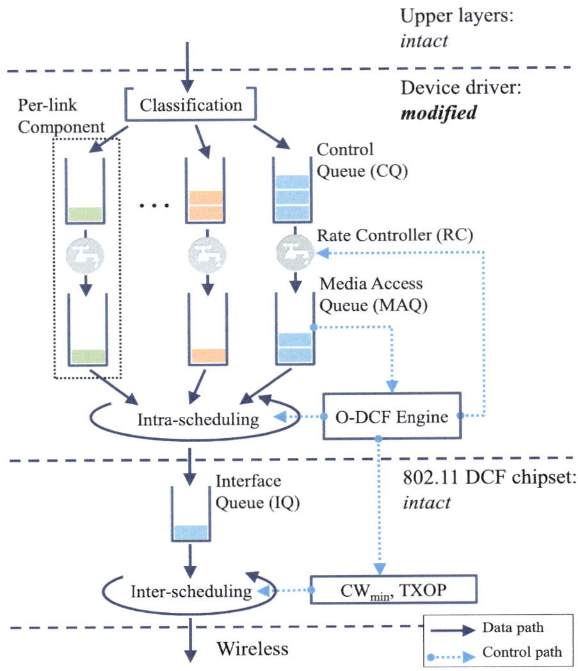

Fig. 4 System architecture of O-DCF

5.2.2 Key Mechanisms of O-DCF

The MAQ maintains the supply–demand differential, and the dequeuing rate and the access aggressiveness are controlled by its queue length. For high performance, O-DCF translates the access aggressiveness into an adaptive combination of access probability and transmission length.

Rate Control

Let $Q_l(t)$ denote the length of MAQ for each link l at time t. O-DCF controls the dequeuing rate from CQ to MAQ as follows:

$$\text{Rate from CQ to MAQ for link } l = \frac{V}{q_l(t)}, \tag{12}$$

where $q_l(t) = bQ_l(t)$, and b and V are some constants. Intuitively, O-DCF decreases the rate for the long MAQ, and increases the rate when the MAQ is well-served. b is a small value that corresponds to a step size, being responsible for slowing down the variations of queue length. V is the constant that controls the sensitivity of dequeuing rate from CQ to MAQ. This form of dequeuing pattern is for achieving proportional fairness, derived from the log utility maximization; the dequeuing rate is $U'^{-1}(q_l(t)/V)$, where $U(\cdot)$ is a utility function, and $U(\cdot) = \log(\cdot)$ thus, $U'^{-1}(q_l(t)/V) = V/q_l(t)$. By suitably choosing the form of the utility function, various fairness criteria can be achieved.

Access Aggressiveness Control

CSMA has two critical parameters for controlling its aggressiveness: (i) access probability and (ii) transmission length. In many practical MACs such as 802.11, access probability is typically controlled by contention window (CW) size, and transmission length corresponds to the number of consecutive transmitted frames without separate media sensing. Aggressiveness simply means the product of access probability and transmission length, which are controlled differently for different neighboring links. Aggressiveness in O-DCF is basically controlled by the following simple rule:

$$\text{Aggressiveness (access prob.} \times \text{trans. length) for link } l = \exp(q_l(t)). \qquad (13)$$

Intuitively, $q_l(t)$ tracks how well a link has been served over time. When a link has not been served for a long time, then it has high access aggressiveness by having either small CW size and/or long transmission length. How to choose the combination of CW size and transmission length is described next.

Adaptive Combination

The key design aspects of O-DCF lies in which combination of access probability and transmission length should be chosen in practice to achieve high performance. When a frame (or a multiple of frames) from a MAQ is moved to IQ by the intra-scheduling for being ready for actual transmission, O-DCF's procedure of setting CSMA parameters is divided into the following three steps:

1. *Initial access probability*: For a frame *f* enqueued to IQ, using its per-neighbor state (i.e., its MAQ's length), an initial CW is smartly selected, where the basic principle is that the frames from under-served MAQs in terms of queue length are assigned smaller CWs. First, in order to effectively prioritize an under-served link, access probability of the link is calculated from a sigmoid function as shown in Fig. 5a. Then, the access probability is converted into CW size conforming to the restriction of the 802.11 chipset[7] as in 5b.
2. *BEB for actual CW*: Once the initial CW size is chosen as a function of MAQ's length, the actual medium access is attempted, allowing BEB (Binary

[7]CW sizes are one of values in $\{2^{i+1} - 1 : i = 0, \ldots, 9\}$.

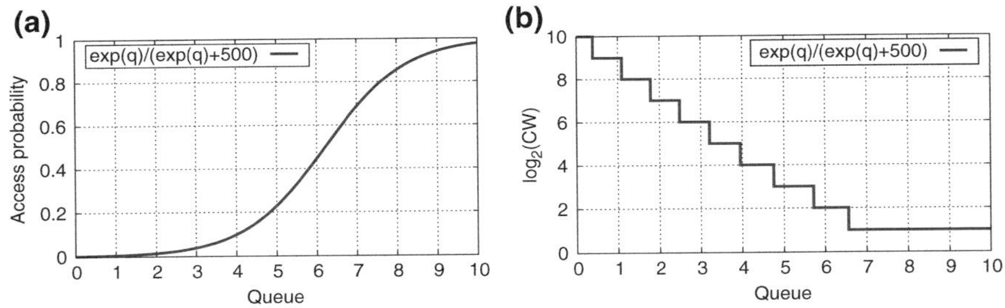

Fig. 5 Illustration of sigmoid function with respect to queue length

Exponential Backoff) to occur, which corresponds to a distributed search of the actual access probability.

3. *Transmission length selection*: Once the actual CW is obtained after BEB, it is converted to an access probability, and then the transmission length is determined from (13) by considering the corresponding MAQ's length and the maximum transmission length specified in the legacy 802.11 chip.

5.2.3 Performance Evaluation

O-DCF is compared with (i) 802.11 DCF, (ii) two versions of optimal CSMA in theory, and (iii) DiffQ [35]. For the standard optimal CSMA, two versions are tested to show the effect of the adaptive CSMA parameter combination in O-DCF: (i) *CW adaptation* in which the transmission length μ is fixed with a single packet and the access probability $p_l(t)$ is controlled, such that $p_l(t) \times \mu = \exp(q_l(t))$ [7], and (ii) *μ adaptation with BEB* (shortly, μ adaptation in this paper) in which the selection of $p_l(t)$ is delegated to 802.11 DCF and $\mu_l(t) = \exp(q_l(t))/p_l(t)$. Note that to understand the effect of different methods for the adaptation of CWs, μ adaptation is evaluated with BEB using 802.11's CW size, and is compared with O-DCF. DiffQ is a *heuristic* harnessing the 802.11e feature, and schedules the interfering links with different priorities based on queue lengths.

For performance comparisons, 16-node testbed is deployed as shown in Fig. 6a. Each node is a netbook platform (1.66 GHz CPU and 1 GB RAM) running Linux kernel 2.6.31 and equipped with a single 802.11a/b/g NIC (Atheros chipset) running the modified MadWiFi driver for O-DCF's operations. To avoid external interference, a 5.805 GHz band in 802.11a is selected. The default link capacity is fixed with 6 Mb/s. In the 16-node testbed topology, two cases of five and seven concurrent flows under the default capacity are tested. This random topology enables to see how the algorithms perform in the mixture of hidden terminals and heavy contention scenarios including flow-in-the-middle (FIM) scenarios. The source and destination of each single-hop flow is chosen randomly. For each case,

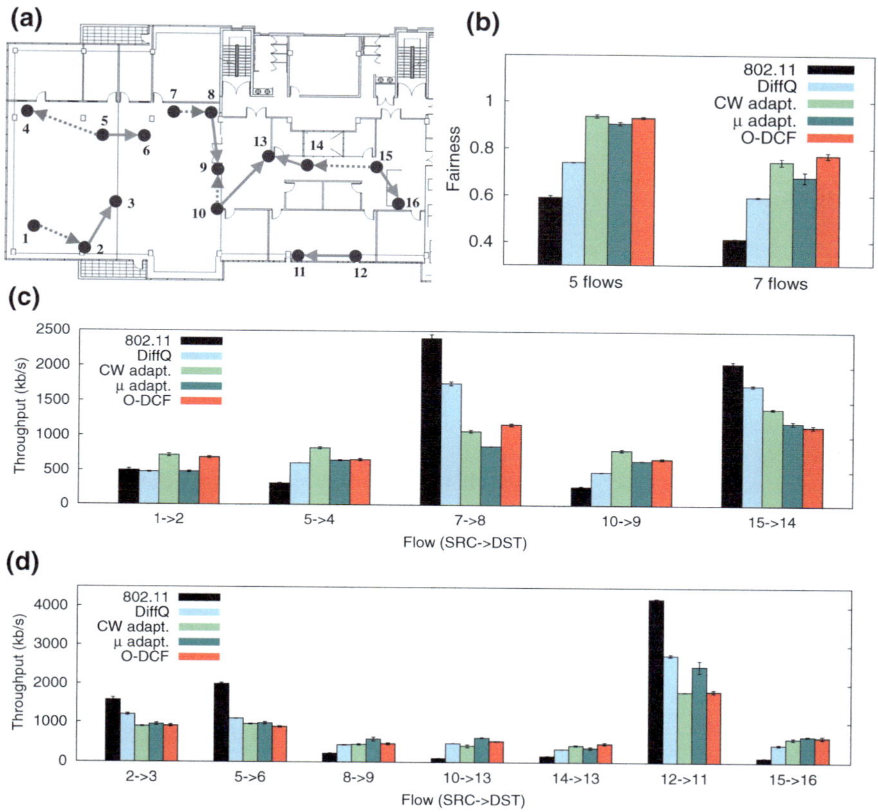

Fig. 6 Tested topology and performance comparison; **a** 16 nodes denoted by *triangles* are distributed in the area of 40 × 20 m; *dotted* (*solid*) *arrows* represent 5 (7) flows for the first (second) scenario. **b** Jain's fairness comparison. **c–d** Per-flow throughput distributions

ten runs are repeated and error bars in all plots represent standard deviation. The duration of each run is 60 s.

Figure 6b compares Jain's fairness achieved by all the algorithms for two scenarios. Over all the scenarios, O-DCF outperforms others in terms of fairness (up to 87.1 % over 802.11 and 30.3 % over DiffQ). The fairness gain can be manifested in the distribution of per-flow throughput, as shown in Fig. 6c, d. O-DCF effectively prioritizes the flows with more contention degree (e.g., flow 10 → 9 forms *flow-in-the-middle* with flows 7 → 8 and 15 → 14) and provides enough transmission chances to highly interfered flows (i.e., 8 → 9, 10 → 13, and 14 → 13), compared with 802.11 DCF and DiffQ. The experimental topology is somewhat limited in size, tending to be full-connected. This leads to a small performance gap between the standard optimal CSMA and O-DCF, but 802.11 DCF yields severe throughput disparities of more than 40 times between flows 12 → 11 and 10 → 13 in the second scenario. Compared with 802.11, DiffQ performs fairly well in the sense that it prioritizes highly interfered flows. However, its access prioritization is heuristic, so there is still room for improvement compared with O-DCF.

6 Summary

An extensive array of analysis and protocols are proposed on what are efficient MAC schemes. Efficiency can be measured by control overhead, throughput, fairness, etc. This survey demonstrates that a simple, fully distributed MAC with no or little message passing, such as CSMA, can be designed to achieve optimality, where various findings have been explored, and people are starting to looking at their practical values by evaluation and implementation in real hardware. Despite a long history of MAC research, there still exist under-explored areas toward simple, yet highly efficient MAC. We hope that this survey paper helps the readers with summarizing the current research progress on optimal CSMA.

Acknowledgments This work is supported by the Center for Integrated Smart Sensors funded by the Ministry of Science, ICT and Future Planning as the Global Frontier Project.

References

1. Akyildiz IF, Melodia T, Chowdhury KR (2007) A survey on wireless multimedia sensor networks. Comput Netw 51(4):921–960
2. Chen W, Wang Y, Chen M, Liew SC (2011) On the performance of TCP over throughput-optimal CSMA. In: Proceedings of IWQoS
3. Hegde N, Proutiere A (2012) Simulation-based optimization algorithms with applications to dynamic spectrum access. In: Proceedings of CISS
4. Jang H, Yun SY, Shin J, Yi Y (2014) Distributed learning for utility maximization over CSMA-based wireless multihop networks. In: Proceedings of INFOCOM
5. Jiang L, Leconte M, Ni J, Srikant R, Walrand J (2011) Fast mixing of parallel Glauber dynamics and low-delay CSMA scheduling. In: Proceedings of INFOCOM
6. Jiang L, Shah D, Shin J, Walrand J (2010) Distributed random access algorithm: scheduling and congestion control. IEEE Trans Inf Theory 56(12):6182–6207
7. Jiang L, Walrand J (2010) A distributed CSMA algorithm for throughput and utility maximization in wireless networks. IEEE/ACM Trans Networking 18(3):960–972
8. Jiang L, Walrand J (2011) Approaching throughput-optimality in distributed CSMA scheduling algorithms with collisions. IEEE/ACM Trans Networking 19(3):816–829
9. Kai CH, Liew SC (2012) Applications of belief propagation in CSMA wireless networks. IEEE/ACM Trans Networking 20(4):1276–1289
10. Kim TH, Ni J, Srikant R, Vaidya N (2011) On the achievable throughput of CSMA under imperfect carrier sensing. In: Proceedings of INFOCOM
11. Kwak J, Lee CH, Eun DY (2014) A high-order markov chain based scheduling algorithm for low delay in CSMA networks. In: Proceedings of IEEE INFOCOM

12. Lee CH, Eun DY, Yun SY, Yi Y (2012) From Glauber dynamics to Metropolis algorithm: smaller delay in optimal CSMA. In: Proceedings of ISIT
13. Lee D, Yun D, Shin J, Yi Y, Yun SY (2014) Provable per-link delay-optimal CSMA for general wireless network topology. In: Proceedings of IEEE INFOCOM
14. Lee H, Moon S, Yi Y (2015) A-DCF: design and implementation of delay and queue length based wireless mac. In: Proceedings of IEEE INFOCOM
15. Lee J, Lee H, Yi Y, Chong S, Nardelli B, Chiang M (2013) Making 802.11 DCF near-optimal: design, implementation, and evaluation. In: Proceedings of IEEE SECON
16. Lee J, Lee HW, Yi Y, Chong S (2012) Improving TCP performance over optimal CSMA in wireless multi-hop networks. IEEE Commun Lett 16(9):1388–1391
17. Lee J, Lee J, Yi Y, Chong S, Proutiere A, Chiang M (2009) Implementing utility-optimal CSMA. In: Proceedings of Allerton (2009)
18. Li B, Eryilmaz A (2012) A fast-CSMA algorithm for deadline-constrained scheduling over wireless fading channels. In: ArXiv e-prints
19. Li H, Vaidya N (2014) Optimal CSMA-based wireless communication with worst-case delay and non-uniform sizes. In: Proceedings of IEEE INFOCOM
20. Liu J, Yi Y, Proutiere A, Chiang M, Poor HV (2008) Maximizing utility via random access without message passing. Tech. rep., Microsoft Research Labs, UK
21. Liu J, Yi Y, Proutiere A, Chiang M, Poor HV (2010) Towards utility-optimal random access without message passing. Wiley J Wirel Commun Mobile Comput 10(1):115–128
22. Lotfinezhad M, Marbach P (2011) Throughput-optimal random access with order-optimal delay. In: Proceedings of INFOCOM
23. Murty RN, Mainland G, Rose I, Chowdhury AR, Gosain A, Bers J, Welsh M (2008) Citysense: an urban-scale wireless sensor network and testbed. In: Proceedings of IEEE HST
24. Nardelli B, Lee J, Lee K, Yi Y, Chong S, Knightly E, Chiang M (2011) Experimental evaluation of optimal CSMA. In: Proceedings of INFOCOM
25. Ni J, Tan B, Srikant R (2010) Q-CSMA: Queue-length based CSMA/CA algorithms for achieving maximum throughput and low delay in wireless networks. In: Proceedings of INFOCOM
26. Proutiere A, Yi Y, Lan T, Chiang M (2010) Resource allocation over network dynamics without timescale separation. In: Proceedings of INFOCOM
27. Qian D, Zheng D, Zhang J, Shroff N (2010) CSMA-based distributed scheduling in multi-hop MIMO networks under SINR model. In: Proceedings of INFOCOM
28. Rajagopalan S, Shah D, Shin J (2009) Network adiabatic theorem: an efficient randomized protocol for contention resolution. In: Proceedings of ACM SIGMETRICS. Seattle, WA
29. Shah D, Shin J (2010) Delay optimal queue-based CSMA. In: Proceedings of ACM SIGMETRICS
30. Shah D, Shin J (2012) Randomized scheduling algorithm for queueing networks. Ann Appl Probab 22:128–171
31. Shah D, Shin J, Tetali P (2011) Medium access using queues. In: Proceedings of IEEE FOCS
32. Shah D, Tse DNC, Tsitsiklis JN (2011) Hardness of low delay network scheduling. IEEE Trans Inf Theory 57(12):7810–7817
33. Subramanian V, Alanyali M (2011) Delay performance of CSMA in networks with bounded degree conflict graphs. In: Proceedings of ISIT
34. Tassiulas L, Ephremides A (1992) Stability properties of constrained queueing systems and scheduling for maximum throughput in multihop radio networks. IEEE Trans Autom Control 37(12):1936–1949
35. Warrier A, Janakiraman S, Ha S, Rhee I (2009) DiffQ: Practical differential backlog congestion control for wireless networks. In: Proceedings of IEEE INFOCOM
36. Wikipedia VSN. http://en.wikipedia.org/wiki/Visual_sensor_network

37. Yi Y, Chiang M (2011) Next-generation internet architectures and protocols. In: Chap. 16: Stochastic network utility maximization and wireless scheduling, Cambridge University Press, Cambridge
38. Yun SY, Shin J, Yi Y (2013) CSMA over time-varying channels: optimality, uniqueness and limited backoff rate. In: Proceedings of ACM MobiHoc
39. Yun SY, Shin J, Yi Y (2013) CSMA using the bethe approximation for utility maximization. In: Proceedings of ISIT
40. Yun SY, Yi Y, Shin J, Eun DY (2012) Optimal CSMA: a survey. In: Proceedings of ICCS

Wireless Sensor Network for Video Sensors

Hyung Won Kim

Abstract This chapter reviews various wireless sensor networks that have been proposed in the literature or have been widely used as experimental or commercial networks. It then analyzes topologies of sensor networks and compares various routing protocols for classical wireline networks and wireless sensor networks. While many types of wireless sensor networks have been developed, most of them are targeted for low data rate sensor devices with sparse events. In such networks, only one RF channel is often used, and they still can find routing solutions that provide data throughput enough for all the sensors and also meet their low power requirements. As the speed and range of wireless networks improve, wireless networks have been adopted for video sensors such as surveillance cameras, and factory or field monitoring cameras. These video sensors usually have much higher data rate and tighter power requirements than the above low rate sensors, and so demand more complex routing schemes. The goal of routing for video sensor network is also different from the low rate sensor network. Its goal is usually the real-time delivery of high data rate bursty video streams from all active video sensors. This chapter introduces new routing and channel allocation methods that use multiple channels and realistic link utilization models. It discusses how to extend the multi-channel routing methods for video sensor networks to various future applications including smart grid and vehicle-to-vehicle wireless networks.

Keywords Wireless sensor network · Routing and channel allocation · Video sensor network

H.W. Kim (✉)
School of Electronics Engineering, Chungbuk National University, Chungdae-ro 1, Seowon-Gu, Cheongju, Chungbuk 362-763, Korea
e-mail: hwkim@cbnu.ac.kr

C. Kyung (ed.), *Theory and Applications of Smart Cameras*,
KAIST Research Series, DOI 10.1007/978-94-017-9987-4_15

1 Introduction to Wireless Sensor Networks

This section reviews the structure and operation of wireless sensor networks, and compares them with other wireless networks. It also describes the demand for new wireless sensor networks.

1.1 Applications of Wireless Sensor Networks

The best method to transfer the data from many sensors over a long distance is widely considered to be wireless sensor network. A variety of sensors are increasingly used to automatically monitor factories and chemical plants, detect wild fire or disaster, measure environmental data such as temperature, atmospheric pressure, and humidity, and also monitor building and home automation, etc. Efficient network protocols that establish and manage the wireless networks to connect all these sensors are the key to a successful deployment of a large number of sensors [1–3].

Lately, surveillance cameras are also often connected through wireless networks. Such wireless networks usually need high data rate and real-time delivery, and so impose a difficult problem that conventional protocols for wireless sensor networks cannot address effectively [4–6].

Other new applications of wireless sensor networks include precision agriculture, which can provide fertilizer, pesticides, irrigation only where and when needed. Wireless sensor networks can also be used for medicine and health care. For example, monitoring sensors on patients can form a wireless sensor network to help intensive care of postoperative patients and long-term monitoring of patients with chronic disease.

1.2 Review of Existing Wireless Networks

To better understand the properties and requirements of wireless sensor networks, we will review various types of wireless networks here. Wireless networks can be categorized into infrastructure networks, ad hoc networks, and sensor networks.

Infrastructure networks have one or multiple central stations, often implemented as base stations or access points, which wirelessly connect with mobile or portable stations and also connect to a wire-line backbone network. In infrastructure networks, all mobile stations can communicate only with central stations which are, in general, stationary. Examples of infrastructure networks are cellular networks such as 3GPP WCDMA and LTE networks for mobile phone services. Wi-Fi networks based on access points are also infrastructure networks commonly found [7, 8].

Ad hoc networks, on the other hand, do not have central stations, instead they allow the mobile or portable stations to communicate directly with other mobile stations.

Examples of ad hoc networks are Wi-Fi networks configured as ad hoc mode. Since there is no central station in ad hoc networks, finding destination node and routing path to the destination is up to each of the stations. Therefore, broadcasting or flooding of packets is often used to blindly forward packets to the destination. Ad hoc networks whose stations move around are called Mobile Ad hoc Network (MANET) [9].

In infrastructure networks and ad hoc networks, the stations are usually human-carried devices such as PCs, laptops, mobile phones, or portable monitoring systems with human interfaces like keyboard or LCD screen [7, 8].

Wireless sensor networks are distinguished from these networks, since in wireless sensor networks the stations are mostly autonomous sensor nodes. These sensor nodes usually have only sensing circuits and wireless transceiver circuit (and sometimes actuator) and usually do not have human interfaces. Due to the nature of applications, the sensor nodes of sensor networks are often battery powered. Since the sensor nodes are expected to operate for a long time unattended, the battery lifetime must last a long period of time. While infrastructure and ad hoc networks may also have battery-powered stations, these nodes can be recharged easily by humans [7, 8].

Wireless sensor networks often have network topology similar to ad hoc networks. Unlink ad hoc networks, however, wireless sensor networks usually have routing protocols that are optimized for long battery lifetime. Wireless sensor networks are often required to cover a large area of building, factory, field, or mountains. Therefore, the research focus for wireless sensor networks has been on efficient network topology and routing protocols to deliver sensor data to the destination node through other nodes, in other words, a multi-hop routing [7, 8].

Other requirements for wireless sensor networks are operation of sleep modes, auto-configuration or self-organization of networks, fault tolerance, data centric network protocol, energy harvesting or scavenging, and in-network processing (or pre-processing of data in intermediate nodes). In most of low rate sensor applications, wireless sensor networks usually do not provide real-time delivery or data processing [7, 8].

Due to the recent surge in demand for wireless sensor networks, the activity of new standards for more efficient wireless sensor networks is increasing. A few examples of such standards are Zigbee, WirelessHART, ISA100A, IETF 6LowPAN, and IEEE 802.11ah. Here Zigbee, WirelessHART, and ISA100A share the IEEE 802.15.4 standard for their MAC and PHY definition, while they use different network layer and application layer profiles targeting different application goals. The IEEE 802.11ah is an extended Wi-Fi standard under development to modify existing IEEE 802.11ac standard to cover a wider range of sensor networks with low data rate [10–17]. Comparison of these wireless sensor network standards is given in Table 1.

Table 1 Comparison of wireless sensor network standards

Standards	IEEE 802.15.4	Zigbee	Wireless HART	ISA 100A	IEEE 802.11ah
Frequency	800 ~ 900 MHz, 2.4 GHz	2.4 GHz	2.4 GHz	800 ~ 900 MHz, 2.4 GHz	800 ~ 900 MHz
Data rate	20 K ~ 250 Kbps	20 K ~ 250 Kbps	250 Kbps	20 K ~ 250 Kbps	0.72 M ~ 64 Mbps for SISO
Bandwidth	2 ~ 5 MHz	2 ~ 5 MHz	2 ~ 5 MHz	2 ~ 5 MHz	1 ~ 16 MHz
Range	10 ~ 100 m				1 Km
Network size		255 nodes	400 nodes	600 nodes	8000 nodes
Target application	WPAN for industrial, residential and medical	Consumer, smart grid	Industrial automation	Industrial automation	Wireless sensor network
Topology and routing	Not defined	Star, tree, mesh, AODV	Star, cluster, mesh, source routing, TDMA	Star, tree, mesh	Tree, 2hops with a relay AP
PHY standards	Direct sequence spread spectrum	IEEE 802.15.4	IEEE802.15.4 Feq. hopping spread spectrum	IEEE802.15.4 Feq. hopping spread spectrum	IEEE802.11ah

1.3 *Demands for New Wireless Sensor Networks*

The wireless sensor networks described above may work effectively for sensor nodes with low data rate and non-real-time data, or networks with relatively small number of sensor nodes.

There is a growing demand for wireless video sensor networks (WVSN) for the purpose of building or facility surveillance, factory or agriculture monitoring, and disaster prevention. These WVSN need new schemes of routing and channel allocations for optimal data delivery with lower power [4–6].

Even for the case of low data rate sensors, if the number of sensor nodes becomes extremely large, a common situation of Internet of Things, their routing problem tends to behave like high data rate networks. In such networks, the classical routing methods developed for low rate sensors often lead to very poor results.

To provide better solutions for WVSN and large-scale sensor networks such as Internet of Things (IoT), new research is demanded in the areas of network topology, better power saving methods, more efficient routing methods, and efficient allocation and scheduling of multi-channels [10].

In the later sections of this chapter will introduce such research effort.

2 Wireless Sensor Network Topology and Routing Protocols

This section describes various topologies and routing protocols of wireless sensor networks. It will begin with well-known topologies and routing protocols of classical Internet, and then extend the subjects to wireless sensor networks.

2.1 *Review of Wireless Sensor Network Topology*

The topology of a wireless sensor network is to allow which nodes of the network to connect with which nodes. In other words, topology control determines the connectivity of each node by turning on and off the activities of certain nodes or the links between the nodes. The topology becomes more important as the size of wireless sensor network increases, as the decision of connecting which node to which node is becoming a very complex problem in a network with a large number of nodes.

Topologies can be categorized into the following three groups:

- **Flat networks**: All nodes in the network are at the same level. Flat networks need a method to determine which link to enable. Some links are removed by lowering transmit power or deliberately turning off certain connections between two nodes [7]. See Fig. 1a for example.

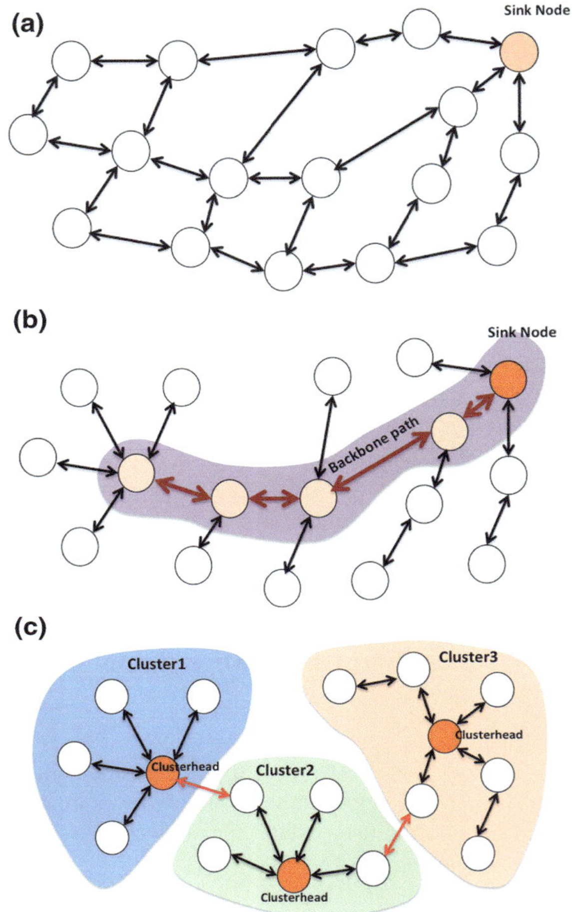

Fig. 1 **a** An example of flat wireless network; **b** Backbone-based hierarchical network; **c** Cluster-based hierarchical network

- **Backbone-based hierarchical networks**: Some nodes are selected as dominating node set, so they control their neighbor nodes. The nodes in the dominating set are connected to form a backbone path. Every other node connects only to one of the backbone nodes. The backbone carries all traffic of non-backbone nodes to the sink node [18]. See Fig. 1b for example.
- **Cluster-based hierarchical networks**: All nodes are partitioned into a set of clusters. Each node belongs to only one cluster, except a bridge node which may belong to multiple clusters to forward traffic between clusters. Each cluster has a central node called a clusterhead. All nodes in a cluster connect with only their clusterhead. Then the clusterhead forwards data to other cluster's clusterhead or a bridge node. Figure 1c gives an example of a cluster-based hierarchical network [19].

In flat networks, the network structure is usually mesh or tree network, which need multiple hops to forward data to the destination or sink node. In backbone-based or cluster-based networks, the connection from regular nodes to a backbone or clusterhead node are often configured as multi-hop to allow a shorter backbone or a larger cluster.

In the following section, therefore, we assume the entire network (flat network case) or a portion of the network (backbone or cluster-based network case) is a multi-hop tree or mesh structure. Given such multi-hop tree or mesh structured networks, we will review routing protocols to find optimal forwarding paths.

2.2 *Routing Protocols of Wireless Networks*

Routing protocols for classical wireline networks are usually categorized into two groups: link state protocols and distance vector routing protocols. In the classic link state protocols, routing is to find a complete path from one node to all possible destination nodes using Dijkstra's shortest path algorithm. Each node of a network constructs complete routing paths from itself to every other node in the network. Example commercial routing protocols based on this category are OSPF (Open Shortest Path First) protocol and IS-IS (Intermediate System to Intermediate System) protocol.

On the other hand, in the distance vector routing category, each node of a network periodically informs its neighbor nodes of topology changes such as direction and distance. These routing protocols often use the Bellman Ford shortest path algorithm. Example Internet protocols based on distance vector routing are RIPv1 (Routing Information Protocol), RIPv2, IGRP (Interior Gateway Routing Protocol).

The classical routing protocols, however, have drawbacks to be applied to wireless sensor networks. They are often too slow to react to the changes in the wireless sensor networks, and their complex protocols impose heavy burden on the low computation power sensor nodes.

Wireless network routing protocols that are enhanced to alleviate these problems can be grouped to proactive routing and on-demand routing protocols. The proactive routing protocols always keep the routing table up-to-date, while the on-demand routing protocols, also called reactive routing protocols, find the routing paths only when needed.

Well-known proactive routing protocols include: DSDV (Destination sequence distance vector) routing [20, 21], OLSR (Optimized Link State Routing) [22], FSR (Fisheye State Routing). DSDV routing improves the classic DV routing by using distributed Bellman-Ford algorithm, therefore it can be applied to ad hoc wireless networks. It also considers aging information by adding a sequence number to routing information, which is propagated through distance vector exchange process. It allows fast routing table updates by sending immediate advertisement upon significant changes in the topology. Another enhancement is introducing a damp function to avoid unnecessary fluctuation of routing paths when unstable changes occur in the topology. The damp function delays new route information until all such information is received, and selects the best route path. OLSR enhances the classic link state routing protocol by limiting the dominating set of each node to its two-hop neighborhood, and thus reducing its flooding overhead.

Popular reactive (on-demand) routing protocols are DSR (Dynamic source routing) [23], AODV (Ad hoc on-demand distance vector) routing [24], TORA (Temporally ordered routing algorithm) [25]. DSR enhances the classic Link State Routing by discovering the complete or partial path from a source node to the destination instead of updating routing table in every node. It also uses Route Request/Route Reply packets to find a path. An intermediate node can send Route Reply if it already has a path. In case an error occurs, the route path can be updated locally instead of redoing the entire route process. See Fig. 2a for an example of DSR.

AODV improves the classic DV routing by controlling Route Request flooding within a region. Each node sends a Route Request only once. Unlike DSR, in AODV each node remembers where a packet came, and records this information in its route table instead of source routing information. It introduces a sequence number to avoid stale cache information and also to break a loop in the route paths. In AODV, when a Route Reply from the destination node is forwarded back to the source node, a forward path is formed. It also allows an intermediate node that knows a route to destination also to send Route Reply to the source, and thus reduce Route Request flood. An example of AODV routing processing is given in Fig. 2b.

3 Low Power Multi-hop Routing Algorithm for Video Sensor Networks

This section highlights a new demand for low power wireless sensor networks for the real-time delivery of high rate video streams. The earlier conventional wireless sensor network protocols are not suitable for such networks. This section later introduces a new approach to solving a more complex and demanding problem of finding an optimal routing and channel allocation for WVSN.

We first describe properties of a WVSN, where each node generates event-driven video sensor data, and forwards the video data toward a designated sink node. Battery powered video camera sensors are often connected wirelessly to cover a large area. Such WVSN are considered as major applications of IoT. We analyze the power consumption model for a WVSN. We then describe an algorithm to route the sensor nodes and allocate channels in a way that minimizes the overall power consumption while satisfying the required data transmission. We then describe a WVSN simulator that proves the performance advantage of the algorithm introduced in this section. Simulation results are provided with wireless sensor networks of various sizes.

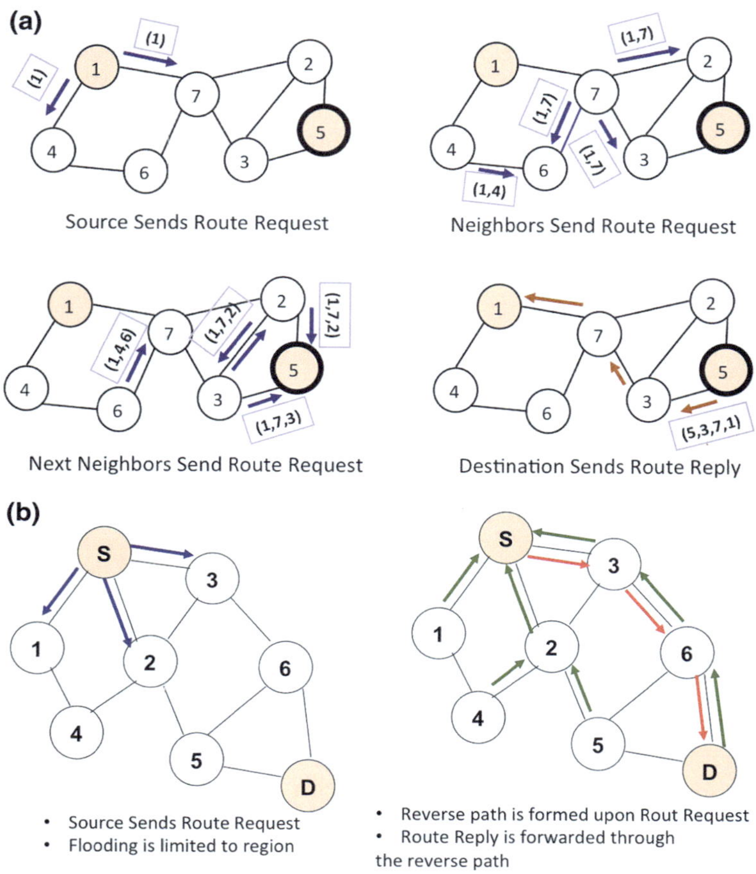

Fig. 2 **a** An example of DSR protocol; **b** AODV routing

3.1 Properties of Wireless Video Sensor Networks

Today, wireless cameras are increasingly adopted in CCTV and home networks, and also for IoT in the near future. Most past wireless cameras use wireless networks such as Wi-Fi based on IEEE 802.11 standards with an access point system operating in infrastructure mode [6]. However, such wireless networks have many restrictions in their data rate, wireless range, and traffic congestion level. When used for WVSNs, their power consumption also becomes a serious issue, because their camera nodes would be battery powered.

A promising solution to these problems is to use a wireless mesh network with each camera acting as each node. Multi-channel routing schemes are often used to reduce the RF interference and traffic congestion, and to enhance the video data rate while minimizing the power consumption [4–6].

In the past, much research has been done to minimize the total energy consumption in wireless sensor networks, to reduce the total delivery time in wireless mesh networks, and to minimize the size of camera data transferred through WVSNs [4–6].

Prior techniques have also attempted to reduce energy consumption by selecting or combining overlapped field of views (FoVs) if such overlapped images are frequently observed [26].

In this section, we describe a new method of multi-channel allocation and routing selection in a wireless mesh network where each node has an event-driven surveillance camera—a camera that operates (active mode) only when it detects a motion, sound, or perturbation, and otherwise stays dormant (sleep mode). We present its simulation results proving that it can minimize the power while allowing maximal data rate using only the active video nodes whenever possible.

Often cameras are battery powered, and so it is required to reduce the energy consumption while delivering their real-time video streams as fast as possible [27–31]. Multi-channel radios can either reduce the interference among neighbor nodes or reduce the data delivery time by concurrently transferring the video streams. It has been proven, however, that the multi-channel allocation and routing problem is NP-hard [6]. In the mesh network of our concern, each node generates intermittent video data (driven by events). Only the nodes currently transmitting data are active while all others are in sleep mode. New constraints are added to minimize the energy consumption. Event-driven video data is delivered through active-mode nodes avoiding sleep-mode nodes whenever possible. This makes it even harder to solve the multi-channel routing problem.

In general wireless networks, the dominant source of the power consumption is the transmit power of each wireless link. Most prior work, however, proposed routing and channel allocation methods assuming each node uses the same unit Tx (transmit) power and the same unit data rate regardless of the channel condition and the distance of the wireless link [4–6, 32, 33]. While this assumption allows simpler optimization formula, it can lead to results drastically different from realistic power consumption. In reality, the path loss increases exponentially along with the distance, and so requiring Tx power to increase in order to keep the equal data rate. In general, since Tx power cannot be increased infinitely, wireless modems also lower their MCS (modulation and code scheme)—in other words lowering the data rate—as the transmit distance increases. In this section, we introduce a routing technique taking into account realistic link data rates and Tx power as a function of the distance between nodes. We introduce a concept of link utilization ratio, which is a succinct but realistic method of calculating Tx power during active time duration of each link.

3.2 Utilization-Based Routing and Channel Allocation

In general, battery powered video camera sensors wake up and transmit video data, only when events are detected, while they are placed in their sleep mode at other times.

A WVSN covers a large area but it has only one or a few data collection gateway nodes, which collect all video data and send it to a data center by a wire-line internet.

A. Consideration of Wi-Fi Networks for WVSN

In this section, we assume without loss of generality that in a WVSN, only a subset of the nodes wake up during a short period of time (e.g., 10–25 % are active among all nodes), and then go back to sleep mode after data transmission. We assume there is only one data collection gateway node, which we call a sink node s. Each active node transmits video data toward the sink node via multi-hop routes.

Each node has multiple IEEE 802.11n (Wi-Fi) modem and transceiver that can connect to multiple neighboring nodes using different channels. Each wireless link between two nodes is called an edge e. Each edge uses one Wi-Fi modem with one channel.

Figure 3 shows an example wireless sensor network. It has a total of 100 nodes, but only 9 nodes are active for a given time duration. Each active node generates its own video data and transmits to the sink node n0 (red circle) through neighboring nodes within its wireless range. Wireless ranges are indicated by dotted circles.

Figure 4 shows potential edges between nodes in their wireless ranges. A routing algorithm, therefore, needs to select a set of active edges from Fig. 4 in a way that the selected edges construct multi-hop paths from every active node to the sink. These paths should deliver all video data to the sink with minimal power consumption. In this example, the potential edges are selected such that the next hop of the edge has shorter distance than the previous hop to prune the edges unlikely to give an optimal routing path.

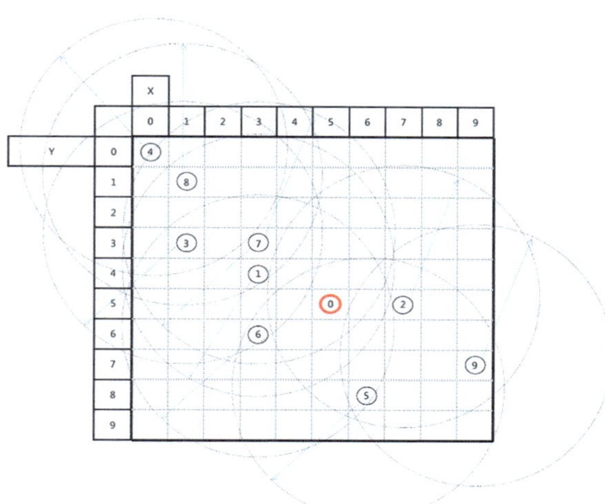

Fig. 3 Example WVSN with an array indicating location of all sensor nodes. *Small circles* indicate nine active nodes. *Large circles* indicate the wireless range of each active node. The *red* node (node 0) is a sink node (data collection gateway node). Reproduced with permission from IEEE

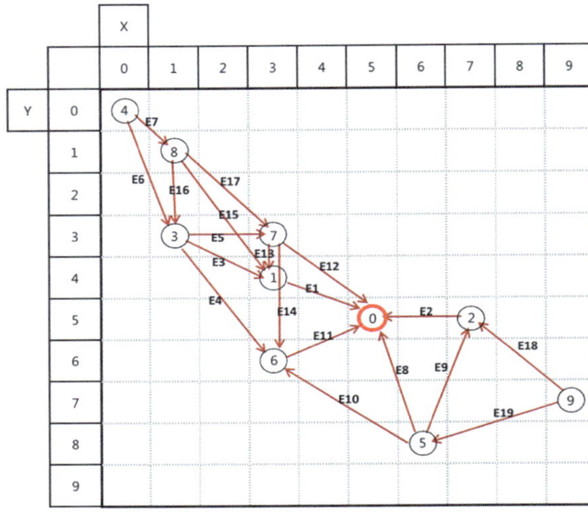

Fig. 4 Wireless video sensor network with its potential active connections (edges) within wireless ranges. Reproduced with permission from IEEE

Since Wi-Fi modem, like other wireless technologies, can interfere with others if operated in the same RF channel, different edges within a wireless range, therefore, must be allocated to different channels. In this section, we use IEEE 802.11n with 5 GHz spectrum, and we use 11 nonoverlapped channels of 40 MHz channel bandwidth. For a single stream Wi-Fi modem configuration, the maximum physical layer data rate is 150 Mbps. Based on the results measured with commercial Wi-Fi modules, the overhead of MAC and IP layer is commonly considered as 1/3 of the overall data rate. We, therefore, assume that the maximum link rate of each edge is 100 Mbps.

B. Link Utilization Analysis and Power Model

Given a wireless network described above, we can fomulate the problem of finding channel allocation and routing as an optimization problem that minimizes the total Tx power. While sleep-mode links have no Tx power, active-mode links consume Tx power only during they transmit actual data. Hence, we calculate the time duration of data transmission by deriving a link utilization rate from a maximum link rate and the total amount of current data rate on each link.

We then introduce a heuristic algorithm that finds a low power routing solution while ensuring delivery of all video data through multi-hop routes. In this way, the proposed method ensures each sensor node's maximal battery life, and avoids data congestion in the sensor network.

We will also present a simulator WiSeSim, which implements the routing and channel allocation algorithm in a C program with network models based on IEEE 802.11n. We present simulation data, which proves that the proposed method gives routing results with lower power compared to conventional routing algorithm.

Each edge's link rate varies depending on the distance between two nodes, and the radio channel condition. From the measurement of wireless modem modules, we can observe the following relations.

Let D be the distance from a source node to a destination node. A general formula for the path loss for a wireless link is given by (in unit of dB):

$$\text{Path Loss, } L_p = 20\log_{10}\left(\frac{4\pi D}{\lambda}\right) \quad (1)$$

Here λ is wavelength of the RF signal. The TX power $P_{\mathrm{TX}}(e)$ for each edge can then be represented as follows (in unit of mW):

$$\text{TX Power } P_{\mathrm{TX}}(e) = \frac{10^{L_P}}{10\alpha} \quad (2)$$

Here α is a channel factor. As described above, the maximum link rate is assumed as 100 Mbps, which is $R_e^{\max}$. Then with distortion factor β, a possible link rate for each edge e can be defined by

$$\text{Link Rate } R(e) = \frac{R_e^{\max}}{P_{TX}(e) \times \beta} \quad (3)$$

In the simulation provided in Sect. 3, α and β are determined impirically using the measurement of Wi-Fi modules.

The total data rate traversing an edge e is defined as $U(e)$. Then the link utilization ratio $\mathrm{UR}(e)$ for edge e is defined by

$$\text{Utilization Ratio } \mathrm{UR}(e) = \frac{U(e)}{R(e)} \quad (4)$$

Here $R(e)$ is a possible link rate for edge e.

For each edge e, the effective TX power $P_{\mathrm{eff}}(e)$ is defined by

$$P_{\mathrm{eff}}(e) = P_{\mathrm{TX}}(e) \times \mathrm{UR}(e) \quad (5)$$

This reflects the important condition that each wireless modem turns on Tx power only when its link transmits data, and otherwise it goes down to power saving mode. Hence, the total effective power consumption $P_{\mathrm{eff}}^{\mathrm{net}}$ of the entire network is defined as

$$P_{\mathrm{eff}}^{\mathrm{net}} = \sum\nolimits_{\forall e} P_{\mathrm{eff}}(e) \quad (6)$$

We use the above analytical model for the following routing and allocation algorithm and its simulator.

C. Routing and Channel Allocation Formulation

For a wireless video sensor network, suppose only a set N of active nodes n_i have video event and send data to the sink node s. We search for an optimal set E of edges e_i (a wireless link between two nodes). The proposed routing and channel allocation method can be formulated as follows.

Minimize (Objective):

$$\sum\nolimits_{\forall e \in E} P_{\mathrm{TX}}(e) \times \mathrm{UR}(e) \quad \text{for all active edges in } E \tag{7}$$

Such that (constraints):

$$\text{For } \forall n \in N \text{ and } \forall e_i \in \mathrm{Path}_{n,s}, \quad \text{satisfy } e_i \in E \tag{8}$$

$$\text{For } \forall n \in N \text{ and } \forall e \in \mathrm{Path}_{n,s}, \quad \text{satisfy } R(e) - U(e) > r_{\mathrm{sensor}} \tag{9}$$

$$\text{For } \forall e \in E \text{ and } \forall f \in V(e), \quad \text{satisfy } C(e) \neq C(f) \tag{10}$$

Formula (7) is an objective to minimize, where E is a set of all active edges e. If we find a set E in a way that minimizes the sum of $P_{\mathrm{TX}}(e) \times \mathrm{UR}(e)$ for all e's in E, then E gives an optimal routing for all active nodes.

Formulas (8)–(10) define the constraints that we must satisfy while minimizing the objective (7). Here $\mathrm{Path}_{n,s}$ is defined as a multi-hop path from node n to the sink node s. Constraint (8) ensures that E contains all the required edges comprising a complete path from n to s. Constraint (9) ensures that the remaining link capacity $(R(e) - U(e))$ on each edge e can still hold new sensor data rate r_{sensor}. In Constraint (10), $V(e)$ is a set of vicinity edges of e, which are within the wireless range of e. Also, $C(f)$ is a channel assigned to edge f. All the edges in $V(e)$, hence, are assigned with channels different from e's. Otherwise, the edges in $V(e)$ may interfere with e causing collision and so loss of data rate.

Since finding an optimum routing and channel allocation is NP-hard, we propose a heuristic approximation to find a near-optimal solution.

D. Routing and Channel Allocation Algorithm

We propose an approximated cost metric $\mathrm{CM}_{n,s}^{\mathrm{path}}$ and a heuristic algorithm based on $\mathrm{CM}_{n,s}^{\mathrm{path}}$. This algorithm alleviates the complexity of the optimization formula (7)–(10) in Sect. 3.2C.

$$\mathrm{CM}_{n,s}^{\mathrm{path}} = \sum\nolimits_{\forall e \in \mathrm{Path}_{n,s}} P_{\mathrm{TX}}(e) \times \mathrm{UR}(e) \tag{11}$$

The proposed heuristic algorithm selects, for each active node n, the best edges in a way that minimizes $\mathrm{CM}_{n,s}^{\mathrm{path}}$. The algorithm starts finding paths for nodes near the sink first, and then paths for nodes farther from the sink. This way it can reuse the cost metric values calculated earlier for the previous paths.

For each active node n, the algorithm forms a graph of edges in E, by searching through all the egress edges from n toward sink node s to select the edge with lowest $CM^{path}_{n,s}$. When evaluating each edge, it ensures that the constraints (8)–(10) are satisfied. If any of the constraints cannot be met, it backs off from the selected egress edge, and searches through other egress paths.

Since the algorithm finds paths for nodes closer to the sink first, the condition in (9) can be easily calculated with prior values of $R(e) - U(e)$ for edges that have been already chosen. This is an important property of the proposed algorithm, which allows its rapid routing speed.

E. Routing and Channel Allocation Example

Figure 5 shows a routing and channel allocation result obtained by the proposed algorithm for the example network in Fig. 4. It first calculates the possible minimum number of hops from each active node to the sink node. For each egress edge e of node n, with distance D, it calculates $U(e), R(e), P_{TX}(e)$, $P_{\text{eff}}(e)$ and then cost metric from them.

For edge e_2 from node $n2$, $D = 2$ (indicating 20 m), $U(e) = 20\,\text{Mbps}$, $R(e) = 90\,\text{Mbps}, P_{\text{TX}}(e) = 276\,\text{mW}$, $P_{\text{eff}}(e) = P_{\text{TX}}(e) \times \text{UR}(e) = 276 \times 20/90 =$ 61.3 mW (indicated by EP). Here edge e_2 is selected and added to E, since it is the only path from $n2$ to $n0$ (sink).

In the same way, for edge e_4 from node $n5$, $P_{\text{eff}}(e) = 56.4$. For edge e_{18} from $n9$, $P_{\text{eff}}(e) = 48.6$, while for edge e_{19} from $n9$ to $n5$ (not shown in Fig. 4), $P_{\text{eff}}(e) = 56.4$. Therefore, cost metric $CM^{path}_{n9,s} = 61.3 + 48.6 = 109.9$ for the path $n9 \rightarrow n2 \rightarrow n0$, while $CM^{path}_{n9,s} = 56.4 + 56.4 = 112.8$ for the path $n9 \rightarrow n5 \rightarrow n0$. The algorithm selects the former path since it has a lower cost metric.

In this fashion, the algorithm selects the best edges of E as shown in Fig. 3. At the same time, it allocates minimal number of channels to the edges, so the edges in vicinity $V(e)$ would not interfere with each other. In Fig. 5, different channels are indicated by different colors and also by channel ID, C0–C10.

The final routing paths in Fig. 5 are:

$<n2 \rightarrow n0>$, $<n5 \rightarrow n0>$, $<n6 \rightarrow n0>$, $<n1 \rightarrow n0>$, $<n7 \rightarrow n0>$,
$<n9 \rightarrow n2 \rightarrow n0>$, $<n3 \rightarrow n1 \rightarrow n0>$, $<n8 \rightarrow n7 \rightarrow n0>$,
$<n4 \rightarrow n3 \rightarrow n1 \rightarrow n0>$

Figure 6 shows a routing result with different routing algorithm that selects paths that maximize each route's data throughput. It produces routing paths different from those of Fig. 5. The path for n4 is changed to:

$$<n4 \rightarrow n3 \rightarrow n7 \rightarrow n0>$$

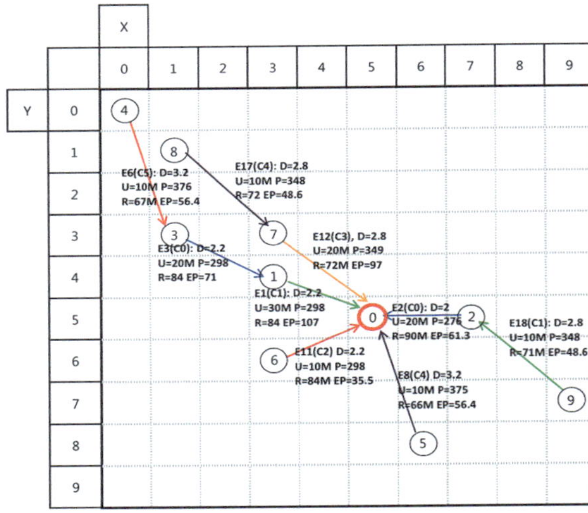

Fig. 5 Example routing and channel allocation result from the proposed algorithm, when applied to the wireless video sensor network of Fig. 4. Reproduced with permission from IEEE

Figure 7 shows another routing result with a routing algorithm that finds each link that has maximal link rate. Its routing results are:

$$<n2 \rightarrow n0>, <n5 \rightarrow n2 \rightarrow n0>, <n6 \rightarrow n0>, <n1 \rightarrow n0>,$$
$$<n7 \rightarrow n1 \rightarrow n0>, <n9 \rightarrow n5 \rightarrow n2 \rightarrow n0>, <n3 \rightarrow n6 \rightarrow n0>,$$
$$<n8 \rightarrow n3 \rightarrow n6 \rightarrow n0>, <n4 \rightarrow n8 \rightarrow n3 \rightarrow n6 \rightarrow n0>$$

It can be observed that Figs. 6 and 7 result in route paths with higher power consumption than those of Fig. 5.

The total Tx power consumption is 1.66 W for Fig. 5 (the proposed algorithm), 1.70 W for Fig. 6, and 2.38 W for Fig. 7. (See the 2nd column of Table 2).

3.3 *Experimental Results of Routing and Channel Allocation*

We implemented a simulator (WiSeSim) based on the the proposed algorithm. We experimented with an extensive set of WVSN. Tables 1 and 2 show the simulation results of 10 networks whose size ranges from 100 to 400 nodes. The number of active nodes ranges from 9 to 100 nodes. (See Table 1). The positions of active nodes are randomly generated.

Table 1 shows the number of routable paths. For some networks of large size, some paths turned out as unroutable owing to its high congestion. WiSeSim produced all paths routable except the two largest network cases. This is better than conventional algorithms.

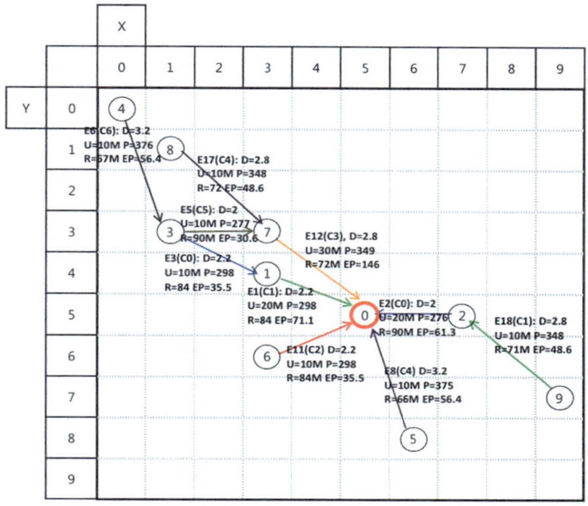

Fig. 6 A result using a routing algorithm that selects paths maximizing each route's data throughput. (Applied to the network in Fig. 4). Reproduced with permission from IEEE

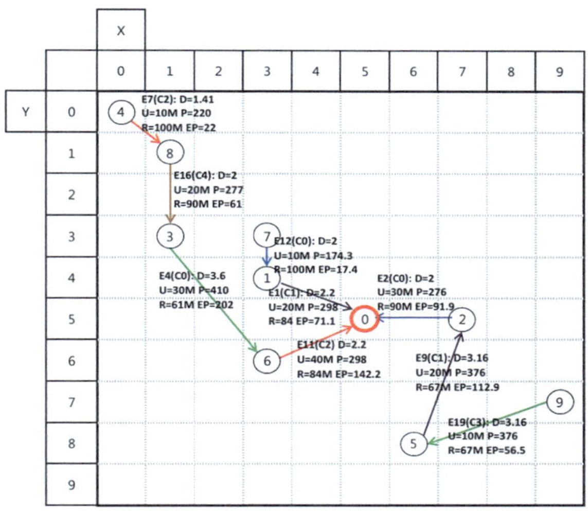

Fig. 7 A result using a routing algorithm that selects links that have maximal link rate. (Applied to the network in Fig. 4). Reproduced with permission from IEEE

Table 2 Simulation result of WiSeSim: Comparison of the number of routable paths for 3 different routing algorithms. Reproduced with permission from IEEE

Number of routable paths										
Network size (Num of nodes)	100	144	169	196	225	256	289	324	361	400
Num of active nodes	9	20	30	40	50	60	70	80	90	100
Routing for high rate link	9	20	30	40	49	53	66	76	83	94
Routing for route throughput	9	20	30	40	50	60	70	79	88	98
Proposed routing for low power	9	20	30	40	50	60	70	80	85	99

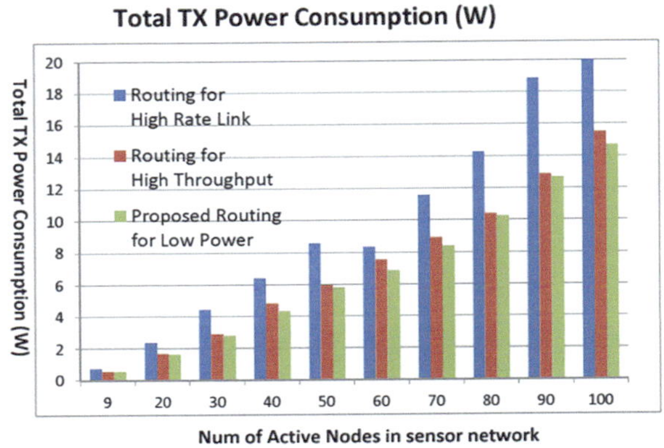

Fig. 8 Simulation result of WiSeSim: comparison of total Tx power consumption for the three different algorithms (Same result as Table 3). Reproduced with permission from IEEE

Table 2 and Fig. 8 compare the total Tx power consumption of the three algorithms described in Sect. 3C. WiSeSim has the lowest Tx power in all cases. WiSeSim has up to 30 % lower power than the routing algorithm for high rate link, and up to 10 % lower power than the routing algorithm for high throughput.

Figure 9 shows the average link utilization ratio for the three algorithms. WiSeSim shows about 10 % higher link utilization than the high throughput algorithm, while 20 % lower link utilization than the high rate link algorithm. This result indicates that the proposed algorithm minimizes Tx power by selecting fewer number of edges, but allocating more data on the selected fewer paths.

In this section, we assume that the routing and channel allocation are processed in a central node. We hence assume that a central node collects other node's information periodically (e.g., wake-up or sleep mode, data rate, and event-detection information), and then broadcasts routing results (Table 3).

The proposed algorithm can also be implemented as a distributed routing method. In this case, a node with an event detected broadcasts its information to its vicinity nodes, so the node and its vicinity nodes can recalculate their routing and channel allocation.

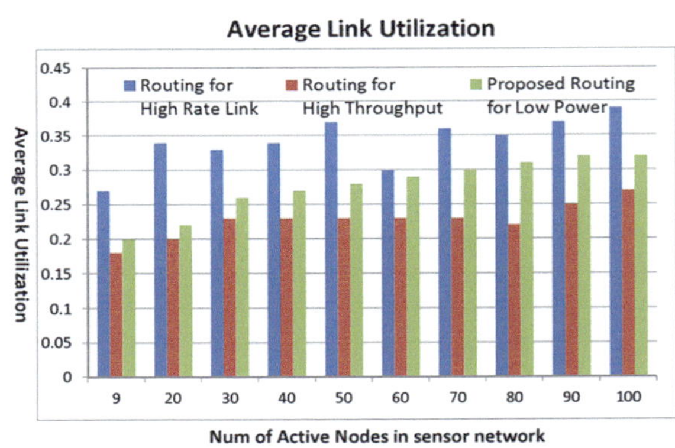

Fig. 9 Simulation result of WiSeSim: Comparison of average link utilization ratio for three different algorithms. Reproduced with permission from IEEE

Table 3 Simulation result of WiSeSim: Comparison of total Tx power consumption for 3 different algorithms. Reproduced with permission from IEEE

Total power consumption (W)										
Network size (Num of nodes)	100	144	169	196	225	256	289	324	361	400
Num of active nodes	9	20	30	40	50	60	70	80	90	100
Routing for high rate link	0.77	2.38	4.48	6.46	8.59	8.36	11.6	14.3	18.9	20.1
Routing for high throughput	0.59	1.7	2.91	4.83	5.96	7.57	8.91	10.4	12.9	15.5
Proposed routing for low power	0.58	1.66	2.82	4.35	5.81	6.89	8.38	10.3	12.7	14.7

3.3.1 Summary of Routing and Channel Allocation of WVSN

Networks of surveillance cameras with an event-driven wake-up function is becoming increasingly important for efficient deployment of low power wireless security networks—one of major research areas for Internet of Things (IoT). We presented a modeling technique for routing wireless sensor network with realistic transmission (Tx) power. We also described a method of formulation for optimal routing and channel allocation with minimal Tx power. We then presented an efficient heuristic algorithm and implemented it in a simulator (WiSeSim). The experimental results have shown Tx power savings up to 30 % compared to a conventional method. The proposed work is expected to contribute to the new IoT research areas of WVSN.

4 Multi-channel Allocation for Low Power Sensor Networks

In the previous section, we described a method of finding routing paths of a WVSN assuming there are a large number of channels available. We allocated different channels to all the selected links within the wireless range to avoid interference between the links. In most real networks, however, the channels are expensive resource and usually limited. In this section, we describe an extended routing and channel allocation method that allows multiple links within the same wireless range to share the same channel. This is possible by allowing the different links to occupy the same channel in different time periods as introduced below.

4.1 Properties of Channel Utilization

We begin by revisiting the notion of utilization ratio of Sect. 3.1, and extend the concept of utilization from link to channel considering the case of wireless links sharing the same channel.

A. Concept of Channel Utilization

The formula of utilization ratio (Eq. (4) in Sect. 3.1) holds when edge e_i has no interference in its wireless range A_{e_i}. This is the case when e_i is allocated a channel different from all other edges in A_{e_i}. When e_i shares a channel c_k with other edge e_j in W_{e_i}, however, the effective link utilization of e_i is increased by the utilization of e_j. This is due to the fact that e_i and e_j compete each other to use the same channel c_k in CSMA fashion. From e_i's perpective, e_i is called a victim, while e_j is called an aggressor. The effective link utilization of a victim edge is the sum of the victim's link utilization and all aggressor's link utilization.

Figure 10 illustrates an example network, where each node has up to 4 Wi-Fi modules. The dotted circle indicates the wireless range W_7 of node n_7. Consider a victim edge e_{13} using channel c_4. The aggressors of e_{13} are the edges that interfere with the received signals of e_{13}'s receiver (destination) node n_7. To determine the aggressors of e_{13}, therefore, we need to use the wireless range W_7 of the receiver node n_7. The aggressors are the edges e_3 and e_5 which use channel c_4 and are within W_7.

Figure 11 compares data transmission of edge e_{13} when there is no aggressor and when there are aggressors e_3 and e_5. Here we assume link rates $R(e_i)$ and utilization $U_L(e_i)$ for the victim and aggressors as indicated in Fig. 11. When e_{13} alone uses channel c_4, the effective utilization for c_4 is 20 % utilized ($U_L(e_{13}) = 0.2$) as shown in Fig. 11(1). On the other hand, when the two aggressor edges share the same channel c_4, the effective utilization for c_4 is the sum of $U_L(e_i)$ of the victim and aggressors. As a result, c_4 is 77 % utilized.

In summary, the formula of calculating effective utilization $UR_C(e_v)$ of a victim edge e_v for channel c_k is given below:

$$\text{For } \forall n_a \in W_v, e_{a,k}^{\text{egress}} \in A_v \text{ for victim } e_v \quad (12)$$

$$\text{For } \forall e_{a,k}^{\text{egress}} \in A_v, \mathrm{UR}_C(e_v) = \sum \mathrm{UR}_L(e_{a,k}^{\text{egress}}) \quad (13)$$

In formula (12), W_v is a set of all nodes within the wireless range from the destination node of victim edge e_v. A_v is a set of all aggressor edges in W_v.$e_{a,k}^{\text{egress}}$ are

Fig. 10 Example wireless video sensor network with channel sharing. Reproduced with permission from IEEE

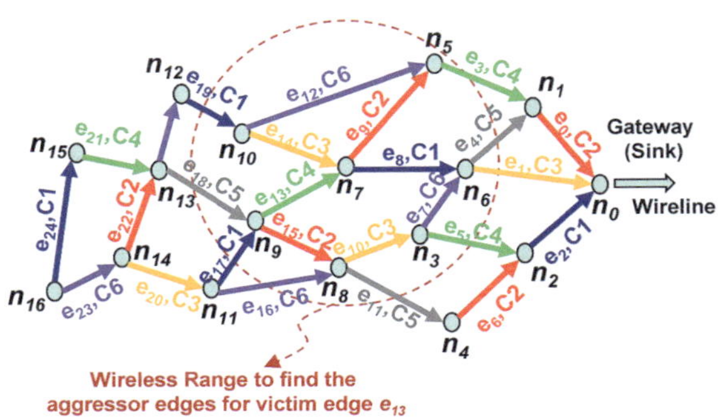

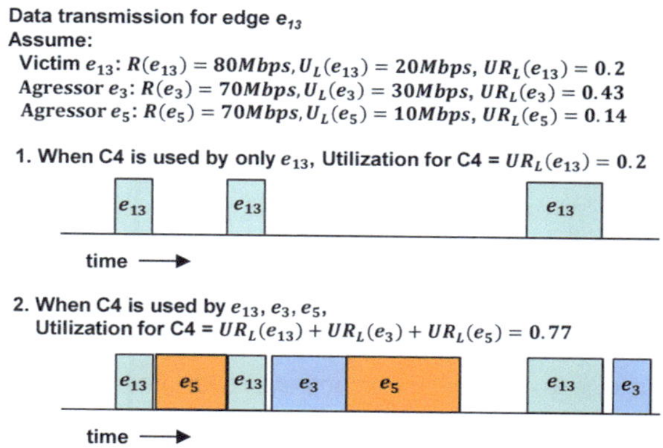

Fig. 11 Example of effective utilization. Data transmission of the victim edge with no aggressors (*1*) and with two aggressors (*2*). Reproduced with permission from IEEE

egress (out-going) edges from node n_a that use the same channel c_k. Formula (13) gives effective utilization $\text{UR}_C(e_v)$ for the victim e_v.

B. Formulation of Multi-channel Allocation

For a wireless video sensor network, suppose only a set N of active nodes n_i have video events and send data to the sink node s. We search for an optimal set E of edges e_i (a wireless link between two nodes) with channel c_i allocated to e_i. The proposed routing and channel allocation method can be formulated as follows.

Minimize (Objective):

$$\sum\nolimits_{\forall e \in E} P_{\text{TX}}(e) \times \text{UR}_C(e) \quad \text{for all active edges in } E \tag{14}$$

Such that (constraints):

$$\text{For } \forall n \in N \text{ and } \forall e_i \in \text{Path}_{n,s}, \quad \text{satisfy } e_i \in E \tag{15}$$

$$\text{For } \forall n \in N \text{ and } \forall e \in \text{Path}_{n,s}, \quad \text{satisfy } R(e) \geq U(e) \tag{16}$$

$$\text{For } \forall e \in E \text{ and } \forall f \in V(e), \quad \text{satisfy } C(e) \neq C(f) \text{ or} \tag{17}$$

$$\text{For } \forall e^{\text{egress}}_{a,C(e)} \in A_v,\ \text{UR}_C(e) = \sum \text{UR}_L(e^{\text{egress}}_{a,C(e)}) \tag{18}$$

Formula (14) is an objective to minimize, where E is a set of all active edges e. If we find a set E in a way that minimizes the sum of $P_{\text{TX}}(e) \times \text{UR}_C(e)$ for all e's in E, then E gives an optimal routing for all active nodes.

Formulas (15)–(18) define the constraints that we must satisfy while minimizing the objective (14). Here $\text{Path}_{n,s}$ is defined as a multi-hop path from node n to the sink node s. Constraint (15) ensures that E contains all the required edges comprising a complete path from n to s. Constraint (16) ensures that $U(e)$ of all the selected edges in E does not overflow after transmitting all data with sensor rate r_{sensor} passing through those edges.

In Constraint (17), $V(e)$ is a set of vicinity edges of e, which are within the wireless range of e. Also $C(f)$ is a channel assigned to edge f. All the edges in $V(e)$, hence, are assigned with channels different from e's, if possible. The edges e that have the same channel as edges in $V(e)$, however, share the channel in a CSMA fashion to avoid collision and so resulting in increased effective utilization. For such edge e with $C(e) = C(f)$, Constraint (18) must be satisfied, where $\text{UR}_C(e)$ gives a formula for the effective utilization of e as in Constraint (13).

Since finding an optimum routing and channel allocation is NP-hard, we propose a heuristic approximation to find a near-optimal solution.

4.2 Multi-channel Allocation Based on Channel Utilization

Cost Metric Calculation: We propose an approximated cost metric $\text{CM}_{n,s}^{\text{path}}$ and a heuristic algorithm based on $\text{CM}_{n,s}^{\text{path}}$. This algorithm alleviates the complexity of the optimization formula in Sect. 4.1B.

$$\text{CM}_{n,s}^{\text{path}} = \sum\nolimits_{\forall e \in \text{Path}_{n,s}} P_{\text{TX}}(e) \times \text{UR}_C(e) \tag{19}$$

Routing Edge Selection: The proposed heuristic algorithm selects, for each active node n, the best edges in a way that minimizes $\text{CM}_{n,s}^{\text{path}}$. The algorithm starts finding paths for nodes near the sink first, and then paths for nodes farther from the sink. This way it can reuse the cost metric values calculated earlier for the previous paths.

For each active node n, the algorithm forms a graph of edges in E, by searching through all the egress edges from n towards sink node s to select the edge with lowest $\text{CM}_{n,s}^{\text{path}}$. When evaluating each edge, it ensures that the constraints (15)–(18) are satisfied. If any of the constraints cannot be met, it backs off from the selected egress edge, and searches through other egress paths.

Figure 12 illustrates a result of the routing and channel allocation algorithm for the network of Fig. 4. On the selected edge, edge ID (channel ID), distance D, used data rate U, link rate R, and effective TX power EP are indicated.

Incremental Utilization Updates: While selecting each edge e as a route, the algorithm allocates a channel that is disjoint from its neighbor edges if possible. If no disjoint channel is left, it reselects a channel in a way that minimizes $\text{UR}_C(e)$.

Each time an edge e is selected as a route, the heuristic algorithm incrementally updates $\mathrm{UR}_C(e)$ by the following formula:

$$\begin{aligned} &\text{For } \forall n_v \in W_a, \text{For } \forall e_v^{\text{ing}} \in V_a \text{ for aggressor } e_a \\ &\mathrm{UR}_C\left(e_v^{\text{ing}}\right)[c_e] = \mathrm{UR}_C\left(e_v^{\text{ing}}\right)[c_e] + \Delta U_L(e_a)/R(e_a) \end{aligned} \tag{20}$$

For edge e selected as a route, formula (20) considers e as an aggressor and finds potential victim edges e_v^{ing} among its neighbor edges within wireless range W_a of e. The algorithm finds potential victim edges e_v^{ing} by selecting all the ingress edges to every victim node n_v in W_a. V_a is a set of all victim edges in W_a. Formula (20) incrementally adds to the effective utilization $\mathrm{UR}_C\left(e_v^{\text{ing}}\right)[c_e]$, the utilization growth due to e_a. Here $[c_e]$ is the index (channel ID) to array $\mathrm{UR}_C[]$. Every edge maintains an array of $\mathrm{UR}_C\left(e_v^{\text{ing}}\right)[c_e]$, $1 \le c_e \le c_{\max}$, where $c_{\max}$ is the max number of channels available.

Channel Allocation for Lowest Utilization: Once $\mathrm{UR}_C\left(e_v^{\text{ing}}\right)[c_e]$ of candidate edges have been calculated by formula (20), the routing and channel allocation algorithm selects a route edge by taking $\mathrm{UR}_C\left(e_v^{\text{ing}}\right)[c_e]$ in place of $\mathrm{UR}_C(e)$ of routing cost metric Eq. (19). This way, it selects edges with lowest effective utilization. It also guarantees that the utilization of the previously selected edges would not overflow. Figures 13, 14 illustrate an example.

In Fig. 13, the algorithm selects e_2 with channel 1. It then finds a victim set V_{e_2} of neighbor edges, which are ingress edges e_v^{ing} to all the nodes within W_a from the source node of e_2. It then updates $\mathrm{UR}_C\left(e_v^{\text{ing}}\right)[1]$ for V_{e_2} with $\Delta U_L(e_2)/R(e_2)$.

In Fig. 14, edges, e_3, e_4, and e_7 are selected with new channels 5, 4, and 6. When edge e_6 is selected, however, no new channel is available, and so channel 2 is selected again. Then it increases $\mathrm{UR}_C\left(e_v^{\text{ing}}\right)[2]$ of all victim edges by adding

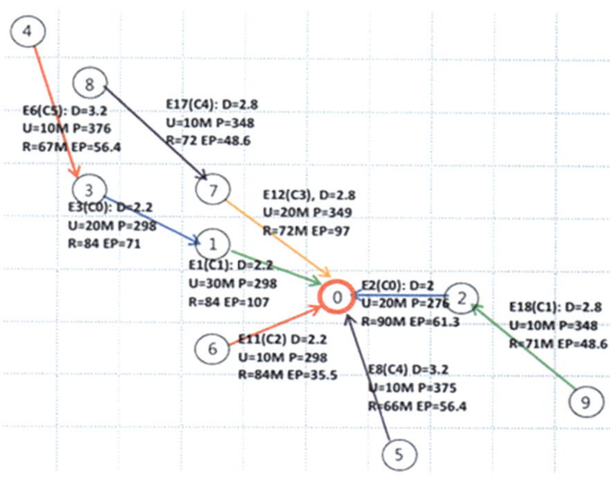

Fig. 12 Example routing and channel allocation result of the proposed algorithm applied to Fig. 4. Reproduced with permission from IEEE

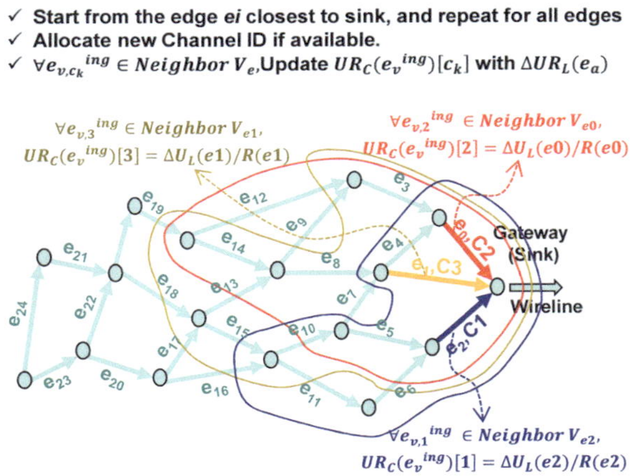

Fig. 13 Channel allocation Example (*Step1*) for the network of Fig. 4. Reproduced with permission from IEEE

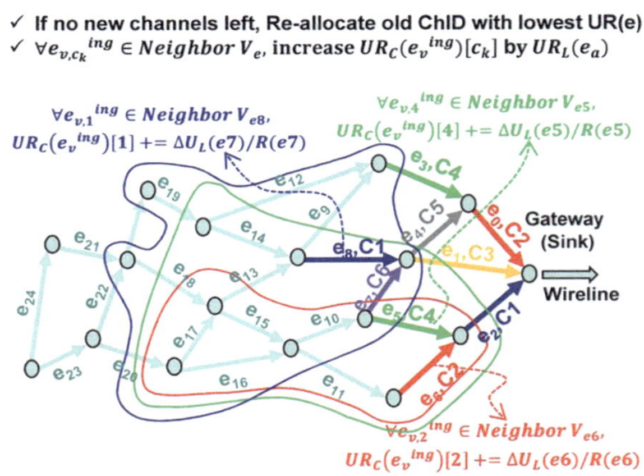

Fig. 14 *Step2*: allocating channels to the next selected edges following *step1* in Fig. 13. Reproduced with permission from IEEE

$\Delta U_L(e_6)/R(e_6)$. Similarly it selects old channels 4 and 1 for selected route edges e_5 and e_8.

Fast but Highly Accurate Algorithm: To avoid an exhaustive search, the proposed algorithm, in each step, selects the best route edge and channel using partial $\mathrm{UR}_C\left(e_v^{\mathrm{ing}}\right)[c_e]$ which has accumulated utilization changes $\Delta U_L(e_a)/R(e_a)$ for partial route edges selected only until the current step. It, however, ensures that $\mathrm{UR}_C\left(e_v^{\mathrm{ing}}\right)[c_e]$ of the previously selected edges would not overflow by the newly selected edge; an important property of the proposed cost-metric-based search algorithm.

This property allows very fast routing and channel allocation, while the recursively calculated effective utilization is accurate enough for the edges within the wireless range.

4.3 Experimental Results of Multi-channel Allocation

We implemented a simulator (**WiSeR**: Wireless Sensor network Router) based on the the proposed algorithm. We experimented with an extensive set of WVSN. Table 4 and Figs. 15, 16 show simulation results of 10 networks whose size ranges from 100 to 400 nodes. The number of active nodes ranges from 9 to 100 nodes. (See Table 4). The positions of active nodes are randomly selected.

In the channel allocation experiment, we used 3 channel allocation methods with different cost metric:

Round-Robin: Select one channel in round-robin fashion among all channels whose $\mathrm{UR}_C\left(e_v^{\mathrm{ing}}\right)[c_e]$ do not overflow for all pre-selected edges.

Minimum Utilization: Select a channel of the lowest $\mathrm{UR}_C\left(e_v^{\mathrm{ing}}\right)[c_e]$ among all channels whose $\mathrm{UR}_C\left(e_v^{\mathrm{ing}}\right)[c_e]$ do not overflow for all pre-selected edges.

Table 4 Routing results of WiSeR under given channel limits. Reproduced with permission from IEEE

Routing under given channel limits										
Network size (Num of nodes)	100	144	169	196	225	256	289	324	361	400
Num of active nodes	9	20	30	40	50	60	70	80	90	100
Sensor data rate	15	15	10	10	5	7.5	7.5	7.5	5	5
Channel limits	4	12	12	16	10	18	28	36	30	32
Round robin channel allocation	9	17	27	33	50	54	67	72	84	93
Min utilization channel allocation	9	19	28	36	49	54	67	73	84	93
Min neighbor util. channel allocation	9	20	30	40	50	60	70	80	90	100

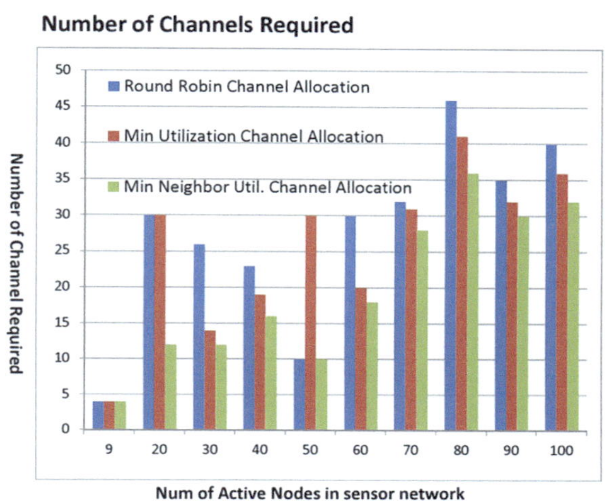

Fig. 15 Comparison of the number of channels required to finish routing and allocation. Reproduced with permission from IEEE

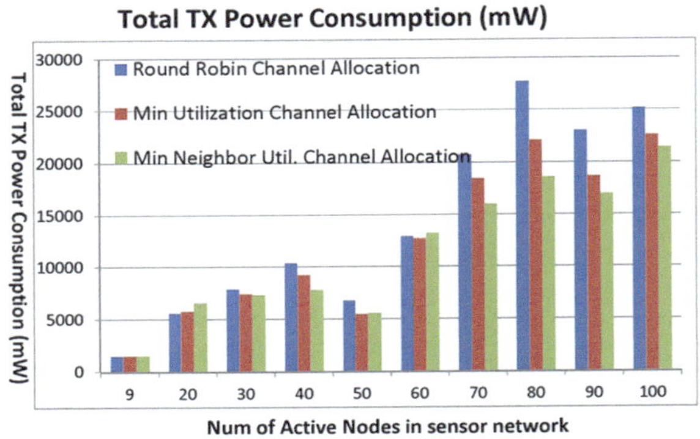

Fig. 16 Comparison of total Tx power consumption for the three different channel selection methods. Reproduced with permission from IEEE

Min. Neighbor Utilization: First find, for each c_e, $\mathrm{UR}_{C,\max}\left(e_v^{\mathrm{ing}}\right) = \mathrm{MAX}(\mathrm{UR}_C\left(e_v^{\mathrm{ing}}\right)[c_e]$, for $\forall e_v^{\mathrm{ing}} \in V_a)$. Then select channel c_e that corresponds to the minimum value of $\mathrm{UR}_{C,\max}\left(e_v^{\mathrm{ing}}\right)$.

Table 4 gives routing results of WiSeR under given channel limits: the number of routable paths for the 3-channel allocation methods. The min. neighbor utilization method finds all routes for all active nodes. For the other two methods, on the other hand, some paths turned out as unroutable owing to high congestion under the limited channel count.

Figure 15 shows the number of channels required to complete the routing and channel allocation for the 10 networks. The min. neighbor utilization method gives the best results for all cases. The other two methods, however, could not finish the channel allocation for some networks.

Figure 16 compares the total Tx power consumption of the 3-channel allocation methods. The min. neighbor utilization method has the lowest Tx power in most of the networks. It has up to 30 % lower power than the round-robin method, and up to 15 % lower power than the min. utilization method.

4.3.1 Summary of Multi-channel Allocation for WVSN

We introduced a multi-channel routing and channel allocation for wireless networks of battery-powered camera nodes with an event-driven wake-up function. We presented a modeling technique for routing and channel allocation of wireless sensor networks with realistic transmission (Tx) power based on link utilization. We then introduced formulation for optimal routing and channel allocation for minimal power. A heuristic utilization-aware algorithm (WiSeR) has been presented. Experimental results have been illustrated with three different channel allocation methods; one of them saved 30 % more TX power than others. More research is demanded in the areas of routing and channel allocation for WVSN to achieve low power network operation.

Acknowledgments This work is supported by the Center for Integrated Smart Sensors funded by the Ministry of Science, ICT and Future Planning as the Global Frontier Project.

References

1. Kim H (2014) Low power routing and channel allocation method of wireless video sensor networks for internet of things (IoT). IEEE World Forum Internet Things. doi:10.1109/WF-IoT.2014.6803208
2. Kim H (2013) Link utilization based routing algorithm for low power wireless visual sensor networks. J Res Inst Comput Inf Commun 21(2)
3. Kim H, Mohamed MGA (2014) Utilization-aware channel allocation and routing for mesh networks for battery-powered surveillance cameras. In: The 28th IEEE international conference on advanced information networking and applications (AINA)
4. Kompella S, Mao S, Hou YT, Sherali HD (2009) On path selection and rate allocation for video in wireless mesh networks. IEEE Trans Netw 17(1):212–224
5. Ding Y, Xiao L (2013) Video on-demand streaming in cognitive wireless mesh networks. IEEE Trans Mob Comput 12(3):412–423
6. Chen L, Zhang Q, Li M, Jia W (2007) Joint topology control and routing in IEEE 802.11-based multiradio multichannel mesh networks. IEEE Trans Veh Technol 56(5):3123–3136
7. Karl H, Willig A (2005) Protocols and architectures for wireless sensor networks. Wiley, New York
8. Akyildiz IF, Su W, Sankasubramaniam Y, Cayirci E (2002) Wireless sensor networks: a survey. Comput Netw 38:393–422
9. Zhou H, Ni LM, Mutka MW (2003) Prophet address allocation for large scale MANETs. In: Proceedings of IEEE INFOCOM, San Francisco, CA, March 2003
10. Ishaq I, Carels D, Teklemariam GK, Hoebeke J, Abeele FVd, De Poorter E, Moerman I, Demeester P (2013) IETF standardization in the field of the internet of things (IoT): a survey. J Sens Actuator Netw 2:235–287
11. Lennvall T, Svensson S, Hekland F (2008) A comparison of WirelessHART and ZigBee for industrial applications. In: IEEE International Workshop on Factory Communication Systems
12. HCF—HART communication foundation (2007) HART7 Specification. Accessed Sept 2007
13. Aakvaag N, Mathiesen M, Thonet G (2005) Timing and power issues in wireless sensor networks—an industrial test case. In: Proceedings of the 2005 international conference on parallel processing workshops, Oslo, Norway, June 2005
14. Society IC (2006) Part 15.4: wireless medium access control (MAC) and physical layer (PHY) specifications for low-rate wireless personal area networks (WPANs). Technical report, IEEE Computer Society, 2006
15. IEEE P802.11ah D3.1 Draft standard for information technology—telecommunications and information exchange between systems local and metropolitan area networks—specific requirements, Part 11: Wireless LAN medium access control (MAC) and physical layer (PHY) specifications, Amendment 6: sub 1 GHz license exempt operation
16. IEEE standard for information technology—telecommunications and information exchange between systems local and metropolitan area networks—specific requirements, Part 11: wireless LAN medium access control (MAC) and physical layer (PHY) specifications, Amendment 4: enhancements for very high throughput for operation in bands below 6 GHz

17. Sun W, Choi M, Choi S (2013) IEEE 802.11ah: A long range 802.11 WLAN at sub 1 GHz. J ICT Stand 1:83–108
18. Liang B, Haas ZJ (2000) Virtual backbone generation and maintenance in ad hoc network mobility management. In: Proceedings IEEE Infocom, Tel-Aviv, Israel, March 2000
19. Dasgupta K, Kalpakis K, Namjoshi P (2003) An efficient clustering-based heuristic for data gathering and aggregation in sensor networks. In: Proceedings of the IEEE wireless communications and networking conference (WCNC), New Orleans, LA, March 2003
20. Perkins C, Bhagwat P (1994) Highly dynamic destination-sequenced distance-vector routing (DSDV) for mobile computers. In: Proceedings of the ACM SIGCOMM, pp 234–244, London, UK
21. Perkins CE, Royer EM (1999) Ad-hoc on-demand distance vector routing. In: Proceedings of the 2nd IEEE workshop on mobile computing systems and applications, pp 90–100, New Orleans, LA, February 1999 (AODV)
22. Dhurandher SK, Obaidat MS, Gupta M (2010) A reactive optimized link state routing protocol for mobile ad hoc networks. In: IEEE international conference on electronics, circuits, and systems (ICECS)
23. Maltz D (2001) On-demand routing in multi-hop wireless ad hoc networks. PhD thesis, Carnegie Mellon University, Pittsburgh, PA (DSR)
24. Perkins CE, Royer EM (1999) Ad-hoc on-demand distance vector routing. In: IEEE workshop on mobile computing systems and applications
25. Park VD, Corson MS (1997) A highly adaptive distributed routing algorithm for mobile wireless networks. In: Proceedings of INFOCOM, pp 1405–1413, Kobe, Japan, April 1997 (TORA)
26. Dai R, Wang P, Akyildiz IF (2012) Correlation-aware QoS routing with differential coding for wireless video sensor networks. IEEE Trans Multimedia 14(5):1469–1479
27. Yu C, Sharma G (2010) Camera scheduling and energy allocation for lifetime maximization in user-centric visual sensor networks. IEEE Trans Image Proc 19(8):2042–2055
28. Li C, Zou J, Xiong H, Chen CW (2011) Joint coding/routing optimization for distributed video sources in wireless visual sensor networks. IEEE Trans Circuits Syst Video Technol 21(2):141-155
29. Pandremmenou K, Kondi LP, Parsopoulos KE (2013) Geometric bargaining approach for optimizing resource allocation in wireless visual sensor networks. IEEE Trans Circuits Syst Video Technol 23(8):1388–1401
30. Akyildiz IF, McNair J, Carrasco L, Puigjaner R (1999) Medium access control protocols for multi-media traffic in wireless networks. IEEE Netw Mag 13(4):39–47
31. Alonso J, Dunkels A, Voigt T (2004) Bounds on the energy consumption of routings in wireless sensor networks. In: International workshop on modeling and optimization in mobile, ad hoc and wireless networks, pp 62–70, Cambridge, UK, March 2004
32. Zhang Y, Zhang Y, Sun S, Qin S, He Z (2010) Multihop packet delay bound violation modeling for resource allocation in video streaming over mesh networks. IEEE Trans Multimedia 12(8):886–900
33. Ho IW-H, Lam PP, Chong PHJ, Liew SC (2014) Harnessing the high bandwidth of multi-radio multi-channel 802.11n mesh networks. IEEE Trans Mob Comput 13(2):448–456

MIX
Papier aus verantwortungsvollen Quellen
Paper from responsible sources
FSC® C105338

If you have any concerns about our products,
you can contact us on
ProductSafety@springernature.com

In case Publisher is established outside the EU,
the EU authorized representative is:
Springer Nature Customer Service Center GmbH
Europaplatz 3, 69115 Heidelberg, Germany

Printed by Libri Plureos GmbH
in Hamburg, Germany